✔ KT-228-645

Resource and Environmental Management

Second Edition

Bruce Mitchell

Professor of Geography and Associate Vice President Academic, University of Waterloo, Canada

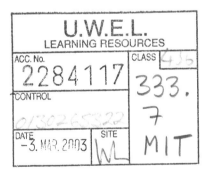

U.W.E.L.
LEARNING RESOURCES

ACC. No. 2284117

CLASS 436

CONTROL 0130265322

333.
7
MIT

DATE -3. MAR. 2003   SITE WL

Prentice Hall

An imprint of Pearson Education

Harlow, England · London · New York · Reading, Massachusetts · San Francisco · Toronto · Don Mills, Ontario · Sydney
Tokyo · Singapore · Hong Kong · Seoul · Taipei · Cape Town · Madrid · Mexico City · Amsterdam · Munich · Paris · Milan

Pearson Education Limited
Edinburgh Gate
Harlow
Essex CM20 2JE

and Associated Companies throughout the world

Visit us on the World Wide Web at:
www.pearsoneduc.com

First published under the Longman imprint 1997
Second edition published 2002

© Pearson Education Limited 1997, 2002

The right of Bruce Mitchell to be identified as author of this work has been asserted
by him in accordance with the Copyright, Designs and Patents Act 1988.

All rights reserved. No part of this publication may be reproduced, stored in a retrieval
system, or transmitted in any form or by any means, electronic, mechanical,
photocopying, recording or otherwise, without either the prior written permission of the
publisher or a licence permitting restricted copying in the United Kingdom issued by the
Copyright Licensing Agency Ltd, 90 Tottenham Court Road, London W1P 0LP.

ISBN 0 130 26532 2

British Library Cataloguing-in-Publication Data
A catalogue record for this book is available from the British Library

Library of Congress Cataloging-in-Publication Data
Mitchell, Bruce, 1944 –
    Resource and environmental management / Bruce Mitchell. – 2nd ed.
        p. cm.
    Includes bibliographical references and index.
    ISBN 0-13-026532-2 (pbk.)
    1. Environmental management.   2. Natural resources – Management.   3. Sustainable
development – Management.   I. Title.
    GE300.M58 2001
    333.7–dc21
                                                                          2001021571

10   9   8   7   6   5   4   3   2
05   04   03   02

Typeset in 11/12 pt Garamond by 35
Produced by Pearson Education Asia Pte Ltd.,
Printed in Singapore

Contents

Contents

Chapter 4 Sustainability — 72

Chapter 5 Ecosystem approach — 96

Chapter 6 Learning organizations and adaptive environmental management — 129

Chapter 7 Assessing alternatives — 155

Chapter 8 Partnerships and Stakeholders — 182

Contents

List of Figures

List of Tables

List of Guest Authors

Preface to the Second Edition

The first edition of *Resource and Environmental Management* has been used in many countries. A Spanish edition, entitled *La Gestion de los Recursos y del Medio Ambiente*, was published in 1999. In 2000 an Indonesian version, *Pengelolaan Sumberdaya dan Lingkungan*, was published, co-authored with Bobi Setiawan and Dwita Rahmi and incorporating Indonesian case study material. This widespread use led to the decision to prepare a second edition.

The second edition contains the updating and different examples usually found in new editions. However, there is much more, with about half of the content of the second edition being "new material". New features of the second edition are:

(1) *Restructuring*, to clarify the overall logic of the focus and the organization. The book begins by highlighting basic concepts of change, complexity, uncertainty and conflict, then considers some general strategies to deal with turbulent decision-making environments full of surprise and uncertainty. Subsequent chapters address developing a *vision* or sense of direction, and using a mix of *processes* to develop a *product* (policy, program, plan) which can be *implemented* and *monitored*.

(2) *New material*, including definitions of resource and environmental management (Chapter 1); future state visioning and appreciative planning (Chapter 3); future search conferences (Chapter 4); factors contributing to success of an ecosystem approach (Chapter 5); learning organizations (Chapter 6); strategic impact assessment, critical learning, and significance of culture in impact assessment (Chapter 7); elaboration on components of partnership programmes (Chapter 8); principles for and limitations of participatory appraisal methods, as well as new insights about factors contributing to effectiveness of co-management (Chapter 9); role of collaborative and stakeholder approaches in implementation, and a checklist of factors for success in implementation (Chapter 12); monitoring and evaluating public involvement in adaptive environmental management (Chapter 13); and environmental justice and interdisciplinary approaches (Chapter 14). Chapters 3 and 4 together provide considerably more coverage of the concept of visioning relative to the first edition.

(3) *Guest statements*, prepared by colleagues from around the world. In these statements, experts from universities, government agencies, consulting firms and non-government organizations in Australia, Canada, China, England, Germany, India, Indonesia, New Zealand, Nigeria, Peru, South Africa, USA and Zimbabwe provide insights based on their experiences.

(4) *Photographs*, incorporated to provide visual images of issues, opportunities and initiatives.

(5) *References*, hundreds of new sources, including those on web sites, have been added for further reading.

I am grateful to various people who provided support and assistance. Alan Diduck, Rodger Schwass, John Sinclair and Philip Xie were generous in allowing me to use photographs taken by them. In addition, Rodger Schwass reviewed Chapter 4, with particular attention to the future search conference concept. Maureen Reed and Bobi Setiawan suggested new concepts to include in the second edition, and their ideas were taken up.

Colleagues at Waterloo and elsewhere continue to stimulate through their own work, and by sharing ideas. Particular appreciation is expressed to Tim Babcock, Rob de Loë, Bob Gibson, Bruce Hooper, Murray Haight, Abdul Manan, Mary Louise McAllister, Geoff McDonald, Ali Memon, Gordon Nelson, Greg Oliver, David Pitts, Dwita Rahmi, Maureen Reed, Rizal, Bobi Setiawan, Dan Shrubsole, Geoff Wall, Susan Wismer and Dave Wood. George Priddle, whose contributions were acknowledged in the first edition, died in September 1998; his enthusiasm and insight are very much missed.

Candace Whitney and Kate Kirvan, graduate students at the University of Waterloo, provided research assistance while the book was being prepared, and their assistance is very much appreciated. In her usual efficient and effective manner, Kate Evans at the University of Waterloo provided technical assistance regarding photographs and figures.

The constructive advice of Louise Lakey, Desk Editor at Pearson Education, and the careful copy-editing and proof reading have been very much appreciated, and have significantly improved the final product.

Joan Mitchell, my wife, continues to be wonderfully supportive, and I would like to thank her for sustained interest, questioning and help.

Bruce Mitchell
Waterloo, Ontario
January 2001

Preface to the First Edition

This book has been written as a replacement for *Geography and Resource Analysis* (first edition, 1979; second edition, 1989). When Longman indicated in 1993 that it would like a third edition of *Geography and Resource Analysis*, an opportunity was created to think about the objectives, structure and content of that book. While the book seemed to have found a useful niche in the literature, and undergraduate and graduate students had indicated that it was helpful for them in their studies, I decided that the field of resource and environmental management had evolved enough that I could not capture it by simply preparing a third edition.

As a result, I explained to Longman that rather than preparing a third edition, a new book would be appropriate. However, because of other commitments, I also indicated that writing could not begin until late in 1995, with completion targeted for mid 1996. Longman accepted the concept of a new book, and the work schedule. This book is the result. Like its predecessor, this book is intended for senior undergraduate students interested in resource and environmental management. It assumes that they have already been exposed to ideas in physical and human geography.

In this Preface, it is appropriate to indicate some of the things that this book is, and what it is not. Reactions to *Geography and Resource Analysis* from students suggested that some key features of that book should be retained. First, providing an extensive list of references was viewed as a strength. As a result, this book also contains numerous references to a wide range of studies. Second, an emphasis on "ways of thinking" and "ways of analysis" was deemed to be useful, and such features are continued. One consequence of this decision is that in places the book is abstract and conceptual. However, my experience is that students appreciate and value these characteristics, as long as their course instructor provides examples or case studies to illustrate practical implications. On the other hand, students commented that *Geography and Resource Analysis* was sometimes "hard reading" because of the compact style and numerous references. A conscious effort has been made here to create a more "user friendly" book, especially for undergraduates.

I also believe it is important to indicate what this book is not, or what aspects have consciously not been covered. First, while *Geography and Resource Analysis* explicitly related work in resource management to enduring themes in geographical investigation, this book does not make such connections. Geography is a well-established discipline with strong traditions. It seems less necessary today to draw out directly the linkages between its traditions and themes, and a field of application such as resource management.

Second, considerable thought was given about whether to include Geographical Information Systems (GIS), and, if so, how best to do that. The challenge was viewed as multiple because: (1) GIS software is evolving quickly, as with most computer software, and therefore material quickly becomes dated, (2) considerable detail must be included if students are to obtain more than a superficial appreciation of GIS, and (3) many books specializing in GIS are available. The decision was to not include GIS in any substantive manner in this book. However, GIS is a very important tool for resource and environmental management, and students aspiring to work in this field should acquire at least a basic understanding of the technique.

The objectives, focus and content of this book have been influenced by many people, and I would like to acknowledge them. First, the opportunity to teach undergraduate and graduate students at the University of Waterloo and elsewhere since 1969 has provided many opportunities to observe the reactions of students to concepts and methods in resource and environmental management. Such reactions and comments from these students have strongly influenced the overall approach to this book. Comments from students who have used *Geography and Resource Analysis* also have been very helpful. Second, jointly teaching a graduate course at Waterloo for 27 years with George Priddle has provided a remarkable opportunity to share ideas and approaches. George's enthusiasm, and ability to identify the value of emerging concepts and approaches before they have become generally accepted, has made working with him a stimulating and enjoyable experience. Third, other colleagues with similar professional interests have provided a supportive atmosphere in which to teach and conduct research. The work and ideas of the following individuals at Waterloo have influenced my thinking, and therefore this book: Jim Bater, M. Chandrashekar, Terry Downey, Len Gertler, Bob Gibson, George Francis, Robbie Keith, Drew Knight, George Mulammottil, Gordon Nelson, and Geoff Wall. Fourth, participation in a major group project in Indonesia, and another team project in Nigeria, exposed me to considerations, concepts and methods not represented in *Geography and Resource Analysis*, but included here. I express my appreciation to Canadian and overseas colleagues (especially Haryadi and Sugeng Martopo in Indonesia; and Peter Adeniyi and Lekan Oyebande in Nigeria) in those two projects for helping both to broaden and to sharpen my thinking.

Tracy Fehl, Gary Loftus and Dave Wood, graduate students at the University of Waterloo, provided research assistance while the book was being prepared, and their work was and is much appreciated. Others who provided information or suggestions were Jim Bauer, Aubrey Diem, Len Eckel, Pauline Peddle, and Lt. Colonel M. Pigeon. Joan Mitchell helped to arrange for permissions to use copyright material. At Addison Wesley Longman, I am particularly grateful for the interest and support of Sally Wilkinson, the editorial support of Tina Cadle and the external freelance copyediting skills of Patrick Bonham and external freelance proofreading skills of Dennis Hodgson.

The first and second editions of *Geography and Resource Analysis* were dedicated to my wife, Joan. It is with pleasure that I dedicate this new book to her.

As stated in the second edition, she has always supported me in my different endeavours but also has helped to keep them in perspective.

Bruce Mitchell
Waterloo, Ontario
May 1996

Acknowledgements

We are grateful to the following for permission to reproduce copyright material:

Figure 1.3 from A Conceptual Model of processes of change in environmental management in *Futures* Vol. 31, p. 586 (Hadfield and Seaton, 1999); Figure 3.3 from The future state visioning process in *Long Range Planning* Vol. 26, p. 92 (Stewart, 1993); Table 5.1 based on General Principles and Characteristics in *Ecological Economics* Vol. 34, p. 10 (Malone, 2000); Figures 5.1 and Table 5.2 from Four systems properties of agroecosystems in Agricultural Administration Vol. 20, pp. 36, 37 (Conway, 1985), reprinted with permission from Elsevier Science; Table 1.1 from Attributes of Change in *Land use change and development in Segara Anakar, Java* published by Indonesia Publications No. 51, p. 21 (Olive, 1998); Figure 1.4 from *Some sources and consequences of renewable resource scarcity* published by Scientific America, 268: 42 (Homer-Dixon *et al.*, 1993); Figure 1.5 from The relationship between political instability, economy and poverty in *Resource Management in Developing Countries*, p. 60, published by Longman/Pearson, Harlow, Essex (Omara-Ojungu, 1992); Figures 1.6 and 1.7 from Coping with uncertainty in planning, reproduced by kind permission of the *Journal of the American Planning Association*, 52, p. 66 (Christensen, 1985); Table 4.1 reprinted with coutesy of Alternatives Journal: Environmental thought, policy and action. Annual subscriptions C$24.00 (plus GST) from *Alternatives Journal*, Faculty of Environmental Studies, University of Waterloo, Waterloo, Ontario, N2L 3G1 *Alternatives*, 17 (2) 36–46 (Robinson *et al.*, 1990); Figure 5.2 from The Baltic Sea ecosystem in *Environment*, 35 (8), p. 9, published by Heldref Publications, Washington DC (Kindler and Lintner, 1993); Figure 5.3 from Environmental Impact assessment in the Himalayas: an ecosystem approach, *Ambio*, 22 (1): 4 Royal Swedish Academy of Science (Ahmad, 1993); Table 6.2 from *Planning Education Reforms in Developing Countries: The Contingency Approach*, Duke University Press. Reprinted with permission (D.A. Rondinelli, 1993); Table 6.3 from Consensus and adaptive management in *Environmental Law* Vol. 16, pp. 448–9 published by the Northwestern School of Law of Lewis and Clark College, Portland, Oregon (Lee and Lawrence, 1986); Table 7.1 from *Education Techniques Identified in the EA Literature* reprinted from *The Canadian Geographer* Vol. 41 No. 3 (Diduck and Sinclair, 1997); Figures 7.2 and 7.3 with kind permission Canadian Standards Association material is reproduced from the CSA Standard: Z760–94 (Life Cycle Assessment), and with kind permission of the Canadian Standards Association, material is reproduced from the CSA Publication: PLUS 1107 (User's guide to Life Cycle Assessment: Conceptual LCA in Practise), which is copyrighted by CSA, 178 Rexdale Blvd, Etobioke, Ontario, M9W 1R3; Table 8.1 from *Arnstein's eight rungs on the ladder of citizen participation* in *Journal of the American Association of Planners* reprinted by permission of the Journal of the American Planning Association Vol. No. 35, issue 35 (Arnstein, 1969); Table 8.2 from Types of Strategic Alliances Table 1 in *Memorandum: MNR Guide to Resource Management Partnerships – Adminstrative Considerations,* Ministry of Natural Resources, Copyright 1995 Queen's Printer, Ontario *Copyright@gov.on.ca* (Ontario Ministry of Natural Resources, 1995); Table 8.3 from *Public participations mechanisms* in Geography and Resource Analysis published by Pearson Education Limited (Mitchell, 1989); Figure 8.1 from Ngamiland Western Communal Remote Zone, Botswana in *Splash* 10 (1), p. 8 (Van der Sluis, 1994); Tables 9.1 and 9.2 reprinted from The Origins and practice of participatory rural appraisal, *World Development* 22 (7), pp. 958–9, with kind permission from Elsevier Science Ltd, The Boulevard, Langford Lane, Kidlington, OX5 1GB (Chambers, 1994); Figure 9.3 from Agroecosystem Transect, Dusur, Natang, South Sulawesi, Indonesia in *Exploring Rapid Rural appraisal for Community-based Watershed Planning: The Bila River Watershed, South Sulawesi, Indonesia,* unpublished Master's thesis, University of Guelph, Ontario NIG 2W1 (Mackenzie, 1993); Table 12.1 from Implementation failure: a suitable case for review? p. 45 in Achieving environmental goals: the concept and practice of environmental performance review, E. Lykke ed. London, Belhaven Press 43–63 (Weale, 1992); Table 12.2 from Why policies Succeed or Fail in *Policy situations,* p. 214, reprinted by permission of Sage Productions Inc. (Berman, 1980); Figure 13.1 from Ironside (1999) Environment Canada, Acid Rain overview SOE Bulletin No. 00–3 Fall. *Framework of acid rain indicators in Canada*; Figure 13.2 (a&b) from Ironside (1999) Environment Canada Acid Rain overview SOE Bulletin No. 00–3. *Indicator: Emissions of sulphur dioxide (a): Indicator: Wet sulphate deposition* (b). Reproduced with the permission of the Minister of Public Works and Government Services, Canada 2001; Figure 13.3 from Dennison and Abel (1999). Moreton Bay Study. *A scientific basis for the healthy waterways campaign 13*; Figure 13.4 from Moreton Bay Catchment Water Quality Management (1998). *The crew members' guide to the Health of our Waterways, 31 and 65.* Please refer to information provided in publication.

Photographs: Figures 1.1 and 1.2 reproduced by kind permission of Hamidur Rahman; Figures 3.1(b) and 3.2 reproduced by kind permission of Philip Xie; Figures 4.1 and 4.2 reproduced by kind permission of Rodger Schwass; Figure 5.4 reproduced by kind permission of Alan Diduck; Figures 5.5 and 10.2 reproduced by kind permission of John Sinclair.

Academic Press for an abridged extract from 'Applied ecology: towards a positive approach to applied ecological analysis' by Walker and Norton, published in *Journal of Environmental Management* 14; Canadian Environmental Assessment Agency (*http://www.ceaa-acee.gcca*) for an abridged extract from 'Common ground: on the relationship of environmental assessment and negotiation' by Sadler and Armour, published in *The Place of Negotiation in Environmental Assessment*; Northwestern School of Law for an abridged extract from 'Adaptive management: learning from the Colombia river basin fish and wildlife program' by K N Lee and J Lawrence, first published in Environmental Law, volume 16: 431, 1986; Oxfam Activities Limited for an abridged extract from

'Half of the world, half of an age: an introduction to gender and development' by J C Mosse; Paper Tiger Enterprises Limited for an abridged extract from 'Net losses: the sorry state of our Atlantic fishery' by Silver Donald Cameron, first published in *Canadian Geographic* magazine, © Paper Tiger Enterprises Ltd, 1990; Policy studies Organization for an abridged extract from *Policy Studies Journal* 1988 by Blackburn; Resources for the Future from an abridged extract from 'Ecosystem management: an uncharted path for public forests' by R A Sedjo, published in *Resources* 19 121: 10, p. 10; and University of New Mexico for an abridged extract from 'The ecosystem as criterion for public land policy' by L K Caldwell, published in *Natural Resources Journal* volume 10, number 2, April 1970, pages 204–21.

Whilst every effort has been made to trace the owners of copyright material, in a few cases this has proved impossible and we take this opportunity to offer our apologies to any copyright holders whose rights we may have unwittingly infringed.

Chapter 1

Change, Complexity, Uncertainty and Conflict

1.1 Introduction

Change, complexity, uncertainty and conflict are encountered in many aspects of life. They are often central in resource and environmental management. They can create opportunities, as well as problems, for analysts, planners, managers, decision makers and members of the public. Two challenges are to recognize their importance and to determine how to function in their presence. Another challenge is to know how to become an agent for positive change.

This chapter begins by outlining an experience that reflects all four of these elements. It also shows how they can affect individuals, and how such individuals are often connected, willingly or unwillingly, to a larger global system which has implications for their life styles and livelihoods. The case study is followed by sections which in turn consider key aspects of each element: change, complexity, uncertainty and conflict. Subsequent chapters focus on various concepts, strategies, methods and techniques which resource and environmental planners are using, or could use more effectively. The over-riding purpose is to familiarize you, the reader, with the strengths, weaknesses, opportunities and threats associated with alternative ways of addressing change, complexity, uncertainty and conflict.

1.2 Arsenic contamination of groundwater in Bangladesh and India

One of the world's potentially most serious environmental problems exists in Bangladesh and India, as a result of use of arsenic-contaminated groundwater for both domestic and agricultural purposes. More than 90 million people in these two countries are at risk from arsenic poisoning. Ironically, this problem has occurred from conscious decisions to avoid using relatively plentiful but contaminated surface water sources, and instead to use groundwater which was considered to be safe for human consumption.

In the year 2000, the population of Bangladesh was between 140 and 145 million people, with 75–80% living in rural areas. In terms of rainfall, the country has distinct seasonal patterns, due to heavy monsoon rains followed by much drier weather. In the wet season, there is abundant surface water. However,

> ## Box 1.1 Arsenic contamination of groundwater
>
> Bangladesh is grappling with the largest mass poisoning of a population in history because groundwater used for drinking has been contaminated with naturally occurring inorganic arsenic. It is estimated that . . . between 35 and 77 million are at risk of drinking contaminated water. The scale of this environmental disaster is greater than any seen before; . . . .
>
> *Source*: Smith, Lingus and Rahman, 2000: 1093.

epidemic levels of faecal bacteria in surface water led to a decision to use groundwater supplies. Supported by overseas aid programmes, during the 1950s tube wells began to be drilled throughout the country. The Department of Public Health and Engineering has reported that 856,000 tube wells have since been drilled in rural areas. By 1999, 97% of Bangladeshis relied on tube-well water for drinking and cooking, while surface water sources continued to be the main source for washing, bathing and other domestic uses.

Recent sampling from the tube wells has shown that many water samples exceed the World Health Organization's recommended maximum permissible limit (0.05 mg/l) for arsenic in drinking water. Since the mid 1990s the effects of arsenic poisoning have been documented, with many deaths being reported as at least partially attributable to this. Naidu and Skinner (1999: 407) explained that by 1999 it was estimated that hospitals and clinics had documented arsenicosis in about 40% of the total population. In other words, about 60 million people in Bangladesh were vulnerable.

In West Bengal, India, contamination of surface water supplies also led to decisions to turn to groundwater for consumptive uses and for irrigation water. In India, newspaper reports have documented contamination of groundwater by arsenic in West Bengal since the early 1970s, but the first major scientific study was not published until 1984. Now, tube-well water in eight districts exceeds the World Health Organization's recommended limits, and it is believed that over 30 million people in West Bengal may be exposed to toxic levels of arsenic. For districts in West Bengal, groundwater is the source of water for virtually 100% of the rural population and for 60% of the urban people. Thus, between Bangladesh and West Bengal, about 90 million people are vulnerable to arsenic toxicity. Arsenic contamination of groundwater has been confirmed in many regions of Bangladesh and eastern India, and has caused serious health problems (Rahman, 1997; Yokota *et al.*, 1997). Figures 1.1 and 1.2 show the damage to feet and hands of arsenic victims from a village in the Chapai Nawalganj area, near the city of Rajshahi in western Bangladesh.

Concern also exists because, in rural areas, groundwater is also used as a source of irrigation water. Thus, not only are people in Bangladesh and West Bengal being exposed to arsenic poisoning through drinking water, but they also are being exposed through consuming food which has been irrigated by groundwater containing arsenic. However, the take up rate of arsenic by different types of crops is not yet well understood, so there is uncertainty about the

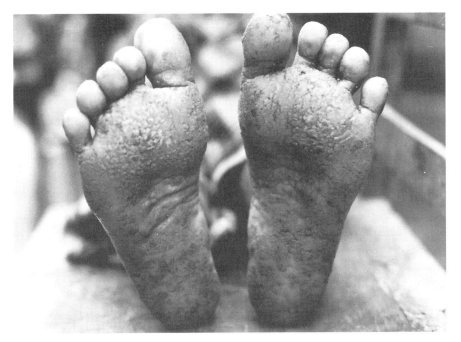

Figure 1.1 Symptoms of arsenic poisoning in soles of feet, male, Bangladesh (Hamidur Rahman)

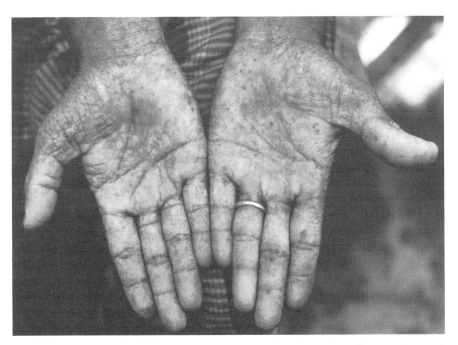

Figure 1.2 Symptoms of arsenic poisoning in hands, female, Bangladesh (Hamidur Rahman)

level of vulnerability through consumption of food (vegetables, meat) exposed to contaminated water.

1.2.1 Much uncertainty and complexity

While the magnitude of the problem has become clearer, much uncertainty concerning the problem of arsenic toxicity exists. Naidu and Skinner (1999: 408–9) have reported that, at an international conference of scientists in Dhaka during 1998, it was concluded that: (1) the original source of the arsenic appearing in groundwater was unclear; (2) the extent of contamination following long-term irrigation with arsenic-tainted groundwater was not understood; (3) the extent of ingestion of arsenic from local food sources, resulting from irrigation using water already contaminated with arsenic, was unclear; (4) the possible significance of additional exposure pathways for arsenic from contaminated groundwater through animals to humans (such as through consumption of milk, meat and other products) had not been investigated systematically; and (5) strategies to minimize exposure to arsenic poisoning for the people of Bangladesh and India were not well developed or available. Thus, much basic research still needs to be completed before the causes of the problem, and possible solutions, will be systematically identified. At the same time, innovative work was underway in Australia to develop low-end technologies to remove arsenic from groundwater (Khoe and Emett, 1999).

1.2.2 Village-level investigations

A research project by a combined team from Japan (Asia Arsenic Network, Research Group for Applied Geology) and Bangladesh (Department of Occupational and Environmental Health, National Institute of Preventive and Social Medicine, as well as the Department of Geology and Mining at the University of Rajshahi) is being conducted in a village in west Bangladesh. The team discovered that in 1983 the first patient with skin infections caused by arsenic was recorded in India, and about 10 years later the first instance was recorded in Bangladesh. The Asia Arsenic Network combined with other Bangladeshi experts and started its joint research project in December 1996. The team subsequently confirmed that groundwater had been contaminated in a large region of Bangladesh, and discovered that many incidences existed in the village of Samta, Jessore District. The team reported that: "we were truly shocked to see many patients suffering with skin problems in the village".

Subsequently, the Samta Village Project was started in 1997 with emphases on three aspects: (1) research, (2) countermeasures and (3) community participation. The ongoing work includes surveys of well water, groundwater and geological characteristics; medical surveys; an applied anthropological survey; and a life-style survey. Measurements of all the wells in the village of Samta during March 1997 revealed that arsenic contamination was above 0.05 mg/l, which is the permissible limit in Bangladesh, in 90% of the wells. Of 135 people examined, 23 had symptoms which suggested skin cancer. In April and

May 1998, detailed research was conducted on groundwater and geological features, and it was discovered that arsenic, with relatively high concentrations, was stored in upper clay layers. Drilling of deeper wells could avoid the arsenic-bearing clays, but the expense associated with drilling for deeper wells is prohibitive for most villagers. During February 1998, all families were surveyed, and it was found that patients showing symptoms of arsenic poisoning were often from low-income families. The work is continuing, and further information can be obtained from the coordinator of the project in Bangladesh: Dr M Hamidur Rahman, Department of Geology and Mining, University of Rajshahi, Rajshahi 6205, Bangladesh, Fax: 880-721-750064 or 880-721-772006.

1.2.3 Implications

Given that the switch from reliance on surface water to groundwater on the Indian subcontinent was supported and aided by international or overseas donor agencies, it is not implausible to suggest that donor agencies share some responsibility for the serious problems associated with arsenic-contaminated water in Bangladesh and India. An obvious question is in what way should international or overseas donor agencies respond? In order to reach an informed and considered decision, overseas donor agencies and water specialists need to consider the moral and other issues arising from such a problem, and the type of response that is appropriate.

Other implications can be identified from the arsenic contamination situation regarding the core themes in this book. Thus:

- **Change**. planners and managers encounter changing conditions, needs and expectations. What might have been acceptable at one time period may not be accepted at a later time period.
- **Complexity**. the ramifications of interactions of human activity with the natural environment are often difficult to understand and predict. Cause-and-effect patterns are difficult to determine due to multiple variables and pathways of interaction. Solving one problem may cause a new problem.
- **Uncertainty**. planners and managers have to make decisions without complete information or understanding of the ecosystems for which their decisions have consequences.
- **Conflict**. different, and often conflicting, values and perspectives are usually involved in resource allocation and use decisions. Such differences frequently reflect different "world views", needs and expectations.

In the remainder of this chapter, some key ideas related to change, complexity, uncertainty and conflict will be introduced. The rest of the book will examine various ways for analysts, planners and managers to address these matters when dealing with environmental and resource management. However, before exploring these concepts, the nature of resource and environmental management is discussed.

1.3 Resource and environmental management

For many decades, Zimmermann's (1933) interpretation of "resources" was generally accepted: he stated that neither the environment, as such, nor parts of the environment are resources until they are, or can be considered to be, capable of satisfying human needs. Thus, coal was not a resource without people whose wants and capabilities allowed them to recognize coal and thereby give it utility. In other words, attributes of nature or the environment are no more than "neutral stuff" until humans are able to perceive their presence, to recognize their capacity to satisfy human wants or needs, and to devise means to utilize them. As a result, in his view, the concept of a resource was subjective, relative and functional.

The interpretation by Zimmermann has been criticized as being too "anthropocentric", i.e. human centred. That is, in his view, aspects of nature are only considered to be resources if they have direct utility to human beings. Critics of Zimmermann's perspective argue that such a view does not recognize that aspects of nature deserve to be recognized as resources simply because they exist, and that they have value even if they do not offer utility to humans. As a result, today many interpret *resources* much more broadly than in a functional or utilitarian sense. In that context, resources are the abiotic, biotic and cultural attributes on, in or above the Earth. *Environment* is broader than resources, as it includes all of the surrounding conditions and influences which affect living and non-living things. In that regard, environment includes the atmosphere, hydrosphere, cryosphere, lithosphere and biosphere. At the same time, we should appreciate that, as Beate Ratter explains in her guest statement, culture is very influential in determining what people consider to be resources, and how they will use them.

Planning and *management* are two other terms that deserve comment. Planning is usually interpreted as a process used to develop a strategy to achieve desired goals or objectives, to resolve problems and to facilitate action. As Brickner and Cope (1977: 203) observed, "Planning is a process by which an individual or organization decides in advance some future course of action. The process consists of a series of steps, . . . to reach desirable ends". In contrast, *management* is normally defined as the capacity to control, handle or direct. In other words, while the role of a *planner* is to identify a desirable future and prepare a course of action to achieve it, a *manager* has the responsibility and authority to allocate capital, technology and human resources to reach the desired future state.

When the concepts of resource, environment, planning and management are combined, the following meanings emerge:

- *Resource and environmental planning*: with regard to resources and/or environment, identification of possible desirable future end states, and development of courses of action to reach such end states.
- *Resource and environmental management*: actual decisions and action concerning policy and practice regarding how resources and the environment are

Guest Statement

The significance of culture for resource and environmental management

Beate M W Ratter, Germany

A shark is a shark is a shark is *not* a shark. . . . When artisanal fishers from the western Caribbean coincidentally get a shark in their nets, their disgust for sharks forces them to return it to the ocean. In contrast, when artisanal fishers from the eastern Caribbean catch a shark, they get excited about a delicacy which will be a treat for their family and friends. Taiwanese fishermen cruising the same waters with their big trawlers catch heaps of the same species, just to cut off their fins and offer them to the huge Asian markets as a soup ingredient – and they throw the still living remains of the sharks back into the Caribbean waters as waste.

A shark – is not a shark. The perceived qualities of the shark as a resource vary widely even in a single geographic area such as the Caribbean Sea. The shark illustrates how culture determines what is viewed to be a resource. This dominance of culture also influences the concept of managing human/nature interactions. The notion of sustainable development is currently defined as development which is ecologically, economically and socially viable. But usually there is a missing link: culture. If we want to protect nature and to manage human/nature interaction, we have to take not only economic, political and social factors into account but also culture, which influences all other societal realms in two ways. First, humans are part of the system. Second, they try to comprehend the whole system and to manage it. This is why culture is significant. Humans live within a specific regional culture shaped by a specific historical context which influence the behaviour of the individual and of society as a whole.

Cultural scientists, human ecologists, as well as cultural geographers, have a considerable amount of experience in analysing regional cultures (Steiner and Nauser, 1993; McNeely and Keeton, 1995). Culture expresses itself in many forms: in stories and texts, in songs and music, in art and pictures, in architecture, in dances and sport. Culture expresses itself, as well, in the perception of space – the definition of distance, of neighbourhood, of centre and periphery. And culture expresses itself in the spatial organization of economic activities.

Understanding regional culture can be undertaken by analysing *Raumlage* (space distribution), *Raumpotential* (space potential), *Raumorganization* (space organization) and *Raumperzeption* (space perception). As Butzer (1990) puts it the starting point could be the analysis of a society's ecohistory. The second step should be to interview the regional population in order to understand their feelings for nature, resources and management. The continuation of the analytical process includes discovering organizational structures as well as describing the specific culture.

This is not to say that global economic networks aren't as important to analyse as the specific local and national situation. Or to say that political as well as legal frameworks and institutional arrangements are not also important. It is only to emphasize the importance of regional cultural analysis as a basis for developing adaptive and adaptable management strategies (Ratter, 2000).

When we understand the people we are dealing with and when the people believe that their way of living is accepted, it is easier to develop and to implement an adaptive management strategy. Participation is not only a democratic cornerstone but it is the basis for public support which is necessary in a sustainable development process. Adaptive management strategies can vary from region to region and among rules, regulations and stewardship programmes. An important instrument which can foster environmental management strategies is the construction and support of a regional identification with the environment as lifeworld – including the environment and natural resources. As we are dealing with people, we are dealing with human beings with all their individuality, subjectivity, pride and prejudices. Environmental management strategies which run against them will fail.

Butzer K 1990 The realm of cultural–human ecology: adaptation and change in historical perspective. In B L Turner *et al.* (eds) *The Earth as transformed by human action – global and regional changes in the biosphere over the past 30 years.* Cambridge, Cambridge University Press, 685–702

McNeely J A and W S Keeton 1995 The interaction between biological and cultural diversity. In B von Droste zu Hülshoff, *et al.* (eds) *Cultural landscapes of universal value.* Jena, Gustav Fischer, 25–37

Ratter B M W 2000 *Natur, Kultur und Komplexität – Adaptives Umwelt-Management am Niagara Escarpment in Ontario, Kanada.* Heidelberg, Springer

Steiner D and M Nauser (eds) 1993 *Human ecology – fragments of anti-fragmentary views of the world.* London/New York, Routledge

Beate M W Ratter is a Lecturer in the Department of Geography, University of Hamburg, Germany. Her research focuses on resource management and cultural geography from theoretical as well as practical perspectives. During the last 15 years, she has spent extensive time in the Caribbean and in Canada. Her current work focuses upon integrated coastal zone management and the integration of regional concepts of public participation. Since February 2000, she has been working with WWF-International on trilateral Wadden Sea policies in Denmark, Germany and the Netherlands.

appraised, protected, allocated, developed, used, rehabilitated, remediated and restored, monitored and evaluated.

1.4 Change

Change is common in natural and human systems, and can take many forms. Hadfield and Seaton (1999) identified numerous kinds of change (Figure 1.3), each with the potential to become a trigger or driver for further change. In dealing with resource and environmental problems, it is useful to keep this "big picture" in mind regarding what kinds of change may occur, and how changes in one variable or area can have a ripple effect, leading to changes in other variables or places. This network of factors which can change, or trigger

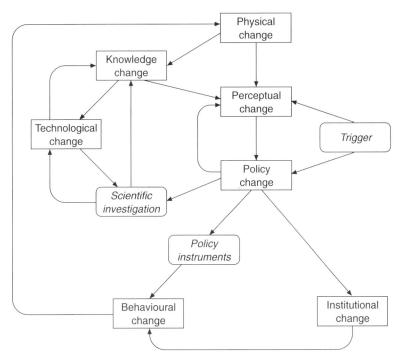

Figure 1.3 A conceptual model of processes of change in environmental management (Hadfield and Seaton, 1999: 586)

change, also highlights the high degree of complexity and uncertainty associated with change, and why there can be conflicting views regarding causes or pathways of change. Such a dilemma appears in the arsenic contamination experience reported earlier in this chapter.

To be able to understand the process of change, it is useful to recognize that change has various attributes (Table 1.1). Public interest and concern often focuses most on change characterized by short duration, high intensity, short magnitude, infrequency, unpredictability and irreversibility. For example, media and public attention focus on events such as earthquakes, floods, hurricanes and tornados, and are less likely to be attracted initially to changes associated with drought.

1.4.1 Environmental change and violent conflict

One implication of environmental change or resource scarcity is the increasing likelihood of conflict among people or nations. As explained by Homer-Dixon, Boutwell and Rathjens (1993), human activity can contribute to environmental change in three ways. First, human actions can result in a *decrease in the quality and/or quantity of resources* if they are used or harvested at a rate faster than they can be renewed. When this occurs, it is said that people are living off their natural capital, rather than from the interest off the natural capital. In some

Table 1.1 Attributes of change

Dimension of Change	Characteristics
Duration	Short to long
Magnitude	Temporary/seasonal to permanent
Intensity	Low to high
Frequency	Rare, seldom, to often
Rate of onset	Slow and gradual to rapid
Spatial dimension/extent	Small to large area
Predictability	Low to high
Reversibility	Low to high

Source: Olive, 1998: 21.

instances, such as use of non-renewable resources like petroleum, natural gas, zinc or silver, human use does deplete the capital, because such resources are renewable on a geologic rather than on a human time scale. However, even renewable resources such as topsoil, forests or wildlife can be degraded if they are used at a rate faster than they can be replenished. A second source of change is from *population growth*. With population growth, a set amount of arable land or water must increasingly be shared among more people, resulting in a steadily reduced amount available per person. Third, *unequal access to resources or the environment* may cause change. Unequal access is usually the result of laws or property rights which lead to or encourage the concentration of supply in the hands of relatively few people, leaving others subject to scarcity. These three activities or factors can act alone or in combination.

As Figure 1.4 shows, decreases in quality and quantity of resources, population growth and unequal access to resources contribute to scarcity or degradation, or both, which in turn become triggers for many other second-order consequences. Thus, worsening environmental conditions may cause people to migrate or to be expelled from a region, giving rise to a phenomenon now referred to as "environmental refugees" (Box 1.2). It can lead to ethnic conflicts, when one group in a society perceives that another has achieved disproportionate ownership or control of crucial resources. Thus, the process of change can have many phases or cycles, some of which may be far enough removed from

---

Box 1.2

Population growth and unequal access to good land force huge numbers of people into cities or onto marginal lands. In the latter case, they cause environmental damage and become chronically poor. Eventually these people may become the source of persistent upheaval, or they may migrate yet again, stimulating ethnic conflicts or urban unrest elsewhere.

Source: Homer-Dixon, Boutwell and Rathjens, 1993: 42.

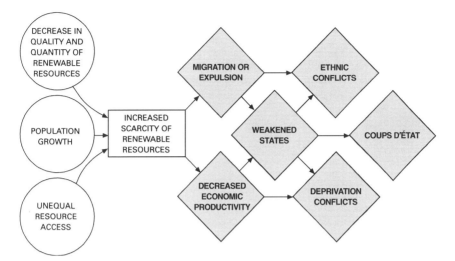

Figure 1.4 Some sources and consequences of renewable resource scarcity (Homer–Dixon *et al.*, 1993: 42)

the scarcity or degradation that most people do not recognize their fundamental causal role. Such a multi-layered process also contributes to significant complexity, making it difficult to know where to draw the boundaries around a resource or environmental "problem" or "issue".

1.4.2 Political ecology

The concept of "political ecology" has been developed to help understand the *political sources*, *conditions* and *ramifications* of environmental change, especially in the Third World (Bryant, 1992, 1997; Bryant and Bailey, 1997). Political ecology requires the analyst, planner or manager to go well beyond the attributes of only a biophysical or natural system when seeking to understand cause-and-effect relationships, or when preparing strategies or plans. In Bryant's view, the three dimensions of political ecology direct attention to:

● **Political sources**: state policies, interstate relations and global capitalism, thereby highlighting the growing importance of national and transnational forces on the environment.
● **Conditions**: conflicts over access, with emphasis upon location-specific struggles. This dimension emphasizes that those with relatively little power will and can fight back to protect environmental conditions which are the basis for their livelihood. Understanding such conditions requires a historical and contemporary appreciation of the dynamics of any conflict. This dimension indicates that understanding of "environmental history" is important to appreciate the broader context of environmental change and problems (Box 1.3).
● **Ramifications**: the political consequences of environmental change, with particular attention given to socio-economic impact and political process issues.

11

Box 1.3

Environmental history was catalyzed by the rise of the environmental movement and the subsequent need to gain greater critical understanding of environmental change and the manner in which nature and human culture have shaped each other over time. In particular, environmental history has explored the impact of various social factors (especially gender, race, ethnicity and class) on our response to the world around us.

Source: Meine, 1999: 1–2.

The issue of *state policies*, within the "political sources" dimension, illustrates how the political ecology framework extends an analyst's perspective regarding environmental changes. State policies have a potentially core role for interactions between people and their environment, as they help to establish priorities and practices for the state, as well as to structure debate about environmental change. Thus, it is important to understand the origins, content, implementation and impact of such policies. Bryant further noted that policies are not developed in a vacuum. Instead, they emerge from interaction and struggle among competing interest groups which strive to influence the development and substance of policy. Furthermore, many policies have implications for the environment and natural resources, ensuring that many groups' – government agencies, national and multinational companies, non-government organizations, multilateral agencies, donor organizations and foreign governments – interests overlap with environmental and resource issues. For example, Black (1990) has outlined how a variety of different and conflicting pressures, some internal and some external, contributed to a major agricultural crisis in the Serra do Alvao in the interior of northern Portugal.

To be effective analysts, it is important to be able to identify and understand the differing and often conflicting pressures placed on policy makers, and the implications for the intent and output of policies. As an example, Bryant suggested that forest policies often attempt to reconcile interests in conservation with pressures from commercial and non-commercial users of forests. In turn, the forest policies which try to accommodate diverse interests are influenced by other considerations ranging from taxation, trade and industrial policies. As a result, analysts often find themselves considering aspects which at first glance appear unconnected to environmental or resource issues.

Conflict regarding access within the "conditions" dimension further illustrates the need to take a broad political ecology perspective. More specifically, Bryant argued that the role of women in conflicts over access to resources in many developing countries is critically important, and yet has mostly been neglected, with a few notable exceptions such as the role of women in their struggle to retain access to common property land, water and forestry resources in India. Only recently have gender roles been examined systematically, a matter to be examined more thoroughly in Chapter 10.

A study of garden orchards in Gambia highlights the importance of considering the role of women in resource allocation conflicts (Schroeder, 1993). Two decades of drought through the 1970s and 1980s stimulated hundreds of women's groups to develop intensive and lucrative fruit and vegetable crops in low-lying communal market gardens. In a separate effort to use tree planting to achieve environmental stability, land developers encouraged male land holders to use female labour, which had been used primarily to irrigate garden plots, to plant orchard trees in the same locations.

One result was that the shade from the orchard trees eventually hindered the production in the garden plots, and led to men taking over control of the plots. The explanation for this shift in control is a gender division of labour for agriculture in the Gambian villages. In the Mandinka villages, men normally grow groundnuts (peanuts) and the coarse grains (millet, sorghum, maize) on higher ground during the rainy season. Groundnut production is the main source of foreign exchange for Gambia, so the male control of this crop gives them control over most of the cash income earned in agriculture. In contrast, women grow rice and vegetables in wet and low-lying areas, and their gardens represent the primary agricultural activity in the dry season. Most of the agricultural produce grown by women is used in home consumption.

The emergence of market-oriented vegetable gardens occurred due to a noticeable decrease in average precipitation. One response was for increased production in shorter duration, drought resistant or cash earning crops. Rice and vegetables, grown in swampy and low-lying areas, were more capable of such production, and women became involved because such production concerned crops and land for which they had traditionally provided the labour. As the men looked for more drought resistant crops from trees, they turned to mangoes, which grow well in wet, low-lying areas. However, mangoes are not compatible with an understorey vegetable or fruit crop. Thus, understanding of different gender roles becomes critical in designing policies to encourage tree planting to avoid environmental degradation.

1.5 Complexity

In considering the relationships between change and conflict, it is necessary to recognize that many other variables or factors besides environmental degradation and resource scarcity contribute to conflicts, as highlighted in the political ecology approach. Other variables include population growth, poverty, inequitable political systems and lack of economic opportunities (Figure 1.5). All of these in combination can contribute to instability within a society, can make it more vulnerable to environmental degradation, and can provide incentives to look outside the society's own borders for solutions.

Notwithstanding the above qualification about the relative significance of environmental degradation, the increasingly important role that environmental change can have in creating conditions which lead to conflict is being recognized. For example, Homer-Dixon, Boutwell and Rathjens (1993: 38)

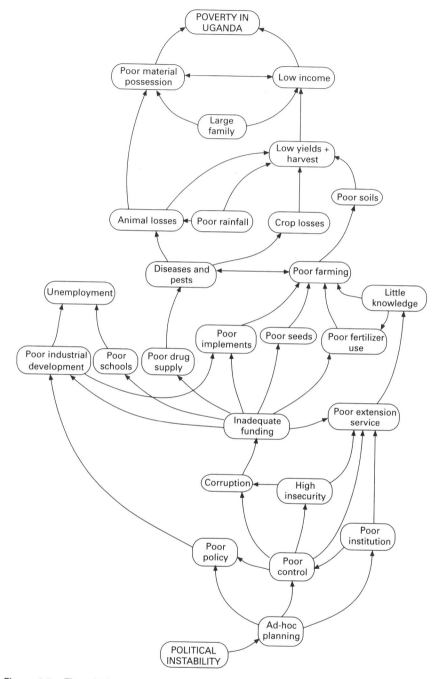

Figure 1.5 The relationship between political instability, economy and poverty (Omara-Ojungu, 1992: 60)

concluded that "scarcities of renewable resources are already contributing to violent conflicts in many parts of the developing world. These conflicts may foreshadow a surge of similar violence in coming decades, particularly in poor countries where shortages of water, forests and, especially, fertile land, coupled with rapidly expanding populations, already cause great hardship". As a result, resource and environmental analysts need to be able to inter-relate environmental changes to other variables which contribute to conflicts and disputes.

As Waldrop (1992: 12) noted, complexity creates major challenges for analysts, planners and decision makers. To illustrate, when dealing with global climate change, attention has to be given to issues as diverse as energy use, food production, forest harvest practices and transportation policies. In his view, complex systems are more spontaneous, more disorderly, and more subject to sudden and unpredictable change. Indeed, complexity is partially the reason why many systems seem chaotic, or discontinuous and erratic, and are best characterized by "jagged edges and sudden leaps" (Gleick, 1987: 5). Being aware of complexity does not make resource and environmental planners' jobs easier, but it does prepare them for inevitable surprises, and to consider how to deal with the uncertainty that it generates (see Chapter 2).

1.6 Uncertainty

The complexity encountered in environmental and resource management contributes to situations in which decisions have to be taken in the face of considerable uncertainty. Our understanding of biophysical systems, of human societies, or of the interactions between natural and social systems is often incomplete and imperfect. Furthermore, we are aware that conditions and circumstances in the future could well change relative to what they are today. And, yet, decisions have to be made because it is not realistic to wait until analysts develop the depth of understanding that we would like to have before committing ourselves to a path of action. Given the reality of change, complexity and uncertainty, environmental and resource managers are increasingly examining approaches which allow for adaptation, and which accept that we should be able to learn from rather than be punished for mistakes. Some of the ideas and strategies for adaptiveness, flexibility and social learning will be examined in Chapters 2 and 6.

Given an uncertain future, it is helpful to be able to recognize different kinds of "uncertainty". Wynne (1992) differentiated among four types of uncertainty, which are described in Box 1.4. By recognizing different kinds of uncertainty, we should be able to focus more readily on the kinds of issues that we need to consider.

Explicit recognition of uncertainty also helps us to identify the kinds of analysis or planning which may be most appropriate. As Christensen (1985) explained, problem-solving situations vary regarding the uncertainty about means and ends. She noted that when people are able to agree on what they want and how to achieve it, then certainty is high and planning can become the

---

Box 1.4 Types of uncertainty

(1) **Risk.** Know the odds.
(2) **Uncertainty.** Do not know the odds. May know the key variables and their parameters.
(3) **Ignorance.** Do not know what we should know. Do not even know what questions we should be posing.
(4) **Indeterminacy.** Causal chains or networks are open. Understanding not possible.

Source: Wynne, 1992: 114.

---

rational application of knowledge. When people agree on what they want but do not know how to achieve it, then planning becomes more of a learning process. If people cannot agree on what they want, but do know which alternative means are preferred, planning becomes a process of bargaining. And finally, if people cannot agree on either means or ends, then planning becomes a search for order in chaos. Figures 1.6 and 1.7 illustrate some of the characteristics of these four conditions, and their implications. The key message is that too often

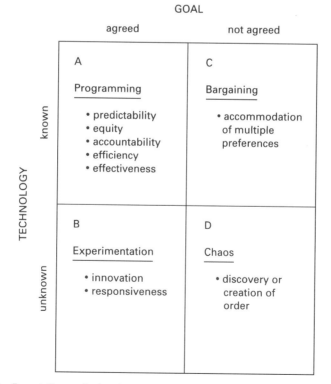

Figure 1.6 Expectations of planning organizations relative to prototype conditions of, and responses to, planning problems (Christensen, 1985: 66)

GOAL

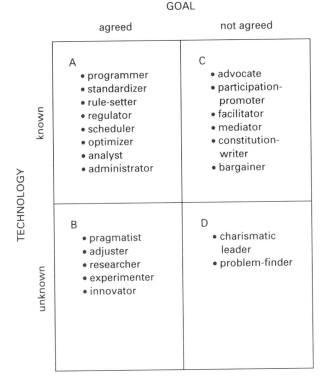

Figure 1.7 Planning roles categorized by planning conditions (Christensen, 1985: 69)

analysts or planners function as if they always are in cell A of Figures 1.6 and 1.7, whereas in reality the conditions are usually more similar to those in cells B, C or D. By recognizing different kinds of uncertainty, we should be able to diagnose more readily the types of conditions which characterize a problem, and thereby be able to handle it more effectively.

1.7 Conflict

It often is suggested that the idea of "managing" the environment or natural resources is a misnomer, given rapidly changing conditions, immense complexity and high uncertainty. Instead, it is argued, we should be focusing upon managing human interaction with the environment and natural resources. If this latter position is accepted, then much of "environmental and resource management" becomes the management of conflict. This situation occurs because it is normal in a society to have individuals or groups with different values, interests, hopes, expectations and priorities. And, often there is at least tension among these different characteristics, if not mutual incompatibility. In the extreme, as outlined above when discussing *environmental change and acute*

Box 1.5 Basic causes for conflict

(1) Differences in knowledge or understanding.
(2) Differences in values.
(3) Differences about distribution of benefits and costs.
(4) Differences due to personalities and circumstances of interested parties.

Source: Dorcey, 1986.

---

conflict, different interests and expectations can lead to armed conflict within or between nations.

If a significant skill for a resource or environmental manager is the capability to deal with conflict, a useful entry point is to be able to recognize some of the causes of conflict (Box 1.5). While this and other aspects of conflict or dispute resolution will be addressed in Chapter 11, at this stage four basic roots of conflict are identified, based on the work of Dorcey (1986). It should be noted that, in many situations, more than one of these causes may exist.

First, *different knowledge or understanding* may lead to conflict. In other words, groups may be using different theories, models, assumptions and information. Thus, one side in a dispute may believe that an aquifer is being over pumped, or that reserves of a mineral are only sufficient for seven to eight more years, whereas another may have concluded that the aquifer is not being over drawn, or that the mineral reserves are sufficient for 30 years. With perspectives about the situation based on different evidence or interpretations of it, the two sides then could be in conflict about whether a problem exists and/or about appropriate solutions to deal with it.

Second, disputes may emerge because of different *values*. For example, there may be agreement about the nature of the problem and the means to resolve it, but sharp divergence regarding the ultimate end point to be sought. Using the aquifer example, some may believe that as much water as possible should be extracted to support economic activities, whether agricultural or industrial. Others may believe that significant quantities of water should not be extracted for human use, in order to ensure that sufficient water can feed into streams and lakes and thereby support fish and other aquatic life, or to nourish wetlands dependent upon the groundwater in the dry season.

Third, even if the various sides in a conflict accept the same evidence and diagnosis, and have similar values, conflict may emerge from *differences in interests*. In other words, conflict arises not because of different understanding (it is agreed that the aquifer is or is not being over pumped) or because of different values (economic development is necessary to provide employment), but because of differences regarding who will be the beneficiaries and who will carry the burden of the costs. For example, one group may believe that the use of water from an aquifer for irrigation, financed through a subsidy from the government for the capital costs, is appropriate because it will lead to food self-sufficiency and the potential for export of food products, both of which could assist in reducing the national debt. Another group could believe that the country should not try to grow food that can be imported from other countries where production costs are lower, and that water should be allocated only to those users, perhaps cities or industries, which can afford to pay a higher price for the water. For each of these groups, the provision of jobs and the support of economic development is important, but the difference is over the allocation of the benefits to rural and urban areas, and among agriculture, industry and service sectors.

Fourth, difficulties may arise because of *personality conflicts*, or *historical circumstances*. Thus, one group may be angry or bitter that at some previous time one of the other groups blocked its attempts to achieve something very important to it. As a result, regarding the current situation, there may be determination to use this occasion to "get even" or to "pay back" the other side. Or, there may be the belief, based on perception or fact, that on a previous occasion another side was dishonest, or at least withheld some key information. Consequently, there may be unwillingness to trust or respect the other side, based on the earlier experience. On that basis, there will be reluctance to reach agreement for fear that yet again it will later turn out that the other side had not been open or honest.

To repeat an observation made near the beginning of this section, it is normal for more than one of these four causes to co-exist in a dispute. Analysts should have the ability to identify which causes are present, and their relative significance. With such a diagnosis, it should be possible to begin to determine how they might be handled as part of the process of resolving or minimizing the conflict.

A final point should also be made. Conflict or disputes are not necessarily bad or undesirable. They can help to identify when a process or procedure is not working effectively, and can remind analysts and managers that there are legitimately different values and views. The challenge is to ensure that conflicts become constructive rather than destructive factors in planning for, and making decisions about, the environment and resources.

1.8 Implications

The purpose of this chapter has been to indicate the pervasiveness of change, complexity, uncertainty and conflict for resource and environmental management. The experience with arsenic contamination of groundwater in west Bangladesh and east India illustrates that resource and environmental management often cannot be isolated from broader issues (such as interests and policies of donor agencies). The discussion of environmental change and violent conflict further highlights the important roles that resource scarcity and environmental degradation can have as triggers for other conflicts, including warfare among nation states.

The concept of *political ecology* encourages us to take a broad perspective on resource issues, and to ensure that we understand their context before developing solutions. In that regard, resource and environmental management is not only a technical exercise. It requires people who can bridge scientific and civic concerns. Such a perspective also highlights the idea that an "expert" in resource and environmental management is not determined only by formal education. Experiential understanding often can be as, or more, significant than insight based on scientific theories and methods.

In the remaining chapters, the themes of change, complexity, uncertainty and conflict provide common threads. If we can improve our capacity to deal with these matters, the likelihood is great that we can improve our capacity for planning and management.

References and further reading

Ali A M S 1995 Population pressure, environmental constraints and agricultural change in Bangladesh: examples from three agroecosystems. *Agriculture, Ecosystems and Environment* 55: 95–109

Ali M J Olley and A Cotton 1998 Agents of change: the case of Karachi City's waste management. *Third World Planning Review* 20: 255–66

Amery H A and A T Wolf (eds) 2000 *Water in the Middle East: A Geography of Peace.* Austin, University of Texas Press

Asia Arsenic Network, Research Group for Applied Geology, and Department of Occupational and Environmental Health, National Institute of Preventive and Social Medicine, Bangladesh 1999 *Arsenic Contamination of Groundwater in Bangladesh: Interim Report of the Research at Samta Village.* Dhaka, National Institute of Preventive and Social Medicine, April, 90 pp.

Asmal K 1998 Water as a metaphor for governance: issues in water resources management in South Africa. *Nature and Resources* 34: 19–25

Atlas R M 1999 Combating the threat of biowarfare and bioterrorism. *BioScience* 49: 465–77

Ayres E 2000 The four spikes. *Futures* 32: 539–54

Baland J-M and J-P Platteau 1999 The ambiguous impact of inequality on local resource management. *World Development* 27: 773–88

Barbier E B and T F Homer-Dixon 1999 Resource scarcity and innovation: can poor countries attain endogenous growth? *Ambio* 28: 144–7

Barrett G W and E P Odum 2000 The twenty-first century: the world at carrying capacity. *BioScience* 50: 363–8

Bauer C H 1997 Bringing water markets down to Earth: the political economy of water rights in Chile, 1976–95. *World Development* 25: 639–56

Beatley T 2000 Preserving biodiversity: challenges for planners. *Journal of the American Planning Association* 66: 5–20

Bennett P 1999 Governing environmental risk: regulation, insurance and moral economy. *Progress in Human Geography* 23: 189–208

Bennett V 1998 Housewives, urban protest and water policy in Monterrey, Mexico. *International Journal of Water Resources Development* 14: 481–97

Black R 1990 "Regional political ecology" in theory and practice: a case study from northern Portugal. *Transactions of the Institute of British Geographers* New Series 15: 35–47

Black R and M F Sessay 1998 Refugees and environmental change in West Africa: the role of institutions. *Journal of International Development* 10: 699–713

Blaikie P 1995 Changing environments or changing views? A political ecology for developing countries. *Geography* 80: 203–14

Blanchard K 2000 Rachel Carson and the human dimensions of fish and wildlife management. *Human Dimensions of Wildlife* 5: 52–66

Brickner W H and D M Cope 1977 *The Planning Process.* Cambridge, Mass., Winthrop Publishers

Bridge G and P McManus 2000 Sticks and stones: environmental narratives and discursive regulation in the forestry and mining sectors. *Antipode* 32: 10–47

Brown G and C C Harris 2000 The US Forest Service: whither the new resource management paradigm. *Journal of Environmental Management* 8: 1–19

Bryant R L 1992 Political ecology: an emerging research agenda in Third World studies. *Political Geography* 11: 12–36

Bryant R L 1997 *The Political Ecology of Forestry in Burma, 1824–1994.* London, C. Hurst & Company

Bryant R L and S Bailey 1997 *Third World Political Ecology.* London and New York, Routledge

Bryant R L and G A Wilson 1998 Rethinking environmental management. *Progress in Human Geography* 22: 321–34

Camacho D E (ed.) 1998 *Environmental Injustices, Political Struggles: Race, Class, and the Environment.* Durham, North Carolina Duke University Press

Campbell D J, H Gichohi, A Mwangi and L Chege 2000 Land use conflict in Kajiado District, Kenya. *Land Use Policy* 17: 337–48

Christensen K S 1985 Coping with uncertainty in planning. *Journal of the American Planning Association* 51: 63–73

Cohen J and I Stewart 1994 *The Collapse of Chaos: Discovering Simplicity in a Complex World.* New York, Penguin Books

Colyvan M, M A Burgman, C R Todd, H R Akçakaya and C Bock 1999 The treatment of uncertainty and the structure of the IUCN threatened species categories. *Biological Conservation* 89: 245–9

Conte C A 1999 Colonial science and ecological change: Tanzania's Mlalo basin, 1888–1946. *Environmental History* 4: 220–44

Cramer J C and R P Cheney 2000 Lost in the ozone: population growth and ozone in California. *Population and Environment* 21: 315–37

Davidson C 2000 Economic growth and the environment: alternatives to the limits paradigm. *BioScience* 50: 433–40

de Loë R 1999 Dam the news: newspapers and water management in Alberta, 1972 to 1992. *Journal of Environmental Management* 55: 219–37

de Loë R and R Kreutzwiser 2000 Climate variability, climate change and water resource management in the Great Lakes. *Climatic Change* 45: 163–79

Denman B and A Ferguson 1995 Human rights, environment and development: the dispossession of fishing communities on Lake Malawi. *Human Ecology* 23: 125–42

Desai U 1998 *Ecological Policy and Politics in Developing Countries.* Albany, State University of New York Press

Diekmann A and A Franzen 1999 The wealth of nations and environmental concern. *Environment and Behavior* 31: 540–9

Döös B R 1994 Environmental degradation, global food production, and risk for large-scale migrations. *Ambio* 23: 124–30

Dorcey A H J 1986 *Bargaining in the Governance of Pacific Coastal Resources: Research and Reform.* Vancouver, B.C., University of British Columbia, Westwater Research Centre

Duffey R 2000 Shadow players: ecotourism development, corruption and state politics in Belize. *Third World Quarterly* 21: 549–65

Ehrlich P R, G Wolff, G C Daily, J B Hughes, S Daily, M Dalton and C Goulder 1999 Knowledge and the environment. *Ecological Economics* 30: 267–84

Farina A 2000 The cultural landscape as a model for the integration of ecology and economics. *BioScience* 50: 313–20

Flynn J E Peters, C K Mertz and P Slovic 1998 Risk, media and stigma at Rocky Flats. *Risk Analysis* 18: 715–27

Gandy M 1999 Rethinking the ecological leviathan: environmental regulation in an age of risk. *Global Environmental Change* 9: 59–69

Gibson R B (ed.) 1999 *Voluntary Initiatives: The New Politics of Corporate Greening.* Peterborough, Ontario, Broadview Press

Gleick P 1987 *Chaos: Making a New Science.* New York, Penguin Books

Gleick P 1991 Environment and security: the clear connections. *Bulletin of the Atomic Scientists* 47: 17–21

Hadfield L and R A F Seaton 1999 A co-evolutionary model of change in environmental management. *Futures* 31: 577–92

Hansis R 1998 A political ecology of picking: non-timber forest products in the Pacific Northwest. *Human Ecology* 26: 67–86

Hefner R 1990 *The Political Economy of Mountain Java: An Interpretive History.* Berkeley and Los Angeles, University of California Press

Hirsch P and C Warren (eds) 1998 *The Politics of Environment in Southeast Asia: Resources and Resistance.* London, Routledge

Homer-Dixon T F 1991 On the threshold: environmental changes as causes of acute conflict. *International Security* 16: 76–116

Homer-Dixon T F 1994 Environmental scarcities and violent conflict: evidence from cases. *International Security* 19: 5–40

Homer-Dixon T F 1999 *Environment, Scarcity, and Violence.* Princeton, NJ, Princeton University Press

Homer-Dixon T F, J H Boutwell and G W Rathjens 1993 Environmental change and violent conflict. *Scientific American* 268: 38–45

Hunter L M 2000 A comparison of the environmental attitudes, concern, and behaviors of native-born and foreign-born U.S. residents. *Population and Environment* 21: 565–80

INFOTERRA (United Nations Environment Programme) and the Institute of Development Studies 1998 *Guide to Environment and Development Sources of Information on CD-ROM and the Internet.* Nairobi, Kenya, United Nations Environment Programme, and Sussex, University of Sussex, Institute of Development Studies

International Committee of the Red Cross 1999 War and Water. *Forum.* Geneva, International Committee of the Red Cross, 112 pp.

International Geosphere–Biosphere Programme 1992 *Global Change: Reducing Uncertainties.* Stockholm, The Royal Swedish Academy of Sciences

Jasanoff S 1999 The songlines of risk. *Environmental Values* 8: 135–42

Jim C Y 2000 Environmental changes associated with mass urban tourism and nature tourism development in Hong Kong. *The Environmentalist* 20: 233–47

Johnston B R 1995 Human rights and the environment. *Human Ecology* 23: 111–23

Kassar M 1999 Rescuing drylands: a project for the world. *Futures* 6: 949–58

Khan N A 1998 *A Political Economy of Forest Resource Use: Case Studies of Social Forestry in Bangladesh.* Aldershot, Ashgate

Khoe G H and M T Emett 1999 *An Interim Report on Field Demonstration of Arsenic Removal Process in Sonargaon Thana*. Sydney, Australia Nuclear Science & Technology Organization and CRC for Waste Management & Pollution Control, 14 pp.

Kwiatowski R E 1998 The role of risk assessment and risk management in environmental assessment. *Environmetrics* 9: 587–98

Lado C 1999 Environmental resources, population and sustainability: evidence from Zimbabwe. *Singapore Journal of Tropical Geography* 20: 148–68

Lawrence W F 1999 Reflections on the tropical deforestation crisis. *Biological Conservation* 91: 109–17

Liddle B 2000 Population growth, age structure, and environmental impact. *Population and Environment* 21: 385–411

Lonergan S C (ed.) 1999 *Environmental Change, Adaptation and Security*. Amsterdam, Kluwer

Lowenthal D 2000 Nature and morality from George Perkins Marsh to the millennium. *Journal of Historical Geography* 26: 3–27

Mabogunje A L 1995 The environmental challenges in Sub-Saharan Africa. *Environment* 37: 4–9, 31–5

Mackey B G 1999 Environmental scientists, advocacy, and the future of Earth. *Environmental Conservation* 26: 245–9

Malhotra K 1999 The political economy of natural resource conflict in the Lower Mekong sub-region. *Development* 42: 20–6

Mehata L, M Leach, P Newell, I Scoones, K Sivaramakrishan and S A Way 1999 *Exploring Understandings of Institutions and Uncertainty: New Directions in Natural Resource Management*. Sussex, Institute of Development Studies

Meine C 1999 It's about time: conservation biology and history. *Conservation Biology* 13: 1–3

Memon P A and H Perkins (eds) 2000 *Environmental Planning and Management in New Zealand*. Palmerston North, Dunmore Press

Miles E L 1998 Personal reflections on an unfinished journey through global environmental problems of long timescale. *Policy Sciences* 31: 1–33

Munn T, A Whyte and P Timmerman 1999 Emerging environmental issues: a global perspective of SCOPE. *Ambio* 28: 464–71

Myers N 1993 *Ultimate Security: The Environmental Basis for Political Stability*. New York, W W Norton

Myers N 1999 Environmental scientists: advocates as well? *Environmental Conservation* 26: 163–5

Naidu R and H C W Skinner 1999 Arsenic poisoning of rural groundwater supplies in Bangladesh and India: implications for soil quality, animal and human health. *Proceedings of the International Conference on Diffuse Pollution*. Wembley, Western Australia, CSIRO Land and Water, pp. 407–18

Norberg-Bohm V 2000 Creating incentives for environmentally enhancing technological change: lessons from 30 years of U.S. energy technology policy. *Technological Forecasting and Social Change* 65: 125–48

Ojima D S, K A Galvin and B L Turner 1995 The global impact of land-use change. *BioScience* 44: 300–4

Olive C A 1998 *Land Use Change and Sustainable Development in Segara Anakan, Java, Indonesia: Interactions among Society, Environment and Development*. Department of Geography Publication Series No 51, Waterloo, Ontario, University of Waterloo

Omara-Ojungu P 1992 *Resource Management in Developing Countries*. Harlow, Longman

O'Riordan T (ed.) 2000 *Environmental Science for Environmental Management*. Second edition. Harlow, Pearson Education

Paul B K and S De 2000 Arsenic poisoning in Bangladesh: a geographic analysis. *Journal of the American Water Resources Association* 36: 799–809

Pelling M 1999 The political ecology of flood hazard in Guyana. *Geoforum* 30: 249–61

Peluso N L 1992 *Rich Forests, Poor People: Resource Control and Resistance in Java.* Berkeley, University of California Press

Peluso N L 1996 Fruit trees and family trees in an anthropogenic forest: ethics of access, property zones, and environmental change in Indonesia. *Comparative Studies in Society and History* 38: 510–48

Peterson G 2000 Political ecology and ecological resilience: an integration of human and ecological dynamics. *Ecological Economics* 35: 323–36

Pleune R 1997 The importance of contexts in strategies of environmental organizations with regard to climate change. *Environmental Management* 21: 733–45

Powell M M 1996 Historical geography and environmental history: an Australian interface. *Journal of Historical Geography* 22: 253–73

Preston-White R 1995 The politics of ecology: dredge-mining in South Africa. *Environmental Conservation* 22: 151–6

Rahman M H 1997 Arsenic hazards in Bangladesh. *Think of Arsenic Contamination Problems in Asia.* Tokyo, The Association for Geological Collaboration in Japan (Chigaku Dantai Kenkyukai), pp. 74–83

Rogers R A and C J A Wilkinson 2000 Policies of extinction: the life and death of Canada's endangered species legislation. *Policy Studies Journal* 28: 190–205

Rowlands D 1999 IMF resource management in a world of mobile capital. *Canadian Journal of Development Studies* 20: 465–89

Rowlands I H 2000 Beauty and the beast?: BP's and Exxon's positions on global climate change. *Environment and Planning C* 18: 339–54

Sanders R 1999 The political economy of Chinese environmental protection: lessons of the Mao and Deng years. *Third World Quarterly* 20: 1202–14

Schroeder R A 1993 Shady practice: gender and political ecology of resource stabilization in Gambian garden/orchards. *Economic Geography* 69: 349–65

Schroeder R A 1999 Geographies of environmental intervention in Africa. *Progress in Human Geography* 23: 359–78

Shindler B 2000 Landscape-level management: it's all about context. *Journal of Forestry* 98(12): 10–14

Sick D 1998 Property, power, and the political economy of farming households in Costa Rica. *Human Ecology* 26: 189–212

Siddiqi T A 2000 The Asian financial crisis – is it good for the global environment? *Global Environmental Change* 10: 1–7

Simmons C and N Chambers 1998 Footprinting UK households: how big is your ecological garden? *Local Environment* 3: 355–62

Singh S P 1998 Chronic disturbance, a principal cause of environmental degradation in developing countries. *Environmental Conservation* 25: 1–2

Slocombe D S 2000 Resources, people and places: resource and environmental geography in Canada 1996–2000. *Canadian Geographer* 44: 56–66

Smith A H, E O Lingus and M Rahman 2000 Contamination of drinking water by arsenic in Bangladesh: a public health emergency. *Bulletin of the World Health Organization* 78: 1093–103

Swaffield S 1998 Contextual meanings in policy discourse: a case study of language use concerning resource policy in the New Zealand high country. *Policy Sciences* 31: 199–224

Trettin L and C Musham 2000 Is trust a realistic goal of environmental risk communication? *Environment and Behavior* 32: 410–26

van Vuuren D P and E N W Smeets 2000 Ecological footprints of Benin, Bhutan, Costa Rica and the Netherlands. *Ecological Economics* 34: 115–30

Vandergeest P, M Flaherty and P Miller 1999 A political ecology of shrimp aquaculture in Thailand. *Rural Sociology* 64: 573–96

Wackernagel M, L Lean and C Borgström Hansson 1999 Evaluating the use of natural capital with the ecological footprint. *Ambio* 28: 604–12

Wackernagel M, L Onisto, P Bello, A Callejas Linares, I S López Falfan, J Méndez Garzia, A I Suárez Guerrero and M G Suárez Guerrero 1999 National natural accounting with the ecological footprint concept. *Ecological Economics* 29: 375–90

Waldrop M M 1992 *Complexity: The Emerging Science at the Edge of Order and Chaos*. New York, Simon and Schuster

Welch A H, D B Westjohn, D R Helsel and R B Wanty 2000 Aresenic in ground water of the USA: occurrence and geochemistry. *Ground Water* 38: 589–604

Westing A H 1994 Population, desertification, and migration. *Environmental Conservation* 21: 110–14 and 109

Wilhite D A, M J Hayes, C Knuston and K H Smith 2000 Planning for drought: moving from crisis to risk management. *Journal of the American Water Resources Association* 36: 697–710

Wilson G A and R L Bryant 1997 *Environmental Management: New Directions for the Twenty-first Century*. London, UCL Press

Wynne B 1992 Uncertainty and environmental learning: reconceiving science and policy in the preventative paradigm. *Global Environmental Change* 2: 111–27

Yohe G 2000 Assessing the role of adaptation in evaluating vulnerability to climate change. *Climatic Change* 46: 371–90

Yokota H, K Tanabe, *et al.* 1997 The largest arsenic pollution of groundwater in Ganges Delta: the report of Samta Village. *Think of Arsenic Contamination Problems in Asia*. Tokyo, The Association for Geological Collaboration in Japan (Chigaku Dantai Kenkyukai), pp. 95–104

Zimmerer K S 2000 The reworking of conservation geographies: nonequilibrium landscapes and nature-society hybrids. *Annals of the Association of American Geographers* 90: 356–69

Zimmermann E W 1933, revised 1951 *World Resources and Industries*. New York, Harper

Ziran Z 1999 Natural resource planning, management and sustainable use in China. *Resources Policy* 25: 211–20

Chapter 2

Turbulence and Planning

2.1 Introduction

In Chapter 1, an argument was made that many environmental and resource management situations are characterized by complexity and uncertainty. Indeed, many environmental and resource problems have been characterized as "wicked", "messes" or "metaproblems", to reflect the high degree of complexity involved. Furthermore, the presence of risk, uncertainty and ignorance ensures that solutions will usually be developed on the basis of incomplete knowledge and understanding of natural and social systems. And yet, life cannot stop until complete understanding is available. If decisions were only to be made after complete understanding were achieved, few decisions would ever be taken. We must be willing to take decisions, accepting that we will make errors. However, acknowledging the presence of complexity and uncertainty may allow us to approach decision making somewhat differently than in the past. It may also lead us to focus and structure our planning and decision-making organizations differently.

In this chapter, consideration is given to the importance of *turbulence* with regard to complexity and uncertainty. Consideration also is given to the concept of *chaos*. Attention is then given to the idea of the *precautionary principle*, as well as the concepts of *hedging* and *flexing*. Then, *alternative approaches to planning* are examined, with particular regard to their assumptions about the best way to address complexity and uncertainty.

2.2 Turbulence

Turbulent conditions seem to be more the norm than the exception. For example, at the beginning of the 1990s, who would have predicted that the

---

Box 2.1 Changing conditions

To be in the midst of change yet oblivious to it is characteristic of the human condition.

Source: Kates, 1995: 627.

Box 2.2 Future shock

. . . I coined the term "future shock" to describe the shattering stress and disorientation that we induce in individuals by subjecting them to too much change in too short a time.

Source: Toffler, 1971: 3.

---

former USSR would break up as a country? Who would have forecast that Nelson Mandela would become the leader of South Africa? While in hindsight such events can be understood, ahead of time or while they are unfolding, they emerge as surprises for many people. They were not predicted much in advance. As a result, turbulence can create bewilderment, anxiety and even suspicion.

Turbulence occurs due to complexity and uncertainty, and due to rapidly changing conditions and associated conflicting interests and positions. In this context, Trist (1980) suggested that, over time, four different kinds of planning or decision-making situations can be identified. Trist's analysis reminds us of the necessity to understand the circumstances or contexts which surround planning and management initiatives.

Type 1: Type 1 is the simplest decision-making environment. It is one in which opportunities and problems are randomly distributed and organization is simple. The best strategy involves each person doing the best as possible on a local scale. A market analogue would be a situation of "perfect competition", represented by a world of small factories, family farms and corner stores. The planning style is *inactive*, or one in which the present is judged to be better than the past or the future. The conventional wisdom is captured by expressions such as it is better to wait and see, look before you leap, or let sleeping dogs lie. This is primarily a non-planning world.

Type 2: In the second decision-making environment, conditions continue to be relatively stable or unchanging. However, opportunities and problems start to group or cluster. The best strategy is to find the optimal location to take advantage of opportunities. While the market analogue for Type 1 was "perfect competition", for Type 2 it is "imperfect competition" with businesses becoming more specialized as they seek a comparative advantage. Here, the planning style is *reactive*, with the past regarded as preferable to the present or the future. In this environment, folk wisdom stresses the good old days, or paradise lost. Planning is used here, as there is a best place to find.

Type 3: At this stage, one which Trist called "disturbed reactive conditions", the environment is dynamic and changing. Other firms or competitors seek and attempt to get the same optimal location. Firms are continually reacting to initiatives of other firms. The best strategy is to amass power, as competitive challenges can best be met through accumulated resources and abundant expertise under the firm's control. Bigger is better. The best market analogy here is "oligopoly" in which there are relatively few but large and specialized firms.

This is the world of large industrial conglomerates or consortia, and of equally large government departments, in which emphasis is placed upon centralized control. In this situation, the future is viewed as better than either the past or the present. Consequently, emphasis is on improving the capacity to predict and to prepare. The art of the calculable becomes valued, with emphasis on econometric models, technical forecasts, operations research and simulations. As in the second type, planning is important. Attention is given to judging where the best waves are going to come from, and then to ensure that you ride the best one and keep others off it. The emphasis is upon technocratic planning.

Type 4: Trist called the fourth type a "turbulent" environment or field. In this type, many large and competing organizations act independently and take actions in diverse directions. The outcome is unanticipated and discordant consequences shared by all. As Trist (1980: 117) remarked, "the result is contextual commotion – as if 'the ground' were moving as well as the organizations. This is what is meant by turbulence". No organization is so big that it can control things, nor is able to do things on its own. A new value becomes collaboration, reflected through consortia-building and sharing, all of which conflict with the basic values found in Type 3 approaches. For Type 4 there is no ready market analogy, other than the concept of *macroregulation*, or increasing involvement by the government in a collaborative manner with the private sector to try and achieve greater stability.

The planning style therefore is best described as *interactive* (Box 2.3). Neither the past, present or future are particularly attractive. Any desirable future will depend on people making it occur, thereby emphasizing the need to make choices among alternative futures. As a result, the interest is less upon forecasting most likely futures, but in creating most desirable futures (a distinction discussed more in Chapters 3 and 4). However, creating a desirable future cannot be done either alone, in isolation from, or in competition with, others. This is why an interactive and collaborative approach to planning and management is so necessary.

Having identified the four types of planning or management situations, Trist then examined the implications, which are several. First, the Type 4 situation requires a different approach to planning and problem solving. Trial-and-error approaches work well in Type 1 conditions. What might be called "craft knowledge" is appropriate in Type 2 situations; the scientific method has most usually been applied in Type 3 circumstances. But the complex

---

Box 2.3 Interactive planning

It [interactive planning] requires the collaboration of interest groups . . . ; the identification of shared values . . . ; continuous learning . . . ; and continuous evaluation and modification. . . . It is an open-ended unfolding process. [It has been called] adaptive planning.

Source: Trist, 1980: 119.

Box 2.4 Institutional gaps

The objective of sustainable development and the integrated nature of the global environment/development challenges raise problems for institutions, national and international, that were established on the basis of narrow preoccupations and compartmentalized concerns. Governments' general response to the speed and scale of global changes has been a reluctance to recognize sufficiently the need to change themselves. The challenges are both interdependent and integrated, requiring comprehensive approaches and popular participation.

Yet most of the institutions facing those challenges tend to be independent, fragmented, working to relatively narrow mandates with closed decision processes. Those responsible for managing natural resources and protecting the environment are institutionally separated from those responsible for managing the economy. The real world of interlocked economic and ecological systems will not change; the policies and institutions concerned must.

Source: World Commission on Environment and Development, 1987: 9.

---

interdependencies of Type 4 situations are usually not amenable to a scientific approach, when that is interpreted as breaking a problem into its component parts. The complex "metaproblems" require a systems approach in which attention is given to relating parts to the whole, and organizations to their larger environments. In Trist's (1980: 119) words, the systems approach is "synthetic, holistic, seeking to capture the Gestalt of system connectedness".

A second implication relates to the nature of public agencies, or the "bureaucracy". The dominant form of bureaucracy in most countries is one which is technocratic, singular (pursues only it own objectives), specialized and centralized (see the comments from the Brundtland Commission in Box 2.4). This approach is best suited to deal with Type 3 conditions. However, as Trist observed, most environmental and resource problems involve Type 4 conditions. The outcome is that we too often rely upon a bureaucracy designed for conditions that no longer exist. And, as we know from military history, being prepared for conditions which no longer exist can be disastrous. For example, after World War I the French built the Maginot Line, an elaborate defensive system of trenches and bunkers, to protect themselves from the Germans. The Maginot Line was based on fighting styles in World War I, when armies occupied trenches opposite from each other, and ground troops skirmished and attacked from the trenches. However, early in World War II, during 1939 and 1940 the Germans used the mobility of their Panzer tanks to sweep through Belgium and into France, and simply outflanked the Maginot Line, making it an anachronism. Resource and environmental management should not be like the Maginot Line – we should be preparing for conditions in the future, not those from the past which may no longer be as relevant.

A third consequence of the Type 4 situation is that the style of comprehensive planning which involves creation of a master plan or blueprint, and which often is very useful in Type 3 situations, will be less appropriate, if appropriate

Guest Statement

Change, turbulence and uncertainty in Nigeria

Peter Adeniyi, Nigeria

Nigeria, located between latitude 4° N to 14° N and longitude 3° E to 15° E, covers a total area of about 923,769 km². The country is characterized by diverse and complex biophysical resources, ethnic groups numbering over 400, and varying socio-economic regimes. Its current estimated population is about 110 million, with urban areas accounting for about 42%. The density of population ranges from 25 to over 2,000 person/km².

Nigeria is generally considered to be endowed with abundant human and material resources. At independence in 1960, agriculture, which accounted for more than 50% of gross domestic product (GDP) and over 75% of export earnings, was the most important sector of the Nigerian economy. The discovery of crude oil in the late 1960s and its exploitation in the early 1970s changed the situation, and Nigeria moved from a position of food self-sufficiency to one of heavy dependence on importation of food and manufactured goods. The importation of food items, for instance, increased from US$353.9 million in 1995 to US$1.6 billion in 1998. Yet only 34 million hectares of the 71.2 million hectares of cultivable land are under cultivation. The cost of imported manufactured goods increased from US$529.4 million in 1995 to US$2.2 billion in 1998.

Inspite of the huge income derived from petroleum resources, which now accounts for over 90% of the total annual national revenue, Nigeria's development has largely remained stunted. Most of the over 100,000 communities in the country lacked basic physical and social infrastructures. In 1992, only about 49% of the communities were served with potable water, while 31, 12, 8, 4 and 2% of the communities were served with educational, health, electricity, communications and banking facilities, respectively. This situation is further compounded by braindrain (i.e. the migration of many capable Nigerians to other countries) and the unprecedented depletion, destruction and degradation of the environmental resource base, leading to increased scarcity; and by unequal access to renewable resources by the rural majority, increasing poverty and frequent violent ethnic conflicts.

The major reasons for this state of affairs include:

(1) Lack of vision and national consensus regarding the role of resource and environmental management.

(2) Political instability. Within 40 years of independence, government has been changed eleven times, eight by the military which governed for 28 years.

(3) Frequent administrative changes. At independence, Nigeria had three regions which were increased to four in 1963. The four regions were divided into 12 states in 1967, 19 in 1976, 21 in 1978, 30 in 1987 and 36 in 1996. The 330 Local Government Areas in 1978 have now become 774. This proliferation led to poor and weak institutions, premature amalgamation of institutions and scrapping of programmes, as well as the instability of programmes and personnel.

(4) Uncritical adoption of external development policy models and programmes which are often uniformly and mechanically implemented mostly by public officials without due consideration of local environmental conditions, local knowledge systems, and the loose structure and capacity of agencies.

(5) Overbearing power of the Federal Government over resources, lack of power at the lower tiers of government, and inadequate involvement of the private sector, non-governmental organizations (NGOs), community leaders and women.

(6) Poor institutional mandate, lack of coordination and cooperation among public organizations at all levels, and lack of respect for the local people by public officials.

(7) Lack of relevant and up-to-date data on all components of natural resources and the environment and the persistent accumulation of disparate data which are inconsistently named, improperly documented; highly fragmented in different locations, very low in integrity; not organized, and cannot be electronically processed or distributed; and generally are unreliable, incomplete and very difficult to identify, access, use and maintain.

(8) Lack of accountability, transparency, proper evaluation of programmes' outcomes and opportunity to learn from past mistakes.

The present civilian administration in Nigeria seems to be aware of the above pitfalls and the national sustainability imperatives of finding lasting solutions to them. The concepts, principles and practices of resource and environmental management, noted in this book, provide a solid foundation for the resolution of these challenging problems. Above all, however, is the need to recognize the significance of the problems, to mount a sustained commitment to addressing them, and to take a long-term perspective by not only the academic community, but also by all those who formulate development policies and those who take development decisions in Nigeria.

Dr Peter Adeniyi is a Professor of Geography at the University of Lagos in Nigeria. He has also been seconded to the federal public service, and has served as National Coordinator for Rural Development Data. A specialist in remote sensing, geographical information systems, and resource appraisal, he has coordinated a collaborative research project in Sokoto State, northwestern Nigeria, with a focus on using remote sensing to improve the capacity to inventory and assess the natural resource potential of that area. He is currently the Managing Director of the University of Lagos Consultancy Services and President of the African Association of Remote Sensing of the Environment.

at all. As Trist argued, the level of change created by an ever-increasing rate of change is too great, and plans too often are outdated before they can be implemented (see Chapter 12 which discusses implementation). Planning for Type 4 conditions requires continuous adaptive planning: *continuous* because frequent modifications will be essential; *participatory* because all stakeholders must have a role (see Chapter 8 about partnerships); *integrated* because various interests must be incorporated; and *coordinated* because the interdependence of issues and decisions must be recognized. A key lesson is that during planning there must

---

Box 2.5 Chaos

. . . chaos is not anarchy or randomness. Chaos is order, but it is order that is "invisible." Nor is chaos merely the result of "noise", or interference, or even insufficient knowledge. What chaos implies is a kind of inherent "uncertainty principle" – not just in how we perceive the world but in how the world actually works.

Source: Cartwright, 1991: 44–5.

---

be equal attention to the *process* as well as to the *product*. Another lesson is that ends need to be agreed upon before means are examined: this is one of the points highlighted in both Chapters 3 and 4. Many of the issues noted above are highlighted in the guest statement by Peter Adeniyi, as he reflects on the experience with resource and environmental management in Nigeria.

2.3 Chaos

If turbulence contributes to complexity and uncertainty, the concept of chaos is even more unsettling. Chaos is order without predictability. In other words, some physical and social systems might be capable of being understood, in the sense that they can be described relative to a set of conditions or rules, but they remain fundamentally unpredictable.

Cartwright (1991) concluded that many of the problems that planners must handle – population growth, land use patterns – may well reflect chaos. In his view, while we may understand the "rules" that influence behaviour at the individual or "local" level, we usually cannot predict the outcomes or impacts at a global scale. The implications of such a conclusion are profound. As Cartwright (1991: 45) remarked, "No matter how much data we gather, no matter how global and complete our models, no matter how rigorously we test them, even so, according to chaos theory, prediction may in some cases be beyond our grasp".

As will be seen later in Section 2.6 on planning models, planners have developed approaches that recognize situations in which there is incomplete understanding. However, these approaches have emerged for pragmatic reasons: there is not always sufficient time to get all the facts; it is often too costly to obtain all the facts; we lack the skill or means to obtain all the facts. In contrast, chaos theory indicates that trying to predict the future is not just impractical in some cases – rather, it is logically impossible.

More attention will be given to predictions or forecasting in Chapter 3. Nevertheless, three conclusions can be reached if it is accepted that chaotic systems are inherently unpredictable and are impossible to understand fully (Cartwright, 1991: 53–4). First, collecting more information and creating more sophisticated models for chaotic systems will likely have little value. Indeed, "research" might be unhelpful if it creates an unrealistic belief about

> ## Box 2.6
>
> While the prediction of chaotic behavior may be impossible, understanding the order that gives rise to it may not be as difficult as we thought. Highly complex and unpredictable behavior, in other words, can be the product of quite simple and accessible rules.
>
> *Source*: Cartwright, 1991: 45.

the capability of planning. Planning strategies that assume the capacity for foresight will be inappropriate and can even be misleading at times.

Alternatively, planners must begin to work with not one or two forecasts about the future, but rather should use an ensemble or suite of forecasts. Planners have normally tested models or forecasts by using different assumptions about future conditions and considering the consequences for the results. It is much less common for planners to begin, or run, their alternative models from different "starting points". It is normally assumed that we know what the present conditions are, and we do not consider alternative perspectives about them. Chaos theory challenges the notion that present conditions are known, and also the idea that even if different starting points were chosen there were would unlikely be major differences in outcomes. As a result, it becomes very important to be willing and able to explore which initial starting points have similar trajectories and which do not. Rather than investing more time and effort in collecting more detailed data, or fine tuning our models, we should look for patterns of behaviour, or points to which systems tend to return, even if we cannot readily predict them.

Second, the concept of chaos requires re-examination of some of our ideas regarding the virtues of order and predictability, and the messiness of chaos and disorder. It is possible that a chaotic environment could be as, if not more, desirable than a neat or orderly one.

Third, we may need to accept that chaotic systems are only predictable at a local scale or on an incremental basis. At a global scale or comprehensive basis, chaotic systems are not predictable due to the cumulative effects of many different kinds of feedback within them. However, at a local or incremental level, the consequences of feedback over time may be identifiable. This conclusion becomes a strong argument for using planning strategies or models which are incremental rather than comprehensive, and based on a capacity to be adaptive rather than structured and "blue printable". Thus, as the comments in Box 2.6 indicate, there are grounds to be optimistic and confident regarding handling chaotic systems.

2.4 The precautionary principle

In Germany, starting in the 1950s, there was discussion focused on the need for *Vorsorge*, or foresight. Of particular interest was recognition of the need for

---

Box 2.7 Statements of the precautionary principle

Bergen Declaration (ECE), 16 May 1990

In order to achieve sustainable development, policies must be based on the Precautionary Principle. Environmental measures must anticipate, prevent and attack the causes of environmental degradation. Where there are threats of serious or irreversible damage, lack of full scientific certainty should not be used as a reason for postponing measures to prevent environmental degradation.

"Maastricht Treaty", Treaty on European Union, February 1992, Article 130r, paragraph 2

Community policy on the environment shall . . . be based on the Precautionary Principle and on the principles that precautionary action should be taken, that environmental damage should as a priority be rectified at source and that the polluter should pay.

Rio Declaration, June 1992, Principle 15

In order to protect the environment, the precautionary approach shall be widely applied by States according to their capabilities. Where there are threats of serious or irreversible damage, lack of full scientific certainty shall not be used as a reason for postponing cost-effective measures to prevent environmental degradation.

---

caution, and to move away from always being reactive in planning and management. Various German governments formally incorporated the need for foresight into their procedures, and this later became known as the *precautionary principle*. This approach soon began to be included in international agreements, especially those regarding marine pollution (Box 2.7). In 1992, Principle Fifteen of the Rio Declaration on Environment and Development from the Earth Summit endorsed the precautionary principle.

The precautionary principle reflects the adage that an ounce of prevention is worth a pound of cure. It also stipulates that rather than waiting for complete understanding, or certainty, managers and decision makers should anticipate potential harmful environmental impacts from actions, and take decisions to avoid such harm. It also recognizes that uncertainty is reality, due to our lack of complete knowledge about ecosystem behaviour; faulty assumptions about ecosystem functions; inability to predict the size, needs and desires of future populations; and difficulty in forecasting future technical innovations. Given such uncertainties, the precautionary principle should be considered when making resource decisions in which (1) the range of possible impacts from one or more uses cannot be predicted, (2) one or more of the outputs or outcomes could have extremely undesirable impacts for future people, and (3) substitutes are not available for the resource to be used.

Young (1993: 15–16) suggested three possible interpretations related to the precautionary principle. These are:

- a conservative interpretation, in which use or activity is approved only if it poses no danger to an ecological system or does not reduce environmental quality, and is confined within boundaries that permit complete reversibility;
- a more liberal interpretation, in which uses or industries judged to be "risky" are required to use the best available technology, and in addition a precautionary safety margin is established which keeps ambient environmental concentrations well below a specified acceptable threshold; and
- a relatively weak interpretation, in which the requirement is the use of best available technology that does not involve undue expense.

These options illustrate that there is as yet no consensus regarding a definition of the precautionary principle. However, this situation is not necessarily bad, as it does provide scope to custom design its application relative to local needs, conditions and circumstances. Young (1993: 14, 16) provided sensible advice regarding such questions. In his view, the principle should be used with actions involving adverse outcomes, particularly when such outcomes are suspected to be of a catastrophic kind. In this manner, its use should force debate about the types and magnitudes of human-induced change to the environment judged to be acceptable by a society. In addition, he commented that application of the principle should never mean that all developments with uncertain environmental impacts should be stopped. What it does mean, however, is that when a possibly irreversible action that could have unpredictable consequences in the future is being contemplated, all alternative options need to be considered before taking a final decision. He also concluded that such irreversible actions should only be undertaken when it is concluded that not doing so could impose significant social costs on the present generation.

Nevertheless, as Bodansky (1991: 43) cautioned, use of the precautionary principle does not provide a guarantee that serious environmental harm will never occur (Box 2.8). Many contemporary environmental problems were not anticipated. In addition, even if regulators had opted for a cautious approach, it is likely that many of the problems would not have been prevented. For example, chlorofluorocarbons (CFCs) and dichlorodiphenyltrichloroethane (DDT) were judged to be environmentally benign when they were first developed. Problems emerged later not because the regulators had approved their use in the context of uncertainties, but because scientists had not tested for the types of environmental impacts that later became a problem. Thus, extensive tests had been

---

Box 2.8 Concerns

Although the precautionary principle provides a general approach to environmental issues, it is too vague to serve as a regulatory standard because it does not specify how much caution should be taken. In particular, it does not directly address two key questions: When is it appropriate to apply the precautionary principle? And what types of precautionary actions are warranted and at what price?

Source: Bodansky, 1991: 5.

conducted on DDT regarding its acute toxicity, but no tests had been completed regarding its chronic effects – and it was the latter that caused problems for health. For CFCs, the properties (persistence and stability) that make it destructive to the ozone layer were judged initially to reduce the likelihood that they would cause negative environmental impacts.

While the precautionary principle is not and cannot be a panacea for complexity and uncertainty regarding environmental problems, it does provide a counterbalance to the "wait and see" attitude that often prevails. For, as Smith (1990: 112) has observed, "Scientific uncertainty is a difficult issue because uncertainty means that it is always possible to argue that better policies can be developed by waiting until a broader scientific consensus emerges".

2.5 Hedging and flexing

Hedging and flexing are strategies for decision making under conditions of uncertainty which are so severe that Collingridge (1983) characterized them as a state of "ignorance". Each is considered here.

2.5.1 Hedging

Hedging is one of the most frequently used strategies when having to make decisions under ignorance. Hedging involves a conscious choice to avoid the worst consequences by comparing all the options against the estimated worst case, and then selecting the alternative with the "least bad worst outcome". Hedging is consistent with the precautionary principle, in that it counsels caution and strives to minimize damage if the worst outcome occurs. However, Collingridge argued that while hedging is a reasonable strategy under extreme uncertainty, it is not the best strategy under conditions of ignorance. His position emphasizes the importance of being able to differentiate among conditions of risk, uncertainty and ignorance, as discussed in Chapter 1.

Collingridge concluded that hedging is not a good approach when ignorance occurs because in such situations no decision can ever be known to be the correct one. Thus, there is always the likelihood that other options than the one taken might be better, if only the planner or decision maker were aware of them.

2.5.2 Flexing

Collingridge maintained that because any decision taken in ignorance may prove to be inappropriate, the decision maker should systematically look for error, or *monitor* any decision, by continuing to scan for other options after the decision has been taken. Furthermore, because there is little point in searching for error unless there is scope for making an adjustment, he recommended that decision makers should always favour flexible options, or ones that can be revised if they are found to be inappropriate. His conclusion was that decision makers should be prepared to take flexible decisions, to monitor them and to modify them. The extreme caution stipulated by hedging becomes less important if there is the willingness and capacity to modify decisions in the light of new knowledge and understanding. In contrast to hedging, flexing involves seeking the best option, risking the worst outcome, but being ready to modify or reverse a decision if the worst outcome should occur.

2.5.3 Relative merits

The relative merits of hedging and flexing are:

- Hedging is an appropriate strategy if it is unlikely that a decision or an action can be reversed or significantly modified. However, if decisions can be modified or reversed, then flexing is the more appropriate approach.
- Flexing is not a prescription for inaction. Flexing provides an opportunity to achieve the best possible outcome; it is not a prescription to do nothing and hope for the best. Hedging tends to lead to preservation of the status quo, or slight variations of it, or to premature adoption of new technologies since they are viewed as the way to resolve problems.
- Flexing increases flexibility by broadening the range of options considered, unlike hedging which limits options.
- Hedging may contribute to the realization of the worst outcome because in situations of ignorance decision makers are often involved in competition with other decision makers whose intents or activities are not easy to predict. For example, if one side develops a weapon system to protect itself from what it thinks the other side might do, the development of such a system might trigger the other side to take the action which was feared as it tries to create a hedge for itself. This pattern leads to the well known phenomenon of self-fulfilling predictions.
- In many decision situations, the worst outcome may be ambiguous because a number of "worst" outcomes may be possible. In such situations it is difficult to know which option to hedge against.
- By seeking to avoid the worst outcome, hedging may also overlook real benefits that could be achieved by attaining some happier state. In contrast, flexing seeks out the best option, but recognizes that monitoring is needed so that if the worst does happen then changes can be made to limit damage.

Box 2.9 Approaches to decision making

Most principles of decision-making under uncertainty are simply common sense. We must consider a variety of plausible hypotheses about the world; consider a variety of possible strategies; favor actions that are robust to uncertainties; hedge, favor actions that are informative; probe and experiment; monitor results; update assessments and modify policy accordingly; and favor actions that are reversible.

Source: Ludwig, Hilborn and Walters, 1993: 36.

Neither hedging nor flexing offers perfect solutions for making decisions in the face of complexity, uncertainty or ignorance. However, the review of them is intended to provide some context for the *precautionary principle*, and to encourage you to think about its strengths and weaknesses at both conceptual and operational levels. Most of the criticisms of the precautionary approach have emphasized operational issues. The above comments highlight that conceptual issues also exist, and require our careful attention.

2.6 Planning models

Different schools or models of planning have been developed. Each reflects different values, assumptions and beliefs about the nature of the world for which planning is done, and about the role of the planner. Several models are considered below, with particular regard to the way in which they address complexity and uncertainty (Box 2.10).

2.6.1 Synoptic planning

Synoptic planning, also referred to as *comprehensive rational planning*, is the dominant planning model. Most other models have been developed as a result of concern about, or criticisms of, the synoptic approach.

Synoptic planning has a number of well established steps or phases, including: (1) defining the problem, (2) establishing goals and objectives, (3) identifying alternative means to achieve the goals and objectives, (4) assessing the options against explicit criteria, (5) choosing a preferred solution and implementing it, and (6) monitoring and evaluation. These phases are linked with feedback loops, creating the possibility to incorporate changes into planning as a result of findings or experience. The assessment of options is often completed

---

Box 2.10 Planning under uncertainty

Where planning for the future is *feasible* (based on good data and analytical skills, continuity in the trends being extrapolated, and effective means to control outcomes), then planning is unnecessary – it is simply redundant to what already goes on. Conversely, where planning is most *needed* (where there is absence of data and skills and controls in the presence of primitive or turbulent social conditions), planning is least feasible.

Source: Hudson, 1979: 393.

. . . planning processes can be understood as addressing different conditions of uncertainty. Thus planners must assess the actual conditions of uncertainty that characterize the particular problem they are confronting and then select a style of planning that suits those conditions.

Source: Christensen, 1985: 69.

---

by using methods such as benefit–cost analysis, operations research and fore-casting. Quantitative analysis is often a central element of analysis.

The comprehensive rational planning model is based on the assumption that the people involved have the characteristics of *economic man* or *economic person*. This individual has the capacity to identify and rank goals, values and object-ives, and can also choose consistently among them after having collected all the necessary data and having evaluated them systematically. This individual also judges alternatives against the criterion of economic efficiency, and seeks to optimize or maximize returns.

A further value associated with comprehensive rational planning is that if the planner or analyst collects enough information, completes enough analyses and studies long enough, she or he will be able to understand the situation, and therefore be able to manage or control it. However, the reality is that in many situations the necessary data are not available. Even when such data can be collected, it is not always easy to analyse them due to the existence of intan-gible attributes. Furthermore, the assumption that the correct problem has been identified is not always verified.

As a result of the characteristics outlined above, and highlighted in Box 2.11, the synoptic or comprehensive rational planning model does not seem well suited to deal with complexity and uncertainty. Its departure point is a belief that with enough effort and work it is possible to achieve understanding and, therefore, to be able to manage or control systems The comprehensive rational approach is unlikely to accept the concept of chaos. It also assumes capability of

---

Box 2.11 Problem solving

There are times, of course, when the models one has cannot accommodate the new information. One feels confused. Confusion is painful, and people strive ardently to quiet it, i.e., to make things understandable. Oftentimes, this striving manifests itself as a tenacious persistence to "get things straight." At other times, however, people jump to conclusions without adequately examining the problem. Their discomfort with the uncertainty is such that any solution will do. Moreover, if these efforts fail to bring cognitive closure, people typically respond emotionally with frustration, anger, helplessness, or apathy. Neither jumping to conclusions nor any of these affective states result in very effective problem-solving. [A study] has provided a summary of problem-solving tendencies, which, even if exagger-ated, suggests the real world implications and frustrations of an inadequate prob-lem exploration. They claim: 90% of problem solving is spent:

- Solving the wrong problem
- Stating the problem so that it can be solved
- Solving a solution
- Stating a problem too generally
- Trying to get agreement on the solution before there is agreement on the problem.

Source: Bardwell, 1991: 605.

participants (economic person) that rarely can be met. As a result, other planning models have emerged. The most frequently considered alternative is *incremental planning*.

2.6.2 Incremental planning

The incremental planning approach has also been referred to as *disjointed incrementalism* or as *muddling through*. Instead of accepting the idea of an economic person, this approach is based on the idea that people are "boundedly rational" and who "satisfices" rather than maximize. In other words, while economic man is able to cope with the complexity of the real world, the boundedly rational person quickly simplifies the buzzing, booming confusion that characterizes the real world into a more simple model. In that sense, the world is "bounded" because not all the detail and complexity is considered. In addition, whereas the economic person strives to maximize, the boundedly rational person searches for a solution which is "good enough" or satisfactory – it does not have to be the optimum.

The incremental approach was developed to describe how things often happen in practice, but over time it also has been interpreted in a normative or prescriptive manner. That is, the model is offered as a preferred way of planning, because it captures much of the reality about the world in which planners function. Thus, in the incremental approach the planner is faced with, and recognizes, multiple problems, goals and values. She or he thus does not try to optimize, but rather to identify practically attainable goals that are generally satisfactory. All alternatives are not known, and no attempt is made to consider a broad range of options. The cost in time, effort and money to obtain additional data and to identify a broader mix of alternatives is recognized to be high.

The incremental approach thus has a number of characteristics, including:

- The problem is not clearly defined. Often, the major task for the policy maker or planner is to determine the nature of the problem to be handled.
- Goals, values and objectives may conflict with one another.
- Only a limited number of options are considered, and those differ only incrementally from each other and from the existing policy or practice.
- For each option, only a restricted number of "significant" impacts are identified.
- The problem is redefined on a regular basis. Normally, it is thought that means are adjusted to ends. Under incrementalism, the reverse often happens. Ends are modified with regard to available means.
- No single correct solution exists. Indeed, the policy maker often does not know what is wanted, but does know what should be avoided. Policies move away from bad or undesirable things, without necessarily moving systematically toward good or desirable outcomes. A satisfactory decision is one for which substantial agreement exists, even though not everyone may believe that the decision is the best relative to a given objective.
- The decision or policy process never ends. The process is viewed as a sequential chain, involving an ongoing series of incremental decisions.

With such attributes, the incremental approach concentrates attention upon familiar and better-known experiences, limits the number of alternatives to be explored, and reduces the number and complexity of variables to be considered. The incremental planner may appear to be tentative, timid, indecisive, hesitant, cautious and narrow. At the same time, such a planner also may be viewed as realistic and pragmatic, and as a shrewd problem solver who is astute enough to recognize and accept the complexity and uncertainty surrounding environmental and resource issues. The incremental planner will deliberately choose a policy or option, knowing very well that it is not quite adequate, in order to leave open a range of other options. This path will be taken rather than selecting a policy that appears to be on target, but difficult to modify. Attention often will be given to solving smaller problems (poor seed) when it is recognized that he or she cannot solve a larger problem (low agricultural productivity). The incrementalist believes that policy making is serial or sequential, and as a result that continual nibbling is as good as taking one large bite. In this manner, the incrementalist approach is compatible with the ideas contained in the precautionary principle and in flexing. As Lindblom (1974: II–34), the architect of the incremental approach, has remarked:

> . . . most of us believe that because we became involved in our environmental difficulties piecemeal, we shall have to get out comprehensively. . . . Clearly the argument contains a fallacy. We did fall into our environmental problems through piecemeal gradualism. That still leaves open the possibility that the same route is the only route out of the problem. . . . Believing that everything is connected, we fall into the logical fallacy of believing the only way to improve those interconnections is to deal with them all at once. . . . But because everything *is* connected, it is beyond our capacity to manipulate variables comprehensively. Because everything is interconnected, the whole of the environmental problem is beyond our capacity to control in one unified policy. We have to find critical points of interventions. . . .

As with all models, the incremental approach has weaknesses. A major criticism is that the incrementalist, in believing that an evolutionary approach is best, will not consider an abrupt or radical (or revolutionary) shift in policy or practice if conditions change markedly. Since the incrementalist only considers options which are marginally different from the status quo, such a planner is unlikely to consider innovative ways significantly different from current practice. There may be situations in which a sharp change in policy or practice is needed, and the incrementalist is unlikely to be prepared to consider a radical shift. As a result, incrementalists are often characterized as being reactive to existing conditions, rather than being proactive in trying to move towards an improved state of affairs.

2.6.3 Mixed scanning

Mixed scanning was developed to capture the strengths of each of the synoptic and incremental approaches, but also to minimize their weaknesses. In particular, mixed scanning rejects the concept of economic person, and also rejects the aspect of incrementalism which creates an inability or unwillingness to consider fundamental changes in policy or practice. A basic idea is that many incremental decisions eventually can lead to a fundamental shift, and that the cumulative effect of many incremental decisions is influenced by fundamental decisions.

The core ideas of mixed scanning are that: (1) the decision maker relies mainly upon a continuous series of incremental decisions, but that (2) the decision maker is also steadily scanning a limited range of other alternatives, each of which represents a major departure from present practice. Thus, unlike the incrementalist, the mixed scanner looks for and considers options which are significantly different from the status quo. On the other hand, unlike the economic person who uses synoptic planning, the mixed scanner restricts attention to a limited number of these markedly different options. As a result, the mixed scanning approach is less cautious than the incremental approach, but it remains pragmatic by recognizing the cost and effort required to examine a wide variety of options.

An analogy has been made with the way in which a captain functions on a large ship, especially as it enters a new port. The attention of the captain and others on the bridge is very focused while approaching the harbour, watching for signs of shoals, small ships and other obstacles to the ship's successful entrance to the port. At the same time, the ship's radar is scanning 360 degrees, and provides information about possible storms or other inbound ships that the captain must consider in making decisions about the most suitable route to enter the port. Thus, the ship's captain is continuously scanning for distant but possibly significant aspects that need to be considered, while still keeping most attention focused on a smaller range of matters which require attention in the short-term future.

2.6.4 Transactive planning

The synoptic approach is often associated with the acceptance of expert input, and of a more centralized form of management or decision making. It usually assumes that the planners are in the best position to define the nature of the issues or problems that require attention, and to develop alternative solutions to them. In the *transactive approach*, the belief is that it is important to consider the experience of people who will be affected by the planning or decisions. Thus, planning is not a technocratic exercise conducted by experts, but rather should involve face-to-face interaction between the planners and those most affected by their activity. In the transactive approach, the key characteristics are inter-personal dialogue and mutual learning.

Transactive planning therefore transforms the role of the planner from one of distant expert to one of facilitator and participant. The planner is not assumed to hold all of the necessary or useful knowledge and wisdom, but rather is one person of many with a constructive contribution to make. Transactive planning also seeks to achieve more decentralized decision making as a means to provide more control over planning by the local people. In this manner, the process is consistent with the ideas of social justice, equity and empowerment advocated by sustainable development (discussed in Chapter 4).

As Hudson (1979) has commented, in contrast to incremental planning, transactive planning gives higher priority to processes of personal and organizational development rather than to the realization of particular functional objectives. As a result, plans are assessed not only in terms of what they provide regarding effective delivery of goods and services, but also to the way in which they affect local people, especially their dignity and sense of effectiveness, their values and behaviour, and their capacity for growth through cooperation. Thus, transactive planning is an approach that places a high value on creation of partnerships (Chapter 8) and on incorporating local knowledge systems into planning (Chapter 9).

2.6.5 Perspective

No single planning approach or model is perfect. Each offers strengths and weaknesses. In addition, each makes different assumptions about the role of planners and analysts, and those for whom planning or analysis is being done. It is important to recognize the merits and shortcomings of various approaches, and not to use a single approach automatically. Depending upon circumstances and conditions, one approach may be more appropriate than another. We need to have a critical appreciation of the various approaches or models, so that we can consciously choose which one is likely to be most effective in a given problem-solving situation. It also is helpful to be able to recognize when viewpoints or comments reflect acceptance of one or another model, so that it is possible to identify why people are disagreeing about a process, and to determine how to resolve such differences.

2.7 Implications

Complexity. Uncertainty. Surprise. Turbulence. All of these conditions or characteristics appear to be more the norm than the exception in resource and environmental management. It appears that resource and environmental managers would be well advised to approach their task with the expectation that surprises will be usual and change will be common. As a result, it is sensible to develop strategies and approaches that allow us to learn and adapt on the basis of experience.

The precautionary principle explicitly recognizes that our information and understanding may be inadequate, and yet that situation is not a reasonable excuse for not making decisions or taking action. In that context, when there is the possibility for reversibility of decisions, *flexing* appears to be a better approach than the more often used *hedging*. Furthermore, *mixed scanning* or *incremental* approaches to planning are more likely to be effective than a *comprehensive rational approach*. In the remaining chapters, attention will be given to various ways to deal with complexity and uncertainty, especially in situations characterized by rapid change and conflict. Interest also will centre on ways to move towards resource and environmental management that will be sustainable in the long term, and that will be consistent with an ecosystem approach.

References and further readings

Adger W N 1999 Social vulnerability to climate change and extremes in coastal Vietnam. *World Development* 27: 249–69

Bardwell L V 1991 Problem-framing: a perspective on environmental problem solving. *Environmental Management* 15: 600–12

Blaikie P, T Cannon, I Davis and B Wisner 1994 *At Risk: Natural Hazards, People's Vulnerability and Disasters.* London, Routledge

Bodansky D 1991 Scientific uncertainty and the precautionary principle. *Environment* 33: 4–5, 43–4

Bolton P A, E B Liebow and J L Olson 1993 Community context and uncertainty following a damaging earthquake: low-income Latinos in Los Angeles, California. *Environmental Professional* 15: 240–7

Briassoulis H 1989 Theoretical orientations in environmental planning: an inquiry into alternative approaches. *Environmental Management* 13: 381–92

Bullock C H 1999 Environmental and strategic uncertainty in common property management: the case of Scottish red deer. *Journal of Environmental Planning and Management* 42: 235–52

Burby R J 1991 *Sharing Environmental Risks: How to Control Governments' Losses in Natural Disasters.* Boulder, Colorado, Westview Press

Burton I, R W Kates and G F White 1993 *The Environment as Hazard.* second edition, New York, Guildford Press

Cameron J 1994 The status of the precautionary principle in international environmental law. In T O'Riordan and J Cameron (eds) *Interpreting the Accountancy Principles.* London, Earthscan, pp. 262–91

Cameron J and J Abouchar 1991 The precautionary principle: a fundamental principle of law and policy for the protection of the global environment. *Boston College International and Comparative Law Review* 14: 1–27

Cardinall D and J C Day 1998 Embracing value and uncertainty in environmental management and planning: a heuristic model. *Environments* 25: 110–25

Cartwright T J 1991 Planning and chaos theory. *Journal of the American Planning Association* 57: 44–56

Christensen K S 1985 Coping with uncertainty in planning. *Journal of the American Planning Association* 51: 63–73

Colglazier E W 1991 Scientific uncertainties, public policy, and global warming: how sure is sure enough? *Policy Studies Journal* 19: 61–72

Collingridge D 1983 Hedging and flexing: two ways of choosing under ignorance. *Technological Forecasting and Social Change* 23: 161–72

Collingridge D, A Genus and P James 1994 Inflexibility in the development of North Sea oil. *Technological Forecasting and Social Change* 45: 169–88

Costanza R and L Cornwell 1992 The 4P approach to dealing with scientific uncertainty. *Environments* 34: 12–20, 42

Costanza R, H Daly, C Folke, P Hawken, C S Holling, A J McMichael, D Pimentel and D Rapport 2000 Managing our environmental portfolio. *BioScience* 50: 149–55

Cutter S L (ed.) 1994 *Environmental Risk and Hazards*. Englewood Cliffs, NJ, Prentice Hall

Dery D and I Salomon 1997 "After me, the deluge": uncertainty and water policy for Israel. *International Journal of Water Resources Development* 13: 93–110

Deville A and R Harding 1997 *Applying the Precautionary Principle*. Sydney, Federation Press

Dickson B 1999 The precautionary principle in CITES: a critical assessment. *Natural Resources Journal* 39: 211–28

Döös B R and R Shaw 1999 Can we predict the future food production? A sensitivity analysis. *Global Environmental Management* 9: 261–83

Dovers S R and J W Handmer 1995 Ignorance, the precautionary principle and sustainability. *Ambio* 24: 92–7

Forsyth T 1999 Environmental activism and the construction of risk: implications for NGO alliances. *Journal of International Development* 11: 687–700

Friedmann J 1987 *Planning in the Public Domain: From Knowledge to Action*. Princeton, New Jersey Princeton University Press

Funtowicz S and J Ravetz 1993 Science for the post-normal age. *Futures* 25: 739–57

Funtowicz S O and J R Ravetz 1994 Uncertainty, complexity and post-normal science. *Environmental Toxicology and Chemistry* 13: 1881–5

Gullett W 1997 Environmental protection and the "precautionary principle": a response to scientific uncertainty in environmental management. *Environmental and Planning Law Journal* 14: 52–69

Handmer J and E Penning-Rowsell (eds) 1990 *Hazards and the Communication of Risk*. Brookfield, Vermont, Bower Technical

Harding R and E Fisher (eds) 1999 *Perspectives on the Precautionary Principle*. Sydney, Federation Press

Hewitt K (ed.) 1983 *Interpretations of Calamity*. Boston, Allen & Unwin

Hewitt K 1997 *Regions of Risk: A Geographical Introduction to Disasters*. Harlow, Longman

Holling C S 1996 Surprise for science, resilience for ecosystems. *Ecological Applications* 6: 773–5

Hudson B 1979 Comparison of current planning theories: counterparts and contradictions. *Journal of the American Planning Association* 45: 387–98

Kadvany J 1996 Taming chance: risk and the quantification of uncertainty. *Policy Sciences* 29: 1–27

Kates R W 1995 Labnotes from the Jeremiah experiment: hope for a sustainable transition. *Annals of the Association of American Geographers* 85: 623–40

Lemons J (ed.) 1996 *Scientific Uncertainty and Environmental Problem Solving*. Oxford, Blackwell

Lindblom C 1974 Incrementalism and environmentalism. In *National Conference on Managing the Environment: Final Report*. Washington, DC, Washington Environmental Research Center, II–32–4

Ludwig D, R Hilborn and C Walters 1993 Uncertainty, resource exploitation, and conservation: lessons from history. *Science* 260(2 April): 17, 36

Mackay B G 1999 Environmental scientists, advocacy and the future of the Earth. *Environmental Conservation* 26: 245–9

McKercher B 1999 A chaos approach to tourism. *Tourism Management* 20: 425–34

Mintzberg H 1994 *The Rise and Fall of Strategic Planning: Reconceiving Roles for Planning, Plans, Planners*. New York, Free Press

Morley D 1986 Approaches to planning in turbulent environments. In D Morley and A Shachar (eds) *Planning in Turbulence*. Jerusalem, Magnes Press, pp. 3–23

Myers N 1993 Biodiversity and the precautionary principle. *Ambio* 22: 74–9

Myers N 1999 Environmental scientists: advocates as well? *Environmental Conservation* 26: 163–5

O'Riordan T 1995 The application of the precautionary principle in the United Kingdom. *Environment and Planning A* 2: 1534–8

O'Riordan T 2000 Environmental science on the move. In T O'Riordan (ed.) *Environmental Science for Environmental Management*. second edition, Harlow, Prentice Hall, pp. 1–27

O'Riordan T and J Cameron (eds) 1994 *Interpreting the Precautionary Principle*. London, Cameron and May

O'Riordan T and A Jordan 1995 The precautionary principle in contemporary environmental politics. *Environmental Values* 4: 191–212

Palm R I 1990 *Natural Hazards: An Integrative Framework for Research and Planning*. Baltimore, Johns Hopkins University Press

Parkes G 2000 Precautionary fisheries management: the CCAMLR approach. *Marine Policy* 24: 83–91

Parsons L S, H Powles and M J Comfort 1998 Science in support of fishery management: new approaches for sustainable fisheries. *Ocean and Coastal Management* 39: 151–66

Pteak W J and A A Atkinson 1982 *Natural Hazard Risk Assessment and Public Policy: Anticipating the Unexpected*. New York, Springer-Verlag

Reckhow K H 1994 Importance of scientific uncertainty in decision making. *Environmental Management* 18: 161–6

Ringland G, M Edwards, L Hammond, B Heinzen, A Rendell, O Sparrow and E White 1999 Shocks and paradigm busters (why do we get surprised?) *Long Range Planning* 32: 403–13

Rogers M G, J A Sindenand and T De Lacy 1997 The precautionary principle for environmental management: a defensive-expenditure application. *Journal of Environmental Management* 51: 343–60

Samset K and T Haavaldsen 1999 Uncertainty in development projects. *Canadian Journal of Development Studies* 20: 383–401

Shackley S and B Wynne 1995 Integrating knowledge for climate change: pyramids, nets and uncertainties. *Global Environmental Change* 5: 113–26

Smith D A 1990 The implementation of Canadian policies to protect the ozone layer. In G B Doern (ed.) *Getting It Green: Case Studies in Canadian Environmental Regulation*. Policy Study 12, Toronto, C D Howe Institute, pp. 111–28

Smith K 1996 *Environmental Hazards: Assessing Risk and Reducing Disaster*. second edition, London and New York, Routledge

Stebbing A R D 1992 Environmental capacity and the precautionary principle. *Marine Pollution Bulletin* 24: 287–95

Streets D G and M H Glantz 2000 Exploring the concept of climate surprise. *Global Environmental Change* 10: 97–107

Strömquist L, P Yanda, P Msernwa, C Lindberg and L Simonsson-Forsberg 1999 Utilizing landscape information to analyze and predict environmental change: the extended baseline perspective, two Tanzanian examples. *Ambio* 28: 436–43

Thapa G B and K E Weber 1995 Natural resource degradation in a small watershed in Nepal. Complex causes and remedial measures. *Natural Resources Forum* 19: 285–96

Toffler A 1971 *Future Shock*. New York, Bantam Books

Trist E 1980 The environment and system-response capability. *Futures* 12: 113–27

Trist E 1983 Referent organizations and the development of inter-organizational domains. *Human Relations* 36: 269–84

Wang M S, J K Fang and W M Bowen 2000 An integrated framework for public sector environmental management in developing countries. *Environmental Management* 25: 463–76

White G F (ed.) 1974 *Natural Hazards: Local, National, Global.* New York, Oxford University Press

Wiman B L B 1991 Implications of environmental complexity for science and policy. *Global Environmental Change* 1: 235–47

Woodward S 1982 The myth of turbulence. *Futures* 12: 266–79

World Commission on Environment and Development 1987 *Our Common Future.* Oxford and New York, Oxford University Press

Yohe G W 1997 Uncertainty, short-term hedging and the tolerable window approach. *Global Environmental Change* 7(4): 303–15

Yohe G, M Jacobsen and T Gaptochenko 1999 Spanning "not-implausible" futures to assess relative vulnerability to climate change and climate variability. *Global Environmental Change* 9: 233–49

Young, M E 1993 *For Our Children's Children: Some Practical Implicatione of Inter-Generational Equity and the Precautionary Principle.* Resource Assessment Commission Occasional Paper 6, Canberra, Australian Government Printing Services

Chapter 3

Looking to the Future

3.1 Introduction

In Lewis Carroll's book, *Alice in Wonderland*, Alice comments that if you do not know where you want to go, any road will get you there. In contrast, Columbus knew where he wanted to go (the Orient) but ended up somewhere else (North America), because he did not know the route to take. The lesson is that it helps to know where you want to go, and also how to get there. Unfortunately, experience suggests that many resource and environmental planners and managers too often are like Alice or Columbus.

In this chapter, attention is directed towards the question of where you want to go, and first consideration focuses upon the concept of a vision. Next, the concepts of forecasting and backcasting, two different ways of viewing the future, are examined. Then, the ideas of future state visioning, appreciative enquiry, scenarios, the Delphi method and analogies are reviewed. Some believe that *sustainable development* or *sustainability* are a good basis for such a vision or sense of direction. In Chapter 4, the nature of sustainability, different views about the concept, the opportunities and problems created if it is used as a vision, and the implications of change, complexity, uncertainty and conflict for it, are considered.

3.2 Visions

3.2.1 Characteristics

Shipley and Newkirk (1998; 1999) have noted that "vision" has been defined in many ways. Here, the concept is interpreted as the capacity to visualize or imagine key attributes of a desired future state. To illustrate, one of the best known visions was that articulated by Martin Luther King during his keynote address before the Lincoln Memorial on 28 August 1963 in Washington DC. During his "I have a dream" speech, King described a USA in which people of different colours lived in respect, trust and harmony (Washington, 1986). Not only did his vision articulate an image of a different future, it was so powerful that it inspired many people to work to achieve that future state. In Box 3.1, various perspectives on a vision are presented, which elaborate on the interpretation used here.

Box 3.1 Perspectives on visions

Quite simply, *a vision is a realistic, credible, attractive future for your organization.* It is your articulation of a destination toward which your organization should aim, a future that in important ways is better, more successful, or more desirable . . . than is the present.

Vision always deals with the future. Indeed, vision is where tomorrow begins, for it expresses what you and others who share the vision will be working hard to create.

A vision is only an idea or an image of a more desirable future for the organization, but the right vision is an idea so energizing that it in effect jump starts the future by calling forth the skills, talents and resources to make it happen.

The right vision attracts commitment and energizes people, . . . creates meaning in workers' lives, ... establishes a standard of excellence, and . . . bridges the present and the future.

Source: Nanus, 1992: 8–17.

A shared vision is . . . a force in people's hearts, a force of impressive power. It may be inspired by an idea, but once it goes further . . . . it is palpable. People begin to see it as if it exists. Few, if any, forces in human affairs are as powerful as a shared vision.

A vision is truly shared when you and I have a similar picture and are committed to one another having it, not just to each of us, individually, having it.

Today, "vision" is a familiar concept . . . . But when you look carefully you find that most "visions" are one person's (or one group's) vision imposed on an organization. Such visions, at best, command compliance – not commitment. A shared vision is a vision that many people are truly committed to, because it reflects their own personal vision.

Source: Senge, 1994: 206.

If a vision is to successfully chart a new future, then, as Senge (1994: 218) explained, it must generate more than compliance, it must generate commitment (Box 3.2). He believed that 90% of the time, what passes for commitment is only compliance. To achieve commitment or "buy in", Senge argued that people have to become enrolled or engaged in the vision, which occurs when individuals believe they are responsible for achieving it. In contrast, compliant followers go along with a vision, mostly doing what is expected of them. Put a different way, whereas compliant people *accept* the vision, enrolled or committed people *want* the vision. Therefore, the committed person brings passion, energy and excitement to the task of achieving it.

Senge (1994: 209) also concluded that a "learning organization" cannot be created without a shared vision. His argument is that without the pull towards some goal in which people truly believe and want to achieve, forces that reinforce the status quo are usually overwhelming. Vision is essential because it establishes an overarching goal, and the power of a well articulated vision drives people to think and act in different ways. The shared vision can also provide a "rudder" to keep learning processes on track at times when stresses emerge and threaten to

---

Box 3.2 Possible attitudes toward a vision

Commitment: Wants it. Will make it happen. Creates whatever "laws" or structures are needed.

Enrollment: Wants it. Will do whatever can be done within the "spirit of the law."

Genuine compliance: Sees the benefits of the vision. Does everything expected and more. Follows the "letter of the law." "Good soldiers."

Formal compliance: On the whole, sees the benefits of the vision. Does what is expected and no more. "Pretty good soldier."

Grudging compliance: Does not see the benefits of the vision. But, also, does not want to lose job. Does enough of what is expected because he or she has to, but also lets it be known that he or she is not really on board.

Noncompliance: Does not see the benefits of vision and will not do what is expected. "I won't do it; you can't make me."

Apathy: Neither for nor against vision. No interest. No energy. "Is it five o'clock yet?"

Source: Senge, 1994: 220–1.

---

derail the common goal. Finally, Senge believed that a powerful vision can encourage people to take risks and to experiment, from which they will learn more.

Visions receive commitment through a reinforcing process of increasing clarity, enthusiasm and communication. As people discuss the vision, it becomes clearer, and as it becomes clearer, enthusiasm builds. Furthermore, people generally are more inspired by a positive vision (moving toward a desirable future) than by a negative vision (moving away from an undesirable future). In contrast, various impediments can cause a vision to wither or fade away. These include: (1) as more people become involved, diverse perspectives dilute the focus and create unmanageable conflicts; (2) people become discouraged due to the perceived difficulty in realizing the vision in practice; (3) people become overwhelmed by the demands and difficulties of the present and lose focus on the future vision – there is not enough time and energy to devote to the future vision; (4) people lose sight of their connections and common links, and focus only on aspects of the vision that directly affect them (Senge, 1994: 227–30). The guest statement by Lin Shunkun, outlines a vision for Hainan Province, southern China, to become a "bioprovince". Figures 3.1 and 3.2 illustrate some of the valued features which are to be protected and enhanced in this vision.

In the following sections, alternative ways of viewing the future are considered. After reviewing them, their implications for developing a vision will be considered.

3.3 Forecasting and backcasting

3.3.1 Forecasts: too often incorrect?

"Forecasting" is usually understood to mean the estimation of probable conditions or events in the future, based on present conditions and trends. For

(a)

(b)

Figure 3.1 Attractive natural environments in Hainan Province, China, such as these beaches (a) on the southeastern coast (Bruce Mitchell) and (b) on the south coast at Yalong Bay (Philip Xie), are to be protected during development of tourist facilities which will diversify the economy

Figure 3.2 The culture of the minority Li people, an important element for tourism in Hainan Province, is to be protected and encouraged under the bioprovince policy. Here, young women are performing a traditional bamboo pole dance (Philip Xie)

environmental and resource management, the emphasis has been upon forecasting the most likely or probable future related to the supply and demand of resources, given assumptions about population growth, changes in economies, technological innovations, consumption patterns and evolving values. With the need to make numerous assumptions, it is not surprising that forecasts often turn out to be wrong.

Forecasts relevant to resource and environmental management have been made for a long time. For instance, in 1798, the Scottish clergyman Robert Malthus published his famous book, *An Essay on the Principle of Population*. Malthus's basic position was that the rate of population growth was much greater than the capacity of the Earth to support humans. This conclusion was based on the "facts" that in Britain, France and America the population had been doubling every 25 years. It was not clear that food production would be able to increase at the same rate indefinitely. Malthus concluded that an inevitable outcome would be starvation and poverty, followed by widespread deaths due to famine and disease. In contemporary terms, such an outcome would occur because humans would exceed the "carrying capacity" of the Earth, humanity's "ecological footprint" would become too large, and the rapidly growing populations would not be "sustainable".

However, as Kennedy (1993) explained, at least three developments allowed the British to avoid the abyss which Malthus thought was inevitable. First, people left Britain to seek better conditions elsewhere. Indeed, between 1815 and 1914, some 20 million people emigrated. Second, significant technological

Guest Statement

Developing an ecological province: an initiative in Hainan Province, China

Lin Shunkun, China

Preserving and improving the ecological environment is a common challenge for humankind, an issue that is drawing great attention from all nations in the world. The Chinese government wants to preserve the ecological environment, and has sought to do this through a basic state policy of family planning and environmental protection, and implementation of a strategy of sustainable development.

Hainan Province is one region in China in which the ecological environment is well preserved. Hainan is a relatively underdeveloped region involving various islands with a total land area of 34,000 km^2 and population of 7.3 million in the year 2000. In 1999 a major policy decision was taken to designate Hainan as an ecological province with a view to exploring a path of sustainable development, which neither seeks development at the expense of the ecological environment nor lays undue emphasis on environmental protection at the expense of development, creating a first-rate ecological environment, promoting sound and rapid social and economic improvement, and facilitating high-level and high-quality development.

The focus of the ecological province includes the prevention and control of environmental pollution, improvement of ecological systems, development of industry, and improvement of human settlements and ecological cultural development. The major tasks of ecological cultural development are to establish a sound system of laws and regulations and a sound management system, spread scientific knowledge about ecology, conduct ecological education, cultivate and guide the modes of production and consumer behaviour which are beneficial to ecology and foster social values for economizing and protecting the environment, and enhance the peoples', enterprises' and decision makers' ecological awareness. And, finally, we will cultivate the ecological view of "who disrupts the ecological environment damages productive forces, who preserves the ecological environment protects productive forces, and who improves the ecological environment develops productive forces".

Designating Hainan as an ecological province is essential to preserve its ecological environment. Because of its geographical location and independent biophysical nature, its ecosystem is fragile. If damaged, this ecosystem will be difficult to restore. Therefore, in developing Hainan, we must attach vital importance to the preservation and improvement of the ecological environment.

Creating an ecological province is a strategic option to accelerate Hainan's economic development. Hainan has favourable conditions for the development of tropical high-efficiency agriculture and tourism. Its strategic plan for making use of the new advantage of being a special economic zone is to combine the improvement of the ecological environment with the development of tourism, to promote the development of agriculture and tourism with a sound ecological environment and the improvement of the ecological environment through the development of ecological agriculture and tourism, and to take advantage of its first-rate

53

environmental quality to attract capital, technology and skilled and talented people, and develop technologically intensive new industries.

The key to developing the ecological province is to implement the strategy of sustainable development in an all-round way, to combine economic and social benefits with environmental benefits, and to consider both short- and long-term benefits. With regard to economic development, an ecological province should skip the traditional stage of industrialized development featuring "high energy consumption, high pollution" and make a break from the traditional model of economic development featuring "treatment after pollution" and "restoration after damage". It should apply modern science and technology, and managerial expertise, and take advantage of the beautiful ecological environment to develop a high-performance ecological economy. In developing the ecological province, we should regard ecological rationality as a criterion; encourage ideas, concepts and social and economic activities beneficial to the preservation of natural resources and the ecological environment; and abandon concepts and behaviours which may lead to the destruction of natural resources and the ecological environment.

In brief, the goal of the people of Hainan Province is to achieve an ecological province which has a sound tropical island ecosystem, a developed ecological industry, and a culture which achieves harmony between nature and humans, and a first-rate living environment.

Lin Shunkun was born in 1944. He graduated from the Department of Chemistry, Nankai University, Tianjing, in 1968. Beginning in September 1968, Lin Shunkun worked successively as a technician, laboratory director, senior engineer and deputy director for the Henan Geology and Mineral Exploration Team in Henan Province. He was promoted to Deputy Director-General of the Department of Geology and Minerals of Henan Province in March 1985. In this position, he initiated and completed a Master Plan for Geological Exploration and Mineral Development for Henan. In May 1992, he moved to Hainan and took the position of Deputy Director-General of the Department of Environment and Resources. He is now the Deputy Director-General of the Department of Land, Environment and Resources of Hainan Province.

Lin Shunkun has devoted himself to the search for a sustainable development path for Hainan Province for many years and has contributed greatly to the initiation and formulation of the vision for developing Hainan into an ecological province.

innovations, in what later was to labelled the Agricultural Revolution, began about the time Malthus completed his analysis: one impact was a significant increase in food production. The third development was the Industrial Revolution, the early stages of which had started in the decade before Malthus presented his ideas. Industrial innovations substituted mechanical devices for human skill and inanimate power (steam, then electricity) for animal and human strength. As Kennedy remarked, the response to the power of population came not from the capacity of the Earth, but from the power of technology which pushed back limits and constraints. As a result, the "carrying capacity", "ecological footprints" or "sustainability thresholds" of the Earth were not set in absolute terms. Nevertheless, while Malthus's conclusions were not accurate for societies which could avoid the trap of food shortages, poverty and pestilence through migration, agricultural innovation and industrialization, for other societies without those options his forecast has been much closer to what has occurred.

Nearly two hundred years later, a book entitled *The Limits to Growth* was published. This book presented the results of a team led by the Meadows (1972). That group developed a model to examine what were considered to be five major trends with global implications: (1) accelerating industrialization, (2) rapid population growth, (3) widespread malnutrition, (4) depletion of non-renewable resources, and (5) a deteriorating environment. The similarity to the concerns of Malthus are striking, although by this time "industrialization" was viewed as part of the problem rather than as part of the solution. The Meadows' team emphasized that their model was imperfect, oversimplified and unfinished. Nevertheless, they reached the following conclusions:

- If growth trends in world population, industrialization, pollution, food production and resource depletion continued, the limits to growth on Earth would be reached sometime within the next 100 years. The most likely outcome would be a fairly sudden and uncontrollable decline in both population and industrial capacity.
- The forecast growth trends could be modified to create a condition of economic and ecological stability that would be sustainable far into the future. It was possible to ensure that the basic human needs of every person on the globe could be satisfied, and that each individual could reach his or her human potential.

The forecasts by the Meadows' team later received sharp criticism. For example, Bailey (1993: 67) noted that their predictions for non-renewable resources "have been proven to be spectacularly wrong". He observed that *The Limits to Growth* had predicted in 1972 that due to exponential population growth rates, the world would run out of gold by 1981, mercury by 1985, tin by 1987, zinc by 1990, petroleum by 1992, and copper, lead and natural gas by 1993. In contrast to these figures, Bailey cited US Bureau of Mines calculations which showed that, at 1990 rates of production, world reserves of gold were adequate for 24 years, mercury for 40 years, tin for 28 years, zinc for 40 years, copper for 65 years and lead for 35 years. From this evidence, Bailey suggested that the forecasts in *The Limits to Growth* had little value, and, in his opinion, were not useful because they were unnecessarily alarmist (Box 3.3).

If people such as Bailey have been critical of environmentalists for being unduly pessimistic in their predictions or forecasts, other people have been remarkably optimistic. To illustrate, Easterbrook (1995) called for a new "eco-realism"

Box 3.3

I hold those environmental alarmists strictly accountable for their faulty analyses, their wildly inaccurate predictions, and their heedless politicization of science, . . .

Source: Bailey, 1993: xi.

which would reject much of the prevailing "environmental orthodoxy". He predicted that pollution in the industrialized West would end within our lifetimes and with little pain or disruption, and also that the anticipated environmental catastrophes from global warming were almost certain to be avoided. Indeed, he argued that adjustments to changing environmental and other circumstances occur virtually automatically through what he termed "organic self-adjustment", as societies react to self correct regarding resource imbalances. Thus, whether considering the prophecies of optimists or pessimists, analysts, planners and managers should be able to identify purposes and assumptions of the forecasts in order to judge what weight to give to them.

3.3.2 Forecasts: uses, assumptions and focus

Uses

Forecasts provide insight regarding what the future could be like if trends continue. Knowing what might be the most probable future allows resource and environmental managers to make choices to ensure that demand and supply are in balance, to try to ensure that they are kept in balance, or to find substitutes if balance cannot be achieved. Forecasts can indicate where shortages may occur, and can help for making decisions regarding how best to deal with anticipated shortages. Or, forecasts which identify resource shortages can be helpful in designing policies to change demand patterns, such as dampening consumption and therefore extending the useful life of existing resources until alternatives can be found or developed. As a result, there can be little doubt that forecasts can be valuable tools. However, there also can be little doubt that forecasts will often be incorrect.

Assumptions

Forecasts often turn out to be wrong because a mix of assumptions normally has to be made. If the assumptions are not met, forecasts cannot be correct. As an example, forecasts regarding adequacy of natural resources and resource products in the USA during the early 1960s included estimates about the future population, labour force and gross national product, all of which were needed to forecast the size and shape of the national economy (Landsberg, 1964). To estimate future population levels, assumptions had to be made about birth, death and emigration rates. Once the national economy was predicted, then other estimates were made regarding basic human needs and wants: food, clothing, shelter, heat and power, transportation, durable goods – all of which in turn required assumptions. The estimates about requirements for end products (food, clothing) were translated into requirements for resource products, such as agricultural raw materials, steel, lumber and textile fibres. In turn, the implications of resource products for demands on land, water, fuels and other minerals were calculated. It is clear that any "forecast" regarding supplies of, or demands for, water, for instance, were built upon a wide array of other assumptions and estimates regarding the nature of the economy, technological change and the capacity for substitution.

Box 3.4

Three basic assumptions, built in from the start, were: continuing gains in technology, improvements in political and social arrangements, and a reasonably free flow of world trade. Two other assumptions on which the whole system of projections rests are that there will be neither a large-scale war nor a widespread economic depression like that of the early 1930s.

Source: Landsberg, 1964: 6.

The above example highlights that assumptions are a key ingredient in any forecast (Box 3.4). To the extent that any of the assumptions turn out not to reflect reality, then our understanding of what is the most probable or likely future may well be inaccurate.

Focus

Forecasts focus upon the most likely or probable future conditions. A reasonable question is whether such a focus is appropriate for environmental and resource management. As Bott, Brooks and Robinson (1983: 11–12) observed, humans cannot know what *will* happen in the future. However, three basic questions can help us to cope with an uncertain future. These questions are:

- What *can* happen? [feasibility]
- What *ought* to happen? [desirability]
- What is *likely* to happen? [probability]

They concluded that too often forecasts focused on the questions of *probability*, and did not consider the questions of *feasibility* and *desirability*. Or, even worse, all three questions are treated as a single question.

For Bott, Brooks and Robinson (1983), recognizing the distinction among the above three questions is essential for dealing with the future. That is, we could spend a disproportionate amount of time and effort attempting to predict the most likely conditions and patterns for resource supply and demand. However, none of the most likely futures might be desirable. As a result, they argued that it is important also to consider at least what are desirable futures. Humans have the capacity to make choices, and to intervene to create change. Thus, it seems logical to work on two parallel tracks. One would be to identify the kind of future we would like to have. The other would be to extrapolate from the present situation to estimate whether current trends and patterns are likely to lead to the desired future.

In dealing with an uncertain and complex future, the above perspective requires at least two considerations. Where would we like to be? And, will the current path get us there? In that regard, we should do forecasting in the traditional way to estimate where the current path could take us. However, it is not sufficient to stop there. We must also consider where we would like to go, and what actions we need to take to reach the desired end point. At this stage of analysis, we must also consider the feasibility of the possible desired futures. Some may be more dreams or wishes than achievable futures. This "reality check" emphasizes that we should consider *feasibility*, *desirability* and *probability* together. In order to do that, we need ways to determine where we want to get to and what needs to be done to make that possible.

In contrast to forecasting, in which the focus is upon likely futures, backcasts focus upon identifying desirable and attainable futures. The general approach is to work backwards from some future end point judged to be desirable in order to determine the feasibility of achieving that end point, and to determine the specific actions required to achieve it. In some ways, backcasting is another name for "critical path analysis" or "programming, planning and budgeting", both of which have been used in business to plan for the future. When President Kennedy announced that the USA would place a man on the moon by 1960, he initiated what today would be termed backcasting. A desirable future end point was identified, and those responsible for achieving it then had to consider what actions had to be taken to overcome the barriers hindering the realization of that vision.

Two important points should be highlighted at this stage. First, identifying a desirable future is challenging. For example, to anticipate the discussion about the nature of *sustainable development* or *sustainability*, in Chapter 4, many people advocate sustainable development as a desired future, and argue that many current policies and practices are unsustainable. However, different interpretations have been developed for sustainable development, making it difficult to agree on just what this ideal future involves. Second, and following from the first point, most societies are not homogenous. That is, different values and interests exist, reflecting differences by age, gender, race and status. At any given time, therefore, different groups may hold visions for the future which may conflict with one another. As a result, obtaining agreement on what constitutes a "common future" is often a formidable challenge (Box 3.5).

Box 3.5 Paradoxes

'A well-known set of paradoxes underlies futures studies. First, as in the absence of time travel, the future is unknowable since it has not happened yet, though predictions can be made about the future with varying degrees of confidence. Second, the study of the future is really the study of data from the past and present, since that is the only data available. Third, what will occur in the future is to a large degree a function of choice and behavior in the present, both of which may be influenced one way or another by predictions and other thinking about the future. Fourth, despite the ostensibly scientific nature of future analysis, it is inherently value-laden, with forecasting models serving mainly to display the implications for the future of the assumptions embedded in the model inputs and structure.

Source: Robinson, 1988: 325.

3.4 Future state visioning

Stewart (1993) has developed what he has termed *future state visioning* (FSV) as an approach to explore what an organization, agency or place could be in the future. In his view, looking forward at what could be, rather than analysing the current situation in order to plan forward from the present, is the key. In the FSV process, outlined in Figure 3.3, several steps are followed:

(1) Stakeholders and participants

The first step involves identifying all stakeholders (see Chapter 8) in order to be able to view the future and present state through their eyes. Identifying the stakeholders is considered to be the essential first step, as stakeholders will be more committed to any outcome if they have been involved from the outset. Even if all stakeholders are not directly involved, they should be identified to ensure that any vision will relate to their interests and views. By considering the views of all stakeholders during the visioning process, those participating will be forced to step beyond only their own views and interests.

Involving stakeholders from the outset does more than build commitment and assist to ensure all views are considered. Such involvement also helps to build support by spreading knowledge regarding the background for the vision, as well as the uncertainties, compromises and values underlying it. Such insight and understanding usually help individuals to regard the vision as a

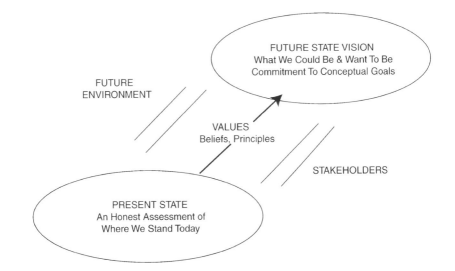

- Future State Vision — what we (our organizations) would like to be in the future
- The Present State — what exists, what we (our organizations) are like in the present
- The Future Environment — in which we must achieve our future state vision
- Values — the beliefs and principles which provide a foundation for our vision
- Stakeholders — those involved in or affected by the actions of the organization

Figure 3.3 The future state visioning process (Stewart, 1993: 92)

flexible guide, and to appreciate that with changing circumstances it may have to be modified. Finally, by participating in the development of a vision, people will better understand the visioning process, and begin to incorporate the vision into day-to-day decision making.

The use of *diagonal slice groups* is suggested to ensure participation from a large organization or region. This approach requires that a visioning group consists of people from at least two and often more parts of an organization or place. The purpose is not to have each individual represent an area or aspect of the organization, but to obtain different perspectives.

(2) The likely future environment

In contrast to many other visioning processes, Stewart strongly recommends that the second step should be to assess the future environment within which the future state vision will have to be achieved. He argued that it is very important to assess the likely future state before considering the present state or working on values. Too early attention to the present state can ensnarl the process with today's issues and barriers, which in turn can cause planning to become little more than divisive complaint sessions. Consideration of what would create an inspiring future has the effect of making participants more open and less defensive when later they turn to describing the reality of the present state.

Stewart suggests that since we usually learn best by doing, rather than by being told, it is important to have the team members identify the likely future environment, rather than relying on external experts. Not only will this approach draw upon the experience and insight of the stakeholders, but it will also sensitize them to uncertainties, threats and opportunities that undoubtedly will emerge in the future. Once the participants conclude that the future will not be the same as the present, and will not be simply the result of present trends, then capacity has been created to be flexible in dealing with unexpected future change.

(3) Creating the future state vision

In this stage, attention should be on developing a comprehensive vision of what could be without regard for present barriers (Box 3.6). Planning usually

---

Box 3.6 Examples of visions

Moreton Bay and its waterways will be a healthy ecosystem supporting the livelihoods and lifestyles of residents and visitors.

Source: Moreton Bay Catchment Water Quality Management Strategy Team, Queensland, Australia, 1998: 9.

In the Wimmera Region, water and land will be nurtured, native flora and fauna will flourish, people will be happy to live, work and visit, and agriculture and other industries will prosper.

Source: Wimmera Catchment Authority, Victoria, Australia, 1998: 6.

---

extrapolates from the present, which results in setting goals based on known issues and then moving to action without a clear vision. Future state visioning focuses on a state of being at a time in the future, far enough into the future so that thinking is less influenced by current conditions, yet also is close enough to maintain a sense of reality. Rodger Schwass, a colleague, has suggested a 15-year time horizon is sensible, especially if stakeholder consultation is being used. In his view, a 15-year time horizon ensures that senior people's influence is reduced, as many of them will be retired within 15 years, and therefore they will not have to live with the consequences of decisions. On the other hand, younger people will have reached positions of influence in 15 years, and then will be in positions to implement and enhance actions needed to achieve the vision. Thus, a 15-year time frame helps to level the playing field, and gives different voices equal weight.

The vision for the desired future state must be imaginative if it is to inspire and motivate, yet also be practicable so that it can be achieved. If it is not achievable, people may become discouraged and unwilling to take necessary actions to realize it. To help monitor progress toward the vision, Stewart urges use of benchmarks against which progress can be tracked. He also reminds us that excellence is a "moving target", so the vision must be conceived and designed to be dynamic, with recognition that continuous improvement should be a central feature.

Consistent with the idea of benchmarks, Stewart advocates that a future state should also be characterized by actionable concepts. In other words, ideals should be expressed in terms which can lead to action and results. As an example, a vision for an organization might include the ideal of trust among people. However, such a statement is not actionable. In contrast, the organization could state that it would strive to have comprehensive and open communication systems so that everyone is aware of what is happening. This will occur by regular face-to-face forums for two-way communication, as well through use of print and electronic bulletins. While enhanced communication is only one dimension to build trust, it is directly actionable.

(4) Contrasting the present state with the future vision

Only after the most likely future environment has been considered, and the future state vision has been created, should the team turn to assessing the present state. As noted already, if this sequence is followed, individuals are less likely to become defensive about their role regarding the present state of affairs, and more likely to understand that the potential future state is considerably more than the present state.

The present state should be thoroughly described, and then assessed with regard to its strengths, weaknesses, opportunities and threats. Systematic description and analysis of the present state will also identify barriers or obstacles to overcome if the desired future state is to be achieved.

(5) Identifying values for the vision

Experience has shown that a vision will not usually be used if it is inconsistent with the values of the people, organization or place, because actions are influenced

---

Box 3.7 A future state vision

The future state visioning process goes well beyond what is normally called visioning. The products – a comprehensive vision of the future in terms of actionable concepts, a contrasting description of the present, a set of values – comprise a fundamental basis for creating change. The FSV process creates a description of a desirable state of being. It also builds understanding, commitment and will to reach the organization's goals.

Source: Stewart, 1993: 97–8.

---

by values. Thus, the most successful visions are those supported by underlying organizational and personal values. However, since values are often not recognized or discussed, it is useful to identify them explicitly. If individuals have difficulty in accepting an emerging vision, it may be because it conflicts with one or more of their basic values. When that situation is recognized, individuals and the organization can decide what might be done either to modify the vision, or determine if the values can be changed. One of the main reasons for developing the future vision before considering values is to ensure that current values do not block consideration of a future state which may be highly desirable, even though it is not consistent with current dominant values.

Stewart suggested that values should be thought of relative to beliefs (what we hold to be true about something), philosophy (the composite of beliefs by which we want to live) and principles (guidelines for translating beliefs and philosophy to action).

(6) Transition to action

Once the five steps have been completed regarding a future state vision, the job is not completed, even though much has been accomplished. The FSV focuses upon the best answers regarding "*what to do?*" or "what we want to be or could be in the future?" FSV does not directly address questions regarding "*how to achieve it?*" Stewart suggests that it is better to focus first upon completing the visioning process before considering questions related to how to achieve it. However, the logical follow up from a visioning process is the preparation of a strategic plan which concentrates upon how the future desired state will be achieved (Box 3.7). As a final observation, however, Stewart remarked that *what is* can be a significant barrier to *what could be*. If bold and effective change is desired, he therefore believes that it is sensible to begin with attention to where you want to be, rather than where you are now.

3.5 Appreciative inquiry

Elliott (1999: 2) has explained that appreciative inquiry uses the energy from the "positive present", develops it to create a vision for a positive, desired future which is based in reality, and assists people to assemble the necessary forces for change to achieve the vision. Appreciative inquiry was developed in the early 1990s to assist corporations and institutions to improve themselves, but it has also been applied at the community level in developing countries. The common thread, wherever it is used, is to shift attention from local problems to local achievements, and from participation to inspiration. Through identification and reinforcement of positive and constructive actions, linkages and visions within a community, appreciative inquiry empowers and fosters local ownership of the inquiry process.

Elliott (1999: 4) noted that appreciative inquiry is distinguished by five attributes: (1) incorporation of a broad range of stakeholders to assess the past and to plan the future; (2) openness or transparency during self-reflection; (3) inclusiveness in determining decision-making criteria which include the items listed in point (4); (4) non-quantifiable and non-financial aspects which may significantly affect the long-term health or well-being of the community or organization; and (5) wide dissemination of power and influence throughout the community, as indicated in both points (2) and (3).

Appreciative inquiry usually involves four steps or stages (Elliot, 1999: 3–4).

Step 1: Discovering excellence and achievement

The analysis, planner or manager, through interviews and storytelling, facilitates local participants to remember significant past achievements and periods of excellence in their community. It allows the community to recognize times during which it was functioning at its best, and what contributed to such excellence. Through sharing memories, the participants are able to identify key variables, such as leadership, values, learning processes, relationships, structures and external relationships, which contributed to achievements.

Step 2: Dreaming an ideal community or organization

Based on identification and appreciation of past achievements, the participants then envisage a desired future. What makes appreciative inquiry distinctive is a process of creating a vision for a desirable future strongly grounded in the history of the community, which is judged to generate possible futures which are both credible and compelling. This approach therefore contrasts sharply with the future state visioning outlined in the previous section, in which the explicit recommendation was to consider the desired future state before assessing past and present conditions. The view here is that, by considering possible futures in the context of the history of past achievements, appreciative inquiry is both practical, through building on what is termed the "positive present", and generative, by seeking to expand and enhance the potential of the community.

Step 3: Designing new structures and processes

In this stage, attention focuses upon creating through consensus specific short- and long-term goals that, if realized, will achieve the dream outlined in the second stage. A key feature is the use of "provocative propositions", or statements declaring what the community will be like in the future, and which are deliberately designed to stretch or extend the community. At the same time, the provocative propositions should be viewed as realistic because they build upon past periods of excellence already identified. A provocative proposition might be that "this community will have a convenient and safe source of potable water within the next two years".

Step 4: Delivering the dream

In this final stage, the community members take action relative to their provocative propositions, and define roles, responsibilities and authority, create strategies, build linkages and networks, and mobilize necessary human and financial resources to achieve the dream. Specific projects are designed and implemented, new plans are developed, old relationships are strengthened and new ones are built. Through the appreciative process, people begin to understand the relevance of each new initiative to the long-term vision being sought.

3.6 Futuristic methods

Various methods have been developed to help explore the future. These can be used as tools in future state visioning or appreciative planning. Three of these are considered here: scenarios, the Delphi method and analogies.

3.6.1 Scenarios

In a scenario, the analyst describes a logical sequence of events which follow from specified assumptions (Jantsch, 1967: 180–1). Scenarios are well suited for incorporating many aspects of a problem, and for allowing the analyst to gain a "feel" for likely future events as well as key decision points. The scenario focuses the investigator's attention on the dynamics and interaction of events. Frequently, several scenarios are prepared. While it is accepted that none will come true as outlined, the general patterns described are viewed as the most likely, if certain assumptions and decisions are made.

Scenarios have often been used to dramatize the nature of possible problems in the future. In *Silent Spring*, Rachel Carson's (1962) first chapter was a scenario. Its purpose was to draw attention to the future consequences of indiscriminate use of chemicals to combat insects and weeds. The chapter describes a hypothetical town in the rural USA which at one time was a prosperous agricultural centre and enjoyed an abundance of wildlife, flowers and other vegetation. Carson then described how the situation changed. Livestock started to die for unexplained reasons. Adults and children became ill, with several sudden and unexplained deaths. Flocks of birds disappeared. Vegetation became withered and brown. Having caught the reader's attention, Carson then went on to explain what had silenced the voices of spring in many American towns – the overuse of biochemicals. She was not predicting that the voices of spring would be eliminated in all small rural towns in the future, but she was able to demonstrate that if current practices of using agrochemicals continued, a strong likelihood existed that there would be some negative impacts such as the ones outlined in her scenario. In this way, scenarios can be used to build a vision, negative or positive, and can also help to identify the actions necessary for such a vision to occur.

3.6.2 Delphi method

The Delphi method is one of many "idea generating strategies". Such strategies can be divided into two types: non-group and group. Non-group processes involve surveys in which participants do not interact with each other. In contrast, group processes incorporate workshops, nominal group techniques and Delphi surveys in which people interact with one another. Evidence confirms that group processes usually are superior to non-group processes in that the former generate a greater number of higher quality ideas.

Group processes can be developed in various ways. The usual brainstorming type involves creating a setting in which people engage in discussion with varying degrees of structure. Nominal group techniques (NGT) involve people interacting in a very structured and controlled environment, such as when a facilitator for a group ensures that all people have an opportunity to comment. Through this approach, NGTs overcome some key problems associated with many group workshops or brainstorming sessions – a small number of people dominating the discussion, many people not contributing, and discussions getting "stuck" on particular topics. Nevertheless, NGTs still face the problems of needing to arrange for people to meet together at a common place and time, which can be both expensive and logistically difficult.

The Delphi method was designed to obtain the benefits of group brainstorming sessions, while avoiding or minimizing the problems noted above. The conventional Delphi involves multiple-round surveys, usually with mailed questionnaires. Participants remain anonymous to one another, but through multiple rounds are exposed to the results of the group thinking. Thus, the Delphi method can be described as involving a series of interactive brainstorming rounds. The Delphi avoids the expense and logistical problems of bringing people to one place, eliminates the possibility of one or more articulate and aggressive people dominating the group, reduces pressure for respondents to conform to the views of senior influential participants, and allows respondents to have more time to reflect before expressing their views. Disadvantages are that each round of a Delphi questionnaire survey can be lengthy to complete, people drop out and stop participating as rounds continue, and the opportunity for spontaneous discussion which occurs at face-to-face meetings is lost. Furthermore, the emphasis on reaching consensus can create its own pressure for conformity.

The conventional Delphi method was developed by the RAND Corporation during research with the objective of identifying the timing and significance of future developments regarding scientific breakthroughs, population growth, automation, space progress, future weapons systems, and probability and prevention of war. RAND proceeded in the following manner. Experts around the world were identified and invited to serve on a panel for one of the topics. For scientific breakthroughs, each panel member was invited to name inventions which were urgently needed and achievable in the next 50 years. Forty-nine items were identified. In a second round, panel members were given a list of the 49 items and asked to indicate the date on which there was a 50:50 probability of the breakthrough occurring. These estimates were combined and displayed as quartiles and medians, and returned to the participants by mail.

Reasonable consensus emerged for 10 items. In a third round, the consensus for the 10 items was presented, and dissenters were asked to elaborate upon why they disagreed. At the same time, 17 items for which agreement had not previously been reached were presented. Panel members were invited to elaborate upon reasons for the timing they had identified. This exercise was repeated in a fourth round. At each stage, the number of items for which consensus was reached went up.

The conventional Delphi was designed to be a decision-helping tool, with the purpose to reach consensus by experts regarding future activities or innovations. Thus, the Delphi has been used as input for forecasts for resource supplies or demands, on the belief that the findings from a Delphi are likely to be more accurate than the opinions of a single expert, or small group of experts. As de Loë (1995) has explained, the conventional Delphi has been used in the following ways related to resource and environmental management: (1) to predict occurrence of events or trends, (2) to rank alternative goals or objectives, (3) to create management strategies, and (4) to allocate scarce resources among competing options.

The output from a conventional Delphi, with its emphasis on consensus and convergence, can be very helpful for resource and environmental managers. However, de Loë (1995) has explained that a derivative, usually termed a *policy Delphi*, also can be useful. In a policy Delphi, the purpose is to generate the strongest and widest possible *different* views about alternative ways to resolve, or deal with, a policy issue. The departure point here is a belief that decision makers do not want analysts to make decisions for them. Instead, they want the analysts to identify all the options, along with supporting evidence. The process for the policy Delphi is the same as for the conventional Delphi. However, the ultimate purposes are quite different. The former emphasizes divergence; the latter, convergence.

Either of the conventional or the policy Delphis can be used by resource and environmental managers as a technique to deal with an uncertain and complex future, and to identify alternative futures which might become candidates for a vision.

3.6.3 Analogies

Analogies also can be used to help understand unfamiliar or uncertain situations (Box 3.8). Creating an analogy generally involves comparing two places or situations. Known information about one is used to make inferences about

---

Box 3.8 Analogies and unfamiliar situations

When confronted by unknown situations, analogies can provide us with a feeling of understanding. They provide a first step toward knowing or, at least, considering the unknown . . . In this regard, analogies are comforting because they provide a bridge between the past (the known) and the future (the unknown).

Source: Glantz, 1991: 14.

---

the yet to be observed, but anticipated, behaviour or structure in the other. The choice of places or experiences for comparison is a major challenge. Should we be more concerned with the *number* of similarities in structure and function, or with the *significance* of the similarities? While the former often seems to influence choices, many believe that significance rather than frequency of similarities is most important.

Familiar analogies can be identified. The *greenhouse effect* has been used to characterize what is happening with regard to the process of climate warming or global change. This analogy has been effective because people know that greenhouses stimulate plant growth by containing warmth in an enclosed environment. The parallel with carbon dioxide released from automobiles and industrial factories, serving as a skin or barrier to trap heat escaping into the atmosphere in a similar way that the panes of glass keep heat in a greenhouse, has helped people to visualize why global warming might occur. An analogy to another *Dust Bowl* in the American Midwest if temperatures increased and precipitation decreased has helped people to imagine what that region might be like if global warming were to occur. And, at the conclusion of a World Conference on the Changing Environment in Toronto, the official concluding statement compared the potential impacts from global warming with those that could be expected from *nuclear war*. As Glantz (1991: 14) explained:

> this analogy was not made by "idealistic, scientifically innocent environmentalists" but was used by the Toronto conference organizers to compare an unknown situation, the consequence of global warming, with a better-known, worst-case scenario, a nuclear holocaust, to capture the attention of policy makers, the public, and especially the media and to urge prompt policy action to address the global warming issue.

As Glantz has explained, analogies can be used for many purposes. They can be used for general education, to identify questions or hypotheses for further research, to forecast future states of systems, and to identify alternative policy options or visions. The education function has already been noted above, when explaining how analogies to the *greenhouse effect*, *Dust Bowl* and *nuclear war* have been made to sensitize the public and policy makers about the processes and potential impacts of climate change. Comparisons have also been made between countries or regions to illustrate what future conditions could be like. Thus, under anticipated climate change scenarios it has been suggested that Iceland

> ## Box 3.9 Caution advised in using analogies for forecasting or generating policy alternatives
>
> The plausibility of a physical or social analogy is not a sufficient reason for it to be used by policymakers because several plausible but contradictory policies could be formulated based on different analogs drawn from the same pool of objective scientific information. But, then, on which historical analog should a decision maker rely?
>
> *Source*: Glantz, 1991: 13.

could have a climate similar to present-day Scotland. Such an analogy helps policy makers to visualize opportunities and problems that could be encountered regarding matters such as agricultural production, and what adaptations might be needed to cope with the new conditions. Such insights can then be used in crafting of a vision for a desirable future.

The use of analogies for forecasting and generating policy options should, in the opinion of Glantz (1991), be treated with great caution (Box 3.9). While analogies can be very useful in stimulating scientific questions or hypotheses about possible future conditions, their implications or adjustments to them, they are less useful as forecasts to identify future states of a climate or societies. In this manner, analogies have the same weaknesses as the forecasts discussed previously – they attempt to converge on most likely conditions, rather than on most desirable conditions. Also, there is always a danger that people who have particular interests or favour certain policies will put forward analogies which support their position, and ignore those which do not.

The above discussion indicates that, as with most things, analogies have strengths and weaknesses. Their careful use can help us to anticipate future conditions, even when we cannot predict exactly what will happen. They can be particularly helpful by providing a first cut at understanding the regional implications of some specified different conditions, such as altered climate. They also can help people in one region to benefit from the experience of people in another region, if there are reasons to believe that conditions in the first region could evolve to approximate those in the second region. Furthermore, analogies can stimulate thinking, as shown by the comments in Box 3.10, as people consider alternative desirable futures in the process of developing a vision.

> ## Box 3.10 Analogies
>
> Analogies can also provide clues, generate ideas, and spark reactions that lead to different searches for new analogies. Each analogy gives additional information about the target problem . . . By providing useful insights, historical analogies may give societies the opportunity to capitalize on their strengths and to minimize their shortcomings.
>
> *Source*: Glantz, 1991: 32–3.

3.7 Implications

People try to see into the future in many ways. Some read tea leaves. Others read palms. Others seek out gurus who are thought to have the ability to anticipate future events. Analysts construct sophisticated models to generate simulations based on different goals, values and assumptions. Whatever approach is used, the future remains uncertain and complex, and the likelihood of change is very high.

For resource and environmental managers, it is important to think about the future for various reasons. One, we want to anticipate possible undesirable conditions so that action can be initiated now to avoid or mitigate them. Two, we would like to act as agents of positive change in order to take decisions that will result in outcomes that will meet the needs and expectations of people in society, as well as consider the needs of non-human living things. As a result, at least three questions deserve our attention regarding the future: (1) what is most likely? (2) what is most desirable? and (3) what is most feasible? Too often, resource and environmental managers have focused primarily upon only the first question.

At the beginning of the chapter, the dilemmas of Alice in Wonderland and Columbus were noted. Alice did not know where she wanted to go; Columbus knew where he wanted to go, but not how to get there. It was suggested that resource and environmental managers need to know where they want to get to and what route to use to get there. However, even knowing both of those things can still lead to problems, as illustrated by Herman Melville's story in *Moby Dick*. Captain Ahab took his crew on the *Pequod* for a well-defined mission (catch and kill the white whale Moby Dick) and knew how he intended to do that. Notwithstanding clear ends and means, the outcome was a disaster for Ahab and the crew – Moby Dick got away, and Ahab and his crew, with the exception of Ishmael, drowned. Captain Ahab's experience is a reminder that a well defined but poorly conceived objective (or vision) pursued with tried and tested methods can still be shipwrecked.

References and further reading

Ache P 2000 Vision and creativity – challenge for city regions. *Futures* 32: 435–49

Ayres R U 1996 Foresight as a survival tactic: when (if ever) does the long view pay? *Technological Forecasting and Social Change* 51: 209–35

Bailey R 1993 *Eco-scam: The False Prophets of Ecological Apocalypse*. New York, St Martin's Press

Barnett H J and C Morse 1963 *Scarcity and Growth: The Economics of Natural Resource Availability*. Baltimore, Johns Hopkins University Press

Bookman C A 2000 Town meeting on America's coastal future: using the Internet to promote coastal stewardship. *Ocean and Coastal Management* 43: 937–51

Bott R, D Brooks and J Robinson 1983 *Life after Oil: A Renewable Energy Policy for Canada*. Edmonton, Hurtig Publishers

Carson R L 1962 *Silent Spring*. Boston, Houghton Mifflin

de Loë R 1995 Exploring complex policy questions using the policy Delphi. *Applied Geography* 15: 53–68 de Loë R and B Mitchell 1993 Policy implications of climate change for water management in the Grand River basin. In M Sanderson (ed.) *The Impact of Climate Change on Water in the Grand River Basin, Ontario*. Department of Geography Publication Series No. 40, Waterloo, University of Waterloo, pp. 189–217

Dewhurst S M and W B Kessler 1999 Scenario planning: wading into the real world. *Journal of Forestry* 97(11): 43–7

Draper D 2000 Toward sustainable mountain communities: balancing tourism development and environmental protection in Banff and Banff National Park, Canada. *Ambio* 29: 408–15

Easterbrook G 1995 *A Moment on the Earth: The Coming Age of Environmental Optimism*. New York, Viking Penguin

Elliott C 1999 *Locating the Energy for Change: An Introduction to Appreciative Inquiry*. Winnipeg, International Institute for Sustainable Development

Floyd, D W, K Alexander, C Bueley, A W Cooper, A DuFault, R W Goret, S G Haines, B B Hronek, C D Oliver and E W Shepard 1997 Choosing a forest vision. *Journal of Forestry* 97(5): 44–6

Glantz M H 1991 The use of analogies in forecasting ecological and social responses to global warming. *Environment* 33: 10–15, 27–33

Godet M 2000 The art of scenarios and strategic planning: tools and pitfalls. *Technological Forecasting and Social Change* 65: 3–22

Gratton L 1996 Implementing a strategic vision – key factors for success. *Long Range Planning* 29: 290–303

Gray B 1989 *Collaboration: Finding Common Ground for Multiparty Problems*. San Francisco, Jossey-Bass

Hailbroner R 1995 *Visions of the Future: The Distant Past, Yesterday, Today, Tomorrow*. New York, Oxford University Press

Höjer M and L-G Mattsson 2000 Determinism and backcasting in future studies. *Futures* 32: 613–34

Jamal B T and D Getz 1997 "Visioning" for sustainable tourism development: community-based collaborations. In P Murphy (ed.) *Quality Management in Urban Tourism*. Chichester, John Wiley, pp. 199–220

Jantsch E 1967 *Technological Forecasting in Perspective*. Paris, Organization for Economic Co-operation and Development

Kennedy P 1993 *Preparing for the Twenty-first Century*. Toronto, HarperCollins

Kuhn R G 1992 Canadian energy futures: policy scenarios and public preferences. *Canadian Geographer* 36: 350–65

Landsberg H H 1964 *Natural Resources for US Growth: A Look Ahead to the Year 2000*. Baltimore, Johns Hopkins University Press

Lazarus M, S Diallo and Y Sokona 1994 Energy and environment scenarios for Senegal. *Natural Resources Forum* 18: 31–47

Linstone H A 1999 Complexity science: implications for forecasting. *Technological Forecasting and Social Change* 62: 79–90

Mannormaa M 1991 In search of an evolutionary paradigm for futures research. *Futures* 23: 349–72

Meadows D H, D L Meadows, J Randers and W W Behrens III 1972 *The Limits to Growth*. second edition, New York, New American Library

Moreton Bay Catchment Water Quality Management Strategy Team 1998 *The Crew Member's Guide to the Health of our Waterways*. Brisbane, Brisbane City Council

Munn R E 1991 A new approach to environmental policy making: the European futures study. *Science of the Total Environment* 108: 163–72

Naisbitt J 1984 *Megatrends*. New York, Warner Books

Nanus B 1992 *Visionary Leadership*. San Francisco, Jossey-Bass

Needham R D and R de Loë 1990 The policy Delphi: purpose, structure, and application. *Canadian Geographer* 34(2): 133–42

Nijkamp P and R Vreeker 2000 Sustainability assessment of development scenarios: methodology and application to Thailand. *Ecological Economics* 33: 7–27

Passig D 1998 An applied social system procedure for generating purposive sound futures. *Systems Research and Behavioral Science* 20: 315–25

Raimond P 1996 Two styles of foresight: are we predicting the future or inventing it? *Long Range Planning* 29: 208–14

Rappert B 1999 Rationalising the future? Foresight in science and technology policy co-ordination. *Futures* 31: 527–45

Raynor M E 1998 That vision thing: do we need it? *Long Range Planning* 31: 368–76

Rhoades R 2000 Integrating local voices and visions into the Global Mountain Agenda. *Mountain Research and Development* 20: 4–9

Robertson M and R Walford 2000 Views and visions of land use in the United Kingdom. *Geographical Journal* 166: 239–54

Robinson J B 1982 Energy backcasting: a proposed method of policy analysis. *Energy Policy* 10: 337–44

Robinson J B 1988 Unlearning and backcasting: rethinking some of the questions we ask about the future. *Technological Forecasting and Social Change* 33: 325–38

Robinson J B 1990 Futures under glass: a recipe for people who hate to predict. *Futures* 22: 820–42

Schwartz P 1996 *The Art of the Long View: Planning for the Future in an Uncertain World*. New York, Currency Doubleday

Senge P M 1994 *The Fifth Discipline: The Art and Practice of the Learning Organization*. New York, Doubleday

Sharmal H D, A D Gupta and Shushil 1995 The objectives of waste management in India: a futures inquiry. *Technological Forecasting and Social Change* 48: 285–309

Shipley R 2000 The origin and development of vision and visioning in planning. *International Planning Studies* 5: 227–38

Shipley R and R T Newkirk 1998 Visioning: did anybody see where it came from? *Journal of Planning Literature* 12: 407–16

Shipley R and R T Newkirk 1999 Vision and visioning: what do these terms really mean? *Environment and Planning B* 26: 573–91

Smill V 2000 Perils of long-range energy forecasting: reflections on looking far ahead. *Technological Forecasting and Social Change* 65: 251–64

Stephenson R B 1999 A vision of green: Lewis Mumford's legacy in Portland, Oregon. *Journal of the American Planning Association* 65: 259–69

Stewart J M 1993 Future state visioning – a powerful leadership process. *Long Range Planning* 26: 89–98

Strong M 2000 *Where on Earth are We Going?* Toronto, Alfred Knopf

Toffler A 1970 *Future Shock*. New York, Random House

Van der Heijden K 2000 Scenarios and forecasting: two perspectives. *Technological Forecasting and Social Change* 65: 31–6

Washington J M (ed.) 1986 *A Treatment of Hope: The Essential Writings of Martin Luther King, Jr.* San Francisco, Harper and Row, pp. 217–20

Weisbord M R 1992 *Discovering Common Ground*. San Francisco, Bennett-Koehler, Jossey Bass

Wimmera Catchment Authority 1998 *Annual Report 1997–1998*. Horsham, Victoria, Australia, PO Box 479

Chapter 4

Sustainability

4.1 Introduction

In Chapter 3, attention focused on the concept of a *vision*, and some of the processes that can be used to identify a desirable future. In resource and environmental management, many people have included *sustainable development*, or, more recently, *sustainability*, as the core or centrepiece for a desirable future state. In this chapter, attention is first given to the emergence of sustainable development (Section 4.2), along with principles and perspectives about the concept. Section 4.3 considers the implications for the concept in developed and developing countries. In Section 4.4, the future search conference process is examined relative to exploring the future, and in Section 4.5 its application to develop a National Conservation Strategy for Pakistan is examined.

4.2 Sustainable development

4.2.1 The concept

Sustainable development was popularized in the report *Our Common Future* prepared by the World Commission on Environment and Development (1987), also referred to as the Brundtland Commission, after its chair (Gro Harlem Brundtland) then the prime minister of Norway. Gro Brundtland explained in the Foreword to *Our Common Future* that she was invited in December 1983 by the Secretary-General of the United Nations to conduct an inquiry and prepare a report to provide a global agenda for change. More specifically, the terms of reference from the General Assembly of the United Nations were: (1) to propose long-term environmental strategies for achieving sustainable development by the year 2000 and beyond; and (2) to identify how relationships among people, resources, environment and development could be incorporated into national and international policies. The Commission included representatives from developed and developing countries, and held public meetings in various countries around the world.

In its report, the Commission was explicit that it had not developed a detailed blueprint for action, but rather a "pathway" through which people in

---

Box 4.1

Those who are poor and hungry will often destroy their immediate environment in order to survive: They will cut down forests, their livestock will overgraze grasslands; they will overuse marginal land; and in growing numbers they will crowd into congested cities.

Source: World Commission on Environment and Development, 1987: 28.

---

different countries could create appropriate policies and practices. Furthermore, the Commission members had quickly agreed that one issue was of primary significance: many development activities were leaving growing numbers of people poor and vulnerable, and at the same time were degrading the environment (Box 4.1). This conclusion convinced the Commission members that a new path for development was needed, one that would sustain human progress not just in a few places for a few years, but for the entire planet into a more distant future. Thus, the planet's main environmental problem was judged also to be its main development problem.

The Commission focused on population, food security, loss of species and genetic resources, energy, industry and human settlements. All of these were deemed to be inter-connected and therefore could not be treated separately. Furthermore, the concept of sustainable development was judged to involve limits. Such limits were not "absolute", but were relative to the state of technology and social organizations, and to the capacity of the biosphere to absorb the effects from human activity.

Perhaps the most frequently quoted statement from the Brundtland Commission is that sustainable development is development that meets the needs of the present without compromising the ability of future generations to meet their own needs. However, less frequently noted has been its associated statement that sustainable development contains two key concepts. These are: (1) *needs*, especially the needs of the poor people in the world, to which over-riding priority was essential; and (2) *limitations* created by technology and social organization regarding the capacity of the environment to satisfy both present and future needs. Thus, sustainable development, as interpreted by the Brundtland Commission, is an *anthropocentric* (human-centred) concept.

The Commission also offered some comments about *growth*. In its view, no set limits could be identified regarding levels of population or resource use beyond which ecological disaster would occur. Different limits existed for use of energy, water, land and materials. Notwithstanding this qualification, the Commission concluded that ultimate limits did exist. Sustainable development required that, well before such limits were reached, the world must have ensured equitable access to constrained resources and have re-oriented technology to relieve pressures. At the same time, it stipulated that every ecosystem in every place could not be preserved intact because economic growth and development inevitably involved changes.

Box 4.2 Perspective on sustainability

Sustainability . . . came into prominence with the publication in 1987 of *Our Common Future* by the World Commission on Environment and Development (Brundtland Commission). It introduced the creatively ambiguous phrase "sustainable development" as an idea to pursue. An intuitively attractive but slippery concept (rather like "democracy" and "justice"), it nevertheless served the intent of the Commission which was to further the debate about what should be the proper relationship between "environment" and "development". No society worth having should be unwittingly undermining the ecological basis of its own continuance, yet the constant litany of environmental problems caused by the degradative impacts of human activities signals that sustainability is indeed in doubt.

The discussion at this point often goes to questions about natural resources and, specifically, questions about how much should be sustained, at what level of quality, for how long a duration, and for whose benefit. This is guaranteed to run in circles. No resource systems, nor the institutions associated with them, can be sustained as is in perpetuity. Changes in both are inevitable. What *must* be sustained however, is the capacity for renewal and evolution in ecosystems, and innovation and creativity in social systems. Sustainability is not some end state to be achieved, but a trajectory to be negotiated continuously as societies learn to recognize the symptoms and evidence of non-sustainability and adjust accordingly. This is much easier said than done.

Source: Francis, 1995: 4.

Having defined sustainable development, and explained what it implied, the Brundtland Commission then identified seven critical objectives for environment and development policies. These were:

- reviving growth;
- changing the quality of growth (emphasizing development rather than growth);
- meeting essential needs for jobs, food, energy, water and sanitation;
- ensuring a sustainable level of population;
- conserving and enhancing the resource base;
- re-orienting technology and managing risk; and
- merging environment and economics in decision making.

Two key points deserve highlighting here. First, the Commission was explicit that while growth is essential to meet basic human needs, sustainable development involves more than growth. It necessitates a change in the nature of growth, to make it less material- and energy-intensive, and to make it more equitable in its impacts. Second, the Commission noted that a common theme in a strategy for sustainable development had to be the integration of economic and ecological considerations in decision making. For this to happen, the Commission concluded that there would have to be changes in attitudes and objectives, and in institutional arrangements and laws at every level. However, the Commission noted that changes in laws alone would not be sufficient to protect common interests. Such protection required community knowledge and support, which in turn necessitated more public participation in decisions about the environment and resources. These aspects are considered in more detail in Chapters 8 (Partnerships and Stakeholders) and 9 (Local knowledge systems).

4.2.2 Principles for sustainable development

Following the publication of *Our Common Future*, considerable effort has been devoted to developing guidelines or principles for sustainable development. The rationale has been that without such guidelines or principles it is not possible to determine if a policy or practice is sustainable, or if initiatives are consistent with sustainable development or with sustainability. Creation of such principles has been a major challenge because, as the Commission recognized, economic and social systems and ecological conditions vary greatly among countries. The result was that no generic model or blueprint could be established, and each nation would have to work out what was appropriate for its context, needs, conditions and opportunities.

Notwithstanding these challenges in developing generic principles, it is helpful to identify general guidelines, which then can be modified for the conditions of a place and time. Table 4.1 provides one set of principles or guidelines. They represent one of the earlier systematic attempts to identify the characteristics of a sustainable society. If the principles are found to be inadequate or incomplete, they challenge the critic to make them more adequate or complete. In addition, for them to be operational or practical, a major task is to develop *indicators* for each principle. In other words, what information or evidence would be required for each principle to allow a decision to be made that a policy or an initiative was consistent with it? Are such data available from the information already being collected in countries as part of their censuses or other monitoring (see Chapter 13)? Or, do new data collecting programmes have to be established?

4.2.3 Perspectives on sustainable development

As a concept, sustainable development has attracted both criticism and support. Sustainable development has been criticized because some view the definitions or interpretations to be too vague or ambiguous, allowing it to be something for everyone, or allowing anyone to use it as a justification for actions, whether those be in the direction of economic growth or environmental protection. Others consider sustainable development to perpetuate the Western capitalist system, so reject it on ideological grounds.

The positive assessment in many ways is the mirror image of the criticisms. Thus, while some see vagueness and ambiguity as a problem, others believe these features provide flexibility and discretion necessary to custom design for the needs of specific places and times. While some view sustainable development as supporting traditional capitalistic systems, others believe its

Table 4.1 Principles of sustainability

A. Environmental/ecological principles

1. Protect life support systems.
2. Protect and enhance biotic diversity.
3. Maintain or enhance integrity of ecosystems, and develop and implement rehabilitative measures for badly degraded ecosystems.
4. Develop and implement preventive and adaptive strategies to respond to the threat of global ecological change.

B. Socio-political principles

B1. From environmental/ecological constraints

1. Keep the physical scale of human activity below the total carrying capacity of the planetary biosphere.
2. Recognize the environmental costs of human activities; develop methods to minimize energy and material use per unit of economic activity; reduced noxious emissions; decontaminate and rehabilitate degraded ecosystems.
3. Ensure socio-political and economic equity in the transition to a more sustainable society.
4. Incorporate environmental concerns more directly and extensively into the political decision-making process.
5. Ensure increased public involvement in the development, interpretation and implementation of sustainable development concepts.
6. Link political activity more directly to actual environmental experience through reallocation of political power to more environmentally meaningful jurisdictions.

B2. From socio-political criteria

1. Establish an open, accessible political process that puts effective decision-making power at the level of the government closest to the situation and the lives of the people affected by a decision.
2. Ensure people are free from extreme want and from vulnerability to economic coercion.
3. Ensure people can participate creatively and self-directedly in the political and economic system.
4. Ensure a minimum level of equality and social justice, including equality to realize one's full human potential, recourse to an open and just legal system, freedom from political repression, access to high-quality education, effective access to information, and freedom of religion, speech and assembly.

Adapted from Robinson *et al.*, 1990: 44.

arguments for including real environmental costs and using environmental pricing, and valuing environmental attributes, are appropriate to modify traditional market thinking which gives more weight to economic than environmental considerations.

While there inevitably will be supporters and detractors for a concept such as sustainable development or sustainability, it is important to recognize that the concept does contain some paradoxes, tensions and conflicts. Dovers and Handmer (1992) identified what they viewed to be eight of the most obvious of these, each of which is considered below.

(1) Technology and culture: cause versus cure

The application of technology has allowed an improvement in the standard of living of many people around the world. It has also led to an increase in resource consumption and in production of wastes. Some societies have become very dependent on technologies, which has been characterized as a "technico addiction".

In some cultures, there is virtually no questioning of the desirability of relying on technology. While some recognition is given to the environmental and social impacts associated with application of technology, there rarely is discussion regarding whether its application is the most appropriate response to a problem, particularly regarding the health of ecosystems. As a result, the mainstream view in many countries has been to apply technology to facilitate resource-intensive growth. Thus, technology has often been the solution to some problems and has created real opportunities, but at the same time has been part of the cause of environmental problems. Sustainable strategies for resource and environmental management will require a re-examination of the role of technology, which for some societies will require a re-examination of fundamental aspects of their culture.

(2) Humility versus arrogance

Consistent with the discussion of complexity and uncertainty in Chapter 1, Dovers and Handmer concluded that despite ever-increasing quantities of information, our understanding of the global environment is characterized by increasing uncertainty. This is troubling for many Western cultures, which have a strong belief that the power of science and technology can allow societies to understand and control nature. Dovers and Handmer conclude that we must be humble, and be able to recognize that our knowledge at best is incomplete, and at worst may be wrong in almost every respect. On the other hand, they argue that we must be arrogant enough to make decisions in the face of inevitable ignorance. The worrisome concern, in their view, is that such humility only seems to occur with regard to the status quo, and arrogance emerges in our willingness to defend the status quo. This situation does not bode well for action required to move societies away from unsustainable activities.

(3) Intergenerational versus intragenerational equity

A key tenet of sustainable development is that meeting basic human needs today should not preclude future generations from being able to meet theirs, often characterized as accepting the desirability of achieving *intergenerational* equity. Some societies address this systematically. For example, native Indian tribes in North America, such as the Algonquins, have traditionally included a person whose task it is to represent the seventh generation in the future during any important group decisions. In the jargon of sustainable development, that person is responsible for thinking about intergenerational equity issues.

However, as Dovers and Handmer comment, if resources are to be preserved or held for future use, how does a society choose how much should be used today and how much should be set aside? This question is more challenging when there are many people today whose basic needs are not being met. And that is the situation in the world, when literally millions of people are unable to meet their basic human needs, let alone decide whether they "need" a second computer or the latest Internet software in their households.

At a simplistic level, the solution for today's problem is either to enlarge the resource supply, which may be achieved through the application of new technologies, or to redistribute the resource supply. The former is likely to pose problems related to environmental degradation, and the latter involves substantial challenges to the privileges and powers of elites in a society. Hence, sustainable development must be able to address both intergenerational and intragenerational equity issues. If the latter are not dealt with, it is unlikely that people who do not have enough to eat each day will be concerned about the needs of future generations.

(4) Growth versus limits

The joining together of "sustainable" and "development" produces a concept which for many people is an *oxymoron* (words with opposite meanings used together, such as "cruel kindness" or "make haste slowly"). From this critical oxymoron perspective, "sustainable" means some activity that can be continued over the long term. In contrast, "development" is interpreted as growth, which implies a primarily physical or material addition to production. The concept of endless and increasing growth is one of the characteristics of a cancerous cell, and that, if left unchecked, often proves fatal. As a result, the idea of having endless growth raises the issue of whether ecological limits exist beyond which irreversible scarcity and/or degradation begin to occur. It is for this reason that the word *sustainability* has become preferred to *sustainable development*, as the former does not become an adjective for development, and therefore can be interpreted more broadly.

The challenge about such limits, which also been debated in the *carrying capacity* and *ecological footprint* literature, is that limits are usually not fixed or absolute (Box 4.3). Rather, they can be variable, and depend upon expectations and objectives. Furthermore, depending upon social values and technological capacity, limits may be stretched or constricted. For example, Stone Age people did not have the technology to take advantage of bronze or other metals. Such "resources" were present, but they were not accessible to those people until

---

Box 4.3 The concept of limits

The concept of sustainable development does imply limits – not absolute limits but limitations imposed by the present state of technology and social organization on environmental resources and by the ability of the biosphere to absorb the effects of human activities.

Source: World Commission on Environment and Development, 1987: 8.

---

technologies were developed that allowed them to transform what, until the arrival of such technology, had only been "neutral stuff".

The Brundtland Commission argued that growth was essential, if basic human needs were to be satisfied. However, the Commission also recognized the existence of constraints or limits. The dilemma is to determine what type of growth is needed to meet human needs, how to sustain such growth, and how to ensure that growth does not unacceptably degrade the environment which provides part of the base for growth.

(5) Individual versus collective interests

The achievement of sustainable development or sustainability requires some trade offs between individual and collective interests. Many Western cultures place a premium on the primacy of individual rights and choice, as reflected in dependency on the private automobile, attitudes to property rights and land tenure, and preference for individual household units. Many have argued that a desirable sustainable future will require much more use of public or mass transport, shifting of values from private land ownership to land stewardship, and acceptance of different sizes and types of housing.

Most environmental issues reflect *collective* problems emerging from many individual decisions which have *cumulative* negative consequences for the environment. Some individual sovereignty may therefore have to be constrained or forgone to achieve desired sustainability. Such a move will generate tension within and among nations, as those required to forfeit some of their individual "rights" are among the most wealthy, powerful and influential people in society, and may not forgo their "rights" (or privileges) with enthusiasm. Nation states also tend to be very protective of their sovereignty and rights. As a result, tension can be expected to occur in defining the appropriate balance between individual and collective rights.

(6) Democracy versus purpose

Sustainable development often is associated with an approach that seeks to empower local people, and to encourage their participation in development and environmental decisions. The rationale for this argument is that people living in an area will have to live with the impacts of development, and therefore are likely to be able to anticipate negative impacts. To achieve such empowerment, it is often argued that there needs to be both decentralization and deconcentration of decision making away from national to local governments.

There is much that is sensible about the arguments for greater local empowerment, including the improvement of ability to take advantage of local knowledge and understanding, as will be discussed in Chapter 9. However, as noted in the discussion about individual and collective interests, many environmental problems occur because of decisions taken by a variety of people in different places. If there is no capacity for overview or oversight, and no capacity for defining a general set of objectives or targets for something such as reducing emission of greenhouse gases, it is unlikely that local governments acting unilaterally will be able to make a significant contribution.

Thus, while there is a need to provide for more participation and local roles in environmental and resource management, there is also a need for the creation of a common purpose or interest, or vision, as discussed in Chapter 3, that people can work to achieve, even if decisions are being implemented at a local level. It is too simplistic to assume that if everything were delegated to the local level that all environmental problems would be resolved. Furthermore, if everything were delegated to the local level, it could be difficult to use a holistic or ecosystem approach which strives to consider an entire system (see Chapter 5).

(7) Adaptability versus resistance

Most societies and their institutions resist change. This resistance can be beneficial through creating stability. However, such resistance can also create severe conservatism and unwillingness to consider new visions, paths or actions. Indeed, often the "gatekeepers" who resist change are those who are best served by the status quo; they are not anxious to see their "comfort zone" affected.

A paradox exists because humans are among the most adaptable beings on Earth. Again and again humankind has demonstrated creativity through technological innovations that often have allowed increased food production from farms, or more fish to be caught from the oceans. Nevertheless, these types of innovations also have contributed to pressure on the environment and resources. Once again, tension and conflict exist regarding the best way to institute change. Change is not always painless. Some people will gain more than others from any new arrangements.

(8) Optimization versus spare capacity

The concept of optimization is based on the idea that it is desirable to achieve the best possible use of resources or the environment. Such a perspective assumes that unused resources are "wasted". It is also a very anthropocentric viewpoint, by implying that unless resources are developed for human benefit, they are not being used optimally. This view does not recognize that other living things are dependent on the environment, and that human interventions sometimes have adverse consequences for them. On the other hand, with steady population growth and the need to satisfy basic human needs, the notion of optimization is very attractive to many people.

The challenge is to determine a credible way to give value to aspects not readily measurable in quantitative or monetary terms. A more basic issue, however, is that when we aspire to use resources and the environment to the fullest possible extent, there will be little or no spare capacity which would be extremely useful to have if and when a decision is made to change direction. If there is no spare capacity, then any change will have to come through redistribution of present use, which will mean that some people will be worse off than they were before the changes. Spare capacity provides the flexibility to consider some changes that could provide gains for some without taking away from others. However, once again it is difficult to defend the protection of spare capacity, when some people's basic needs are not being satisfied.

The eight contradictions presented by Dovers and Handmer (1992) deserve attention if sustainable development or sustainability is to be transformed from a concept to action. In that regard, we should consider key issues, questions and opportunities associated with:

- the paradox of technology,
- humility or arrogance in the face of uncertainty,
- intergenerational and intragenerational equity,
- economic growth versus ecological limits,
- reconciliation of individual and collective interests,
- balance between democracy and purposeful action,
- different styles of resilience,
- the role of optimization.

These eight issues provide the start of an agenda for anyone contemplating how to create a sustainable development strategy.

4.3 Sustainable development in developed and developing countries

The Brundtland Commission was emphatic that it did not have a blueprint for sustainable development, and argued that each country or region would have to develop its own approach. In that context, it is not surprising that there have been different interpretations and emphases in developed and developing countries. In developed countries, the primary interest regarding sustainable development has been to integrate environmental and economic considerations into decision making. Considerable attention also has been focused upon intergenerational equity issues. Furthermore, developed countries have been concerned that in incorporating environmental issues they do not jeopardize their economic competitiveness, particularly given the low wage advantages of the developing countries. The developed countries also have argued that developing countries should modify their economic activities to avoid destruction of rain forests and other resources with global value.

In contrast, for developing countries the priority regarding sustainable development has been to meet the basic human needs of its present citizens and to ensure economic development. Thus, the focus has been more upon intragenerational than intergenerational issues. There has been understandable resentment from the developing countries when industrialized nations suggest that they should forgo development opportunities by harvesting rain forests in order to protect the global environment. The leaders of developing countries believe their citizens have the same right to have basic needs met, and that they should not be told not to do what all the developed countries did to achieve their high level of economic development. Indeed, at the Earth Summit in Rio de Janeiro during June 1992, many disagreements were based on the fundamentally different interpretation between developed and developing countries regarding what sustainable development should mean.

Box 4.4 Characteristics of future search conferences

Typically, 30 to 65 people meet for up to two-and-a-half days. We do five tasks of about three hours each. We explore in turn the past, present and future – of the world, ourselves, our institution. Everybody puts in information, discusses it, and decides what to do. The "technique" is a series of semi-structured dialogues. They take place in mixed, voluntary, and/or "stakeholder" groups, usually of eight people. Small groups report their conclusions to the whole.

We explore and validate differences, but we don't work them. . . . the task is finding common ground and future aspirations. As we discover them, that is where we plant our action flags. When we work on common ground and common futures, we tap deep wells of creativity and commitment.

Source: Weisbord, 1992: 6.

4.4 Future search conferences

Related to the ideas discussed in Chapter 3, Eric Trist and Fred Emery, working at the Tavistock Institute of Human Relations in England, developed a five-step procedure which became known as the *future search conference*, the purpose of which is to identify a desirable future. Marvin Weisbord and others began to use this concept in the late 1980s, and Weisbord popularized it in the USA. The characteristics of future search conferences are summarized in Box 4.4.

Weisbord suggested that future search conferences have four distinctive characteristics. First, a broad cross-section of stakeholders is involved. These people all affect each other, but they rarely if ever meet face-to-face. Second, the participants become involved in a self-managed process of discovery, dialogue, learning and planning. Third, the participants explore together an entire system with regard to its history, ideals, constraints, opportunities and global trends.

The fourth characteristic, related to conflict, is the most radical. In a future search conference, it is accepted that assembling a wide cross-section of participants will lead to disagreements and differences. The approach neither avoids nor confronts the differences. Instead, during a future search conference, the focus is on finding the widest common ground on which all participants can agree without having to be forced or compromised. From such a base, new actions are identified to achieve the desired future. As Weisbord (1992: 7) remarked, "we seek to hear and appreciate differences, not reconcile them. We seek to validate polarities, not reduce the distance between them. We learn, innovate and act from a mutual base of discovered ideals, world views, and future goals. Above all, we stick to business." Thus, a conscious decision is not to get trapped in irreconcilable differences.

Combining the above characteristics, a future search conference is designed to facilitate change through focusing attention on ideal futures rather than past problems, on common ground rather than conflict, and on shared assumptions rather than differences, as well as evolving from providing new input into

---

Box 4.5 Core values underlying future search conferences

1. The real world is knowable to ordinary people and this knowledge can be collectively and meaningfully organized.
2. People can create their own future.
3. People want opportunities to engage their heads and hearts as well as their hands.
4. Everyone participates as an equal.
5. Given the chance, people are much more likely to cooperate than fight.
6. The process should empower people to feel more knowledgeable about and in control of the future.
7. Diversity should be appreciated and valued.

Source: Weisbord, 1992: 13.

---

people toward self-management. All of these characteristics create a useful process for identifying a desirable future, *if* people are willing to participate in open dialogue, to accept their differences, and to be open to new outcomes. If that assumption is met, then the future search conference can be based on the core values shown in Box 4.5.

In the next section, the application of the future search conference process to develop a National Conservation Strategy for Pakistan is reviewed.

4.5 Conservation strategy for Pakistan

More than 40 countries have conservation strategies, many of which were developed to achieve sustainable development, as illustrated in Box 4.6. The conservation strategy for Pakistan was developed using the future search conference process. That process is presented here to illustrate how the future search conference or other visioning processes discussed in Chapter 3 can be used to identify, and move toward, a desirable future state. The experience from Pakistan is based upon conversations with, and a report prepared by, Rodger Schwass (1992), who facilitated the use of a future search conference in Pakistan.

When Pakistan became independent in 1947, it had a population of 34 million people, which grew to 110 million in the early 1990s. The World Bank forecast that the population would increase to 280 million by 2025. Until the early 1980s, economic growth had been strong, but at the cost of significant environmental degradation, ranging from over-irrigation associated with deforestation and overgrazing, as well as inadequate drainage, along with poor capacity for treatment or disposal of wastes. As Schwass (1992: 160) observed, "the present development process in Pakistan is clearly unsustainable".

During 1984, the International Union for Conservation of Nature and Natural Resources (IUCN) decided to support the World Wildlife Fund and other groups in Pakistan to prepare a national conservation strategy whose goal would

---

Box 4.6 National conservation strategy for Zambia

The theme of the strategy is that conservation and development are two sides of the same coin; *conservation can aid development* because it nurtures the productive capacities of natural resources and sustains the environment in which people live and work; *development can help conservation* by ensuring that people's needs are adequately supplied, so that they are not obliged to over exploit and damage soils, forests, fisheries, etc. in an effort to survive.

The goal of the strategy is to satisfy the basic needs of all the people of Zambia, both present and future generations, through the wise management of natural resources.

The objectives of conservation which must be fulfilled by future development activity are:

to ensure the *sustainable use* of Zambia's renewable natural resources;

to maintain Zambia's *biological diversity* (the range of biological material governing the quality and productivity of plants and animals, as well as the rich diversity of wild species);

to maintain essential *ecological processes and life-support systems* (soil regeneration and protection, nutrient recycling, protection and cleansing of waters, etc).

Source: Government of the Republic of Zambia, 1985: 8.

---

be, in the words of the IUCN Mission which identified the first necessary steps, to "use natural resources to satisfy the material, spiritual and cultural needs of all the people of Pakistan, both present and future generations" (Halle and Johnson, 1984: 20). Or, in the words of Schwass (1992: 161), the purpose would be to formulate a design for the future which would be sustainable in both environmental and economic terms.

4.5.1 Application of the future search conference process

The future search conference process involved various steps, each of which is described below.

Step 1: Appointment of facilitators

Schwass and the Director of the World Wildlife Fund in Pakistan were designated to organize the process.

Step 2: Interviews with key stakeholders, and preparation of overview report, late 1985

Interviews were conducted with about 80 key stakeholders, with particular emphasis on government agencies, in order to identify key issues. Following the interviews, a report was prepared. In terms of the present situation in Pakistan, the report noted that many people believed they had little opportunity or influence regarding the issues, and no mechanism existed for sharing concerns. The boundaries between government agencies were well established and discouraged communication and collaboration. The dominant priority was economic growth, with little concern for resource depletion or environmental degradation. Finally, the military control of the government inhibited open dialogue related to allocation of resources.

Step 3: Preparation of background papers focused on issues, 1986

A technical steering committee, consisting of the most senior government officials and private sector people, was established. Using the results of the interviews and the summary report, they identified 30 issues requiring attention, and arranged for 30 experts to prepare background papers on these issues. The issues ranged from population and waste management to resource depletion. Each paper described the present situation and what would likely happen if it were allowed to continue, identified desirable goals for the future, considered the main constraints and obstacles for change and ways to overcome them, and the most desirable directions for action. The 30 papers were completed within a three-month period, and were used as background information for the fourth stage, the actual search conference.

Step 4: Workshop, August 1986

Senior government ministers for key departments were briefed about the process and the purpose of the four-day August workshop, and were requested to attend the opening session to show their support for the conservation strategy, and to use the results in their ministries.

The intent had been to have the background papers distributed to all participants in advance of the conference, but this did not happen so the papers were given to people when they arrived at the conference. Schwass later remarked that this arrangement became a serious limitation on the depth of discussions during the workshop.

During the first day of the workshop, the 30 experts each made short presentations in which they highlighted the main points from their analyses. Emphasis was placed upon identifying key issues and possible actions. No detailed discussion was scheduled on this day, but after each paper there was opportunity for those in the audience to ask questions for clarification. It took 10 hours to get through all the presentations, and over 100 people, including some senior ministers, stayed throughout (Figure 4.1). The purpose of the first day was achieved: participants became aware of the wide array of problems and their interconnections. Most participants had arrived being only familiar with a small subset of the issues, and, perhaps for the first time, the environmental problems at a national level came into focus.

On the second day, about 85 people started the core of the future search conference process. Five groups were created, and each examined the current situation and where it might lead if no changes were made. In reporting back to the entire workshop, these groups highlighted the evidence which had been presented on the previous day, and emphasized the seriousness of the interlocked problems.

Figure 4.1   Salim Saifullah Khan, Minister of Environment and Urban Affairs (fifth person from left, front row) and Rahim Mahsud, Permanent Secretary of the Ministry (fourth person from left, front row), flanked by Ministry staff, at the future search conference for the National Conservation Strategy of Pakistan, National Conference Centre, Islamabad, 25–28 August 1986 (Rodger Schwass)

During day three, the focus shifted to develop scenarios (see Chapter 3) related to a more desirable future. It was recognized that some fundamental aspects of a more desirable future, such as re-emergence of democracy, were beyond the scope or power of the participants to effect, but other goals were identified that could be achieved if people and agencies were able to collaborate. The group concluded that it was possible to improve environmental conditions, while still achieving strong economic development. Later in the same day, the working groups identified the constraints which would hinder achieving the more desirable future, as well as listed the obstacles that would have to be overcome.

On the morning of day four, the groups concentrated on determining ways by which the constraints and obstacles could be overcome, as well as identifying opportunities that could be pursued in the immediate future. The rest of the day was spent in developing an action plan, to present toward the end of the day when the ministers returned. The final session of the workshop was used to present the action plan to the ministers (Figure 4.2). A key recommendation was creation of a new senior steering group at the political level to be chaired by the Prime Minister.

Step 5: Commitment at the senior political level

The ministers responded positively to the recommendations in the action plan, and two senior groups were created. One was a high-level committee, chaired by the person responsible for national planning in Pakistan, to follow up on the recommendations of the action plan at the working level of the government

Figure 4.2 The final plenary session of the future search conference for the National Conservation Strategy of Pakistan, was attended by nine cabinet ministers to hear the results and recommendations from the workshop (Rodger Schwass)

agencies. A second was a political steering committee, chaired by the Prime Minister. The second committee provided credibility or legitimacy to the conservation strategy.

The team that had facilitated the stages that led to the action plan was asked to prepare a report in which the recommendations were presented, and this was done later in 1986. While this work was progressing, other initiatives were taken to raise funds to allow the action plan to be implemented, with regard to new legislation, research and a media campaign to raise the awareness of Pakistanis to the threats to and opportunities for their country.

Step 6: Fund raising

After about one year, the Canadian International Development Agency agreed to provide financial support of more than one million dollars. This funding allowed IUCN to recruit a team to oversee implementation of the recommendations, and to create a Secretariat to assist in the necessary coordination and facilitation. The National Conservation Strategy team established working relationships with the key ministries and individuals responsible for the five-year planning process in Pakistan, to ensure that the concept of sustainable development would be included into the plan at all levels. The team also started publishing a quarterly newsletter to provide information to the media and the public about the situation in Pakistan, and about the National Conservation Strategy.

Step 7: Decentralizing activity

It was recognized that if significant progress were to be achieved, much of the work and action would have to occur at the state and community levels. As a result, in December 1988 and January 1989 more focused search workshops were held in four regional cities, and were attended by about 300 state and local representatives, business people, media representatives and public officials. Funding was provided for 40 demonstration projects, with financial support being provided from aid organizations in a variety of countries.

Step 8: Maintaining momentum

Conscious decisions were taken to build on the momentum established during the search conference process in August 1986, reminding us that creating a plan is important but not sufficient to create change (see Chapter 12 on implementation). The Pakistan National Conservation Strategy (NCS) was approved by the Pakistan Cabinet in March 1991, and systematic action was taken to obtain financial support from donors. An Implementation Conference was organized in October 1992 to bring together major donors and senior government officials to discuss the financial needs for the following five years. During the 1990s, support was obtained for specific aspects of the NCS from the World Bank and the Asia Development Bank, as well as from aid agencies in the USA, Canada, the UK, the Netherlands, Norway, Denmark and Germany. A NCS Division was created within the existing Ministry of Environment and Urban Affairs to provide an institutional home for the NCS work. Furthermore, a business–industry Round Table was created to engage the private sector in ways to implement stricter environmental regulations. The guest statement by Rodger Schwass provides an update of the evolution of the National Conservation Strategy initiative.

The experience in Pakistan illustrates how it is possible to use visioning as one element in a process to move towards a more desirable future. This experience also highlights, however, that developing the vision and the plan, while critically important, is not sufficient to facilitate change. Attention also has to be directed to and sustained regarding implementation. Box 4.7 summarizes the view of Rodger Schwass relative to the role of the future search conference.

4.6 Implications

Sustainability encourages attention to both the present and the future, to incorporate economic, environmental and social considerations into planning and management; to recognize that policies, laws, regulations and institutions must evolve to deal with the linkages and complexity of the world; to combine technical and local knowledge systems; and to seek to change underlying values, beliefs and attitudes. No single path is appropriate to achieve sustainability, as the ends and means towards it should reflect the specific needs, opportunities and obstacles existing at a specific place and time. Thus, the ability to understand

Guest Statement

Update on the Pakistan National Conservation
Strategy and its evolution

Rodger Schwass, Canada

In 1986, Pakistan had many excellent research institutes which covered the "environmental" field, (population, water resources, energy, forestry, tourism, range
management, etc.) but these seldom communicated with
each other. The media generally were silent on environmental issues, since there was little material for them
to work with. Governmental and quasi-governmental agencies such as the Ministry of Environment and Urban Affairs, the Ministry of Planning, the Pakistan
Council for Scientific and Industrial Research, and the Water and Power Development Authority rarely shared information or planned together for the future.
Pakistan's five-year plans were an aggregation of ideas from departments of government, without any effort to define a long-term vision. The Search Conference
was designed to achieve some common understanding of the environmental issues
facing the country and to develop some agreement on an action plan.

The Pakistan office of the International Union for the Conservation of Nature
(IUCN, also known as the World Conservation Union) has become the main
agency for implementing the NCS. It has grown from one half-time person, Ms
Aban Kabraji, to a staff of nearly 300. Her leadership has led directly to the
creation of IUCN organizations in many other countries in the region, including
Nepal, Sri Lanka, Thailand and China.

Provincial conservation strategies have been completed in all provinces. These
have benefited from the original work at the national level, including the original
vision, but they have incorporated many new elements. They also reflect growing
population and resource pressures and, because of the failure of the Government of
Pakistan to develop a sound tax base, the increasingly desperate shortage of funds
for essential social and environmental programmes.

IUCN Pakistan has created or encouraged dozens of NGOs at local, provincial
and national levels, which deal with some aspect of sustainability. The Sustainable
Development Policy Institute (SDPI) was created by IUCN Pakistan, with the en-
couragement of the Government of Pakistan, as an independent, non-governmental
body, to provide policy advice to government and to promote the development of
civil society. It now has more than 50 staff members and plays an important role
in policy development, training and communication for government, the private
sector and for NGOs. Through its training programmes, it also provides a mechanism for inculcating the values of sustainable development in civil servants, NGO
and media leaders. Other national organizations developed or encouraged by IUCN
to promote sustainable development include the Teachers' Resource Centre, the
Journalists' Resource Centre, SUNGHI (labour) and Shirkit Gah (women). The
leader of SUNGHI, Omar Asghar Khan, was appointed Minister for Environment,
Small Business, Local Development by the current interim (military) government.

In the absence of a stable government willing to providing funding, international donors have worked closely with IUCN to provide funding to implement

the National Conservation Strategy. A Donor Committee was established to review the NCS and provincial conservation strategies, to provide support for high priority activities which met each donor's criteria.

The original Search Conference occurred 15 years ago. Pakistan's population has almost doubled since then and the National Conservation Strategy developed in the late 1980s has been overtaken by events. Nevertheless, the NCS has generated a powerful network of environmental NGOs which have made a significant contribution to cleaning up Pakistan's industries, to protecting the remaining natural environment and to affecting personal behaviour. The media in Pakistan now give extensive coverage to environmental issues. Teachers have materials for environmental training in the schools. The link between better environmental management and a more productive economy has been established. Pakistan still requires massive land reform and the creation of a proper tax system to support local government, education, training, social equity and improved environmental management. These are matters which cannot be resolved by outside donors, but require action within Pakistan itself. The effort to bring ideas from the NGO sector into formal government structures may well be the key to reform in Pakistan.

It may be time to convene a new Search Conference in Pakistan, to build on the experience of the first and to aim more broadly at governmental reform.

Rodger Schwass is Professor Emeritus and Senior Scholar at the Faculty of Environmental Studies, York University, Toronto, Canada. He has been Dean of the Faculty and Director of York International and has been involved in environmental, social and communications projects in over 30 countries since 1967. Prior to joining York University in 1976, he was Vice President of Acres International and Hedlin-Menzies and Associates, two Canadian consulting companies and was for seven years, Manager of National Farm Radio Forum, a rural adult education programme in Canada. He is currently a member of the Board of the International Development Research Centre, Ottawa.

Box 4.7 The role of a future search conference

The search process itself is not sufficient to ensure action. It must be embedded in a setting that permits an action team to form. There must be common ground on which various groups can stand together and there must be some common elements in their visions of the future. Committed people, with enough resources at their disposal, are the essential factor, if the plans that are made are to be translated into action.

Ideally, all relevant stakeholders should be present and capable of taking action, the environment should be open, and senior authorities should be receptive. At least some resources need to be available if there is to be hope of reform and new investment.

Source: Schwass, 1992: 166.

the context of a problem-solving situation, and to custom design a solution, should be key elements for a successful approach.

References and further reading

Albrecht S L 1995 Equity and justice in environmental decision-making: a proposed research agenda. *Society and Natural Resources* 8: 67–72

Alexander G 2000 Information-based tools for building community and sustainability. *Futures* 32: 317–37

Amalric F 1998 Sustainable livelihoods: entrepreneurship, political strategies and governance. *Development* 41(30): 31–41

Atkinson G 2000 Measuring corporate sustainability. *Journal of Environmental Planning and Management* 43: 235–52

Barbier E B 1987 The concept of sustainable development. *Environmental Conservation* 14: 101–10

Barrett C B and R Grizzle 1999 A holistic approach to sustainability based on pluralism stewardship. *Environmental Ethics* 21: 23–42

Bartelmus P 1994 *Environment, Growth and Development: The Concepts and Strategies of Sustainability*. London, Routledge

Berke P and M M Conroy 2000 Are we planning for sustainable development? An evaluation of 30 comprehensive plans. *Journal of the American Planning Association* 66: 21–33

Bloomfield P 1998 The challenge of Agenda 21 at the key stage 1, 2 and 3: practical ideas on environment, values and citizenship. *Geography* 83: 97–104

Bowen W M, M J Salling, K E Haynes and E J Cryan 1995 Toward environmental justice: spatial equity in Ohio and Cleveland. *Annals of the Association of American Geographers* 85: 641–63

Boyce J 1995 Equity and the environment: social justice today as a prerequisite for sustainability in the future. *Alternatives* 21: 12–24

Briassoulis H 1999 *Who* plans *whose* sustainability: alternative roles for planners. *Journal of Environmental Planning and Management* 42: 889–902

Briggs J, G Dickinson, K Murphy, I Pulford, A E Belal, S Moalla, I Springuel, S J Ghabbour and A-M Mekki 1993 Sustainable development and resource management in marginal environments: natural resources and their use in the Wadi Allaqui region of Egypt. *Applied Geography* 13: 259–84

Budhya G and S Benjamin 2000 The politics of sustainable cities: the case of Bengare, Mangalore in coastal India. *Environment and Urbanization* 12: 27–36

Bulkeley H 2000 Down to Earth: local government and greenhouse policy in Australia. *Australian Geographer* 31: 289–308

Cairns J 1998 The Zen of sustainable use of the Planet: steps on the path to enlightenment. *Population and Environment* 20: 99–108

Cobb D, P Dolman and T O'Riordan 1999 Interpretations of sustainable agriculture in the UK. *Progress in Human Geography* 23: 209–35

Cutter S L 1995 Race, class and environmental justice. *Progress in Human Geography* 19: 111–22

Daly H E and J B Cobb (eds) 1989 *For the Common Good*. Boston, Beacon Press

Danaher M 1998 Towards sustainable development in Japanese environmental policy-making. *Sustainable Development* 6: 101–10

Dias A K and M Begg 1994 Environmental policy for sustainable development of natural resources. Mechanisms for implementation and enforcement. *Natural Resources Forum* 18: 275–86

Dovers S R 1993 Contradictions in sustainability. *Environmental Conservation* 20: 217–22

Dovers S R and J W Handmer 1992 Uncertainty, sustainability and change. *Global Environmental Change* 2: 262–76

Ellery W N and T S McCarthy 1994 Principles for the sustainable utilization of the Okavango Delta ecosystem. *Biological Conservation* 70: 159–68

Elliott J 1999 *An Introduction to Sustainable Development.* second edition, London and New York, Routledge

Emery M and R E Purser 1995 *The Search Conference: A Comprehensive Guide to Theory and Practice*, San Francisco, Jossey-Bass

Emery M, R E Purser and F Emery 1996 *The Search Conference: A Powerful Method for Planning Organizational Change and Community Action.* San Francisco, Jossey-Bass

Englehardt J D 1998 Ecological and economic risk analysis of Everglades: phase 1 restoration alternatives. *Risk Analysis* 18: 755–71

Fagin A and P Jehlicka 1998 Sustainable development in the Czech Republic: a doomed process? *Environmental Politics* 7: 113–28

Floyd D W, S L Vonhof and H E Sefang 2001 Forest sustainability: a discussion guide for professional resource managers. *Journal of Forestry* 99(2): 8–27

Francis G 1995 PI's perspective on sustainability. *Eco-Nexus* Eco-Research Project Newsletter, Waterloo, Ontario, University of Waterloo: 4

Franks T 1994 Managing sustainable development: Abdul Karim's dilemma. *Project Appraisal* 9: 205–10

Frazier J G 1997 Sustainable development: modern elixir or sack dress? *Environmental Conservation* 24: 182–93

Furuseth O and C Cocklin 1995a An institutional framework for sustainable resource management: the New Zealand model. *Natural Resources Journal* 35: 243–73

Furuseth O and C Cocklin 1995b Regional perspectives on resource policy: implementing sustainable development in New Zealand. *Journal of Environmental Planning and Management* 38: 181–200

Gibbs D 2000 Ecological modernization, regional economic development and regional development agencies. *Geoforum* 31: 9–19

Government of the Republic of Zambia and International Union for Conservation of Nature and Natural Resources 1985 *The National Conservation Strategy for Zambia.* Gland, Switzerland, IUCN

Gow D D 1992 Poverty and natural resources: principles for environmental management and sustainable development. *Environmental Impact Assessment Review* 12: 49–65

Grubb M, M Koch, A Munson, F Sullivan and R Thomson 1993 *The Earth Summit Agreements: A Guide and Assessment.* London, Earthscan

Grumbine R E 1994 Wildness, wise use and sustainable development. *Environmental Ethics* 16: 227–49

Halle M and B Johnson 1984 *A National Conservation Strategy for Pakistan: First Steps.* Gland, Switzerland, IUCN

Holland M M 1996 Ensuring sustainability of natural resources: focus on institutional arrangements. *Canadian Journal of Fisheries and Aquatic Sciences* 53, Supplement 1: 432–39

Howarth R B 1997 Sustainability as opportunity. *Land Economics* 73: 569–79

Infield M and W M Adams 1999 Institutional sustainability and community conservation: a case study from Uganda. *Journal of International Development* 11: 305–15

International Union for the Conservation of Nature, United Nations Environment Programme and World Wildlife Fund 1980 *World Conservation Strategy: Living Resource Conservation for Sustainable Development*. Gland, Switzerland, IUCN

International Union for the Conservation of Nature, United Nations Environment Programme and World Wildlife Fund 1991 *Caring for the Earth: A Strategy for Sustainable Living*. Gland, Switzerland, IUCN, UNEP and WWF

Jamieson D 1998 Sustainability and beyond. *Ecological Economics* 24: 183–92

Jones T 1996 Local authorities and sustainable development: turning policies into practical action through performance review – a case study of the London Borough of Hackney. *Local Environment* 1: 87–106

Karshenas M 1994 Environment, technology and employment: towards a new definition of sustainable development. *Development and Change* 25: 723–56

Khan N A and S K Khisa 2000 Sustainable land management with rubber-based agroforestry: a Bangladeshi example of uplands community development. *Sustainable Development* 8: 1–10

Kleinman P J A, D Pimentel and R B Bryant 1995 The ecological sustainability of slash-and-burn agriculture. *Agriculture, Ecosystems and Environment* 52: 235–49

Knight D, B Mitchell and G Wall 1997 Bali: sustainable development, tourism and coastal management. *Ambio* 26: 90–6

Loucks D P 2000 Sustainable water resources management. *Water International* 25: 3–10

Lundqvist L J 2000 Capacity-building or social construction? Explaining Sweden's shift towards ecological modernization. *Geoforum* 30: 21–30

Martopo S and B Mitchell (eds) 1995 *Bali: Balancing Environment, Economy and Culture*. Department of Geography Publication Series No. 44, Waterloo, Ontario, University of Waterloo

McDonald G T 1996 Planning as sustainable development. *Journal of Planning Education and Research* 15: 225–36

McManus P A 2000 Beyond Kyoto? Media representation of an environmental issue. *Australian Geographical Studies* 38: 306–19

Mercer D and B Jotkowitz 2000 Local Agenda 21 and barriers to sustainability at the local government level in Victoria, Australia. *Australian Geographer* 31: 163–81

Mitchell B 1994 Institutional obstacles to sustainable development in Bali, Indonesia. *Singapore Journal of Tropical Geography* 15(2): 145–56

Mitchell B 1998 *Sustainability: A Search for Balance*. Faculty of Environmental Studies Research Lecture, Waterloo, Ontario, University of Waterloo

Murphy J and A Gouldson 2000 Environmental policy and industrial innovation: integrating environment and economy through ecological modernization. *Geoforum* 30: 33–44

Nagpal T 1995 Voices from the developing world: progress toward sustainable development. *Environment* 37: 10–15, 30–5

O'Riordan T 1999 From environmentalism to sustainability. *Scottish Geographical Journal* 115: 151–65

O'Riordan T and H Voisey (eds) 1998 *The Transition to Sustainability: The Politics of Agenda 21 in Europe*. London, Earthscan

Overton J 1993 Fiji: options for sustainable development. *Scottish Geographical Journal* 109: 164–70

Penning-Rowsell E, P Winchester and J Gardiner 1998 New approaches to sustainable hazard management for Venice. *Geographical Journal* 164: 1–18

Pezzey J C V 1997 Sustainability constraints versus "optimality" versus intertemporal concern, and axioms versus data. *Land Economics* 73: 448–66

Podobnick B 1999 Toward a sustainable energy regime: a long-wave interpretation of global energy shifts. *Technological Forecasting and Social Change* 62: 155–72

Price J 1999 Barriers to the development of sustainable waste management in New York city. *Environments* 27: 15–24

Reed M G 1999 "Jobs talk": retreating from the social sustainability of forestry communities. *Forestry Chronicle* 75: 755–63

Reed M G and O Slaymaker 1993 Ethics and sustainability: a preliminary perspective. *Environment and Planning A* 25: 723–39

Robertson W A 1993 New Zealand's new legislation for sustainable resource management: the Resource Management Act 1991. *Land Use Policy* 10: 303–11

Robinson J, G Francis, R Legge and S Lerner 1990 Defining a sustainable society: values, principles and definitions. *Alternatives* 17: 36–46

Sabin P 1998 Searching for middle ground: native communities and oil extraction in the Northern and central Ecuadorian Amazon, 1967–1993. *Environmental History* 3: 144–68

Salim E 1991 Towards a sustainable future. *Development* 2: 61–3

Schwass R 1992 A conservation strategy for Pakistan. In M R Weisbord *et al.* (eds) *Discovering Common Ground*. San Francisco, Berrett-Koehler, pp. 158–69

Serageldin I 1995 *Toward Sustainable Management of Water Resources*. Washington, DC, World Bank

Stackhouse J 2000 *Out of Poverty: and into Something More Comfortable*. Toronto, Random House Canada

Thompson I B 1999 Sustainable rural development in the context of a high mountain national park: the Parc National de la Vanoise, France. *Scottish Geographical Journal* 115: 297–318

UNESCO Working Group M.IV. 1998 *Sustainability Criteria for Water Resource Systems*. Cambridge, Cambridge University Press

United Nations Conference on Environment and Development 1992 *The Rio Declaration on Environment and Development*. Geneva, Switzerland, UNCED Secretariat

van der Walls J F M 2000 The compact city and the environment: a review. *Tijdschrift voor Economische en Sociale Geografie* 91: 111–21

Vargas C M 2000 Community development and micro-enterprises: fostering sustainable development. *Sustainable Development* 8: 11–26

Velasquez L S 1998 Agenda 21; a form of joint environmental management in Manizales, Colombia. *Environment and Urbanization* 10: 9–36

Voisey H, C Beuermann, L A Sverdrup and T O'Riordan 1996 The political significance of Local Agenda 21: the early stages of some European experience. *Local Environment* 1: 33–50

Weisbord M R 1992 Applied common sense. In M R Weisbord *et al.* (eds) *Discovering Common Ground*. San Francisco, Berrett-Koehler, pp. 3–17

Weisbord M B and S Janoff 2000 *Future Search: An Action Guide to Finding Common Ground in Organizations and Communities*. San Francisco, Berrett-Koehler

White R R 1992 The road to Rio or the global environmental crisis and the emergence of different agendas for rich and poor countries. *International Journal of Environmental Studies, A* 41: 187–201

World Commission on Environment and Development 1987 *Our Common Future*. Oxford and New York, Oxford University Press

Young M D 1992 *Sustainable Investment and Resource Use: Equity, Environmental Integrity and Economic Efficiency*. Paris, UNESCO; and Carnforth, England and Park Ridge NJ, Parthenon Publishing Group

Young M E 1993 *For our Children's Children: Some Practical Implications of Inter-generational Equity and the Precautionary Principle*. Resource Assessment Commission Occasional Paper Number 6, Canberra, Australian Government Printing Service

Chapter 5

Ecosystem Approach

5.1 Introduction

In Chapter 4, attention focused on the concept of *sustainability*, which often is held up as a vision or an ends for resource and environmental management. The *ecosystem approach* can be viewed as one means to achieve sustainability or sustainable development, and it is in that context that the ecosystem approach is considered in this chapter. The distinction between ends and means is important, as too often in resource and environmental management the ecosystem approach is treated as an end in itself, rather than as a means to an end. In the next section, various views regarding the ecosystem approach are considered. That will be followed by examination of comprehensive and integrated interpretations of the ecosystem approach, agroecosystem analysis and some applications of the concept.

5.2 Nature of the ecosystem approach

As the comments in Boxes 5.1 and 5.2 highlight, there are many challenges to develop and implement an ecosystem approach. The very complexity that the ecosystem attempts to capture can be overwhelming, and leave the manager

---

Box 5.1 Complexity and ecosystems

"Ecosystem management" increasingly provides the goals and framework for land, wildlife and protected area management. Broadly speaking, ecosystem management is the process of managing and understanding the interaction of the biophysical and socio-economic environments within a self-similar, self-maintaining regional or larger system. Ecosystem management involves finding institutional and administrative, as well as scientific, ways of managing *whole* ecosystems instead of the often small, arbitrary management units that are found almost everywhere. This is not, in practice, an easy task, and certainly it is easier said than done. Moving ecosystem management beyond the rhetoric and empty adherence that have been the fate of many great new ideas in resource management is crucial.

Source: Slocombe, 1998b: 31.

Box 5.2 Paradox and challenge

As environmental degradation and change continues, decision makers and managers feel significant pressure to rectify the situation. Scientists, in turn, find themselves under pressure to set out simple and clear rules for proper ecosystem management.

. . . However, systems theory suggests that ecosystems are inherently complex, that there may be no simple answers, and that our traditional managerial approaches, which presume a world of simple rules, are wrong-headed and likely to be dangerous. In order for the scientific method to work, an artificial situation of consistent reproducibility must be created. This requires simplification of the situation to the point where it's controllable and predictable. But the very nature of this act removes the complexity that leads to emergence of the new phenomena which makes complex systems interesting. If we are going to deal successfully with our biosphere, we are going to have to change how we do science and management. We will have to learn that we don't manage ecosystems, we manage our interaction with them. Furthermore, the search for simple rules of ecosystem behaviour is futile.

Take for example the diversity–stability hypothesis. This is a classic example of the kind of simple rule people are looking for. . . . [However] ecosystems are dynamic and constantly changing. Stability gives way to the notion of a shifting steady mosaic. Thus, the diversity–stability hypothesis evaporates because the basic concepts of diversity and stability are just too simple to describe the complex reality of ecological phenomena.

Source: Kay and Schneider, 1994: 33–4.

---

unsatisfied because answers are not forthcoming, or are provided after too long a period of time. Advocates of an ecosystem approach need to be sensitive to the needs of managers. Otherwise, a concept which might be conceptually sound may not be accepted or used because it does not meet their practical needs. It is not enough to be convinced about the conceptual value of an ecosystem approach. We also must consider how it can be applied to solve real-world problems in a timely manner. In his guest statement, David Pitts highlights some of the opportunities and challenges in applying an ecosystem approach.

In this section, attention focuses upon the nature of an ecosystem approach, and on some of the ideas or guidelines associated with it. Bocking (1994: 12) concluded that during the 1990s the ecosystem concept signified "the study of living species and their physical environment as an integrated whole. In environmental management, its significance is understood to lie in a comprehensive, holistic, integrated approach." This definition or interpretation captures the essence of what many people associate with an ecosystem approach – the concept of a system, as well as its component parts and the linkages among those parts. However, critics worry that if everything is connected to everything else, then the ecosystem approach can expand the scope of any problem to unmanageable proportions, and thus lead to analyses and planning processes becoming impractical.

Guest Statement

Reflections on experience in working with an
ecosystem approach in Australia

David Pitts, Australia

Integrated resource management is a complex process
that involves multiple stakeholders from both government and non-government sectors; multiple and
sometimes competing objectives; multiple agencies
from different tiers of government; and multiple and
sometimes overlapping jurisdictions. In many cases there
is competition for scarce resources and different perceptions of values among
different interest groups.

Ecosystem approaches to management involve a holistic, ecological view of
natural resources and the environment. They recognize that human activity and
production take place within, and not outside of, ecosystems. Furthermore, ecosystem approaches are not just altruistic. The maintenance of ecosystem health
and viability is a necessary condition for economic sustainability in industries as
diverse as fishing, nature-based tourism and various forms of agriculture.

Ecosystem approaches strive for outcomes that involve long-term economic
benefit rather than short-term financial gain. The desired outcomes of ecosystem
approaches typically include:

- the protection of natural capital;
- the long-term protection of ecosystems and ecological processes;
- the maintenance of biological diversity;
- the sustainable use and harvesting of resources; and
- the recognition and protection of the traditional knowledge, customs and
  practices of indigenous peoples.

From a planning and management point of view, the important point is that
these outcomes are not sector dependent, nor are they confined to a single jurisdiction. They are concerned with ecosystems, ecological processes and combinations
of uses (both multiple and sequential) that are in the best overall, long-term
community interest.

The dilemma, of course, is that the majority of legislative and administrative
arrangements related to resource planning are directed at the management of
specific sectors by means of relatively traditional governance structures based on
areas of functional responsibility. How then, can planning approaches based on
sectors deliver outcomes that are based on ecosystems, ecological processes and the
optimum long-term use of resources across all sectors?

In Australia, some resource planning and management initiatives have moved to
encompass a broader range of environmental and community objectives. In many
cases, however, such initiatives are project driven or they tend to arise as solutions
to particular local issues rather than as part of a broader policy planning framework. Nevertheless, instances of systematic and comprehensive regional planning
initiatives increasingly are being explicitly underpinned by an ecosystem approach
to resource and environmental management. These instances include:

- Regional Forest Agreement processes in several Australian States (www.rfa.gov.au);
- Marine Protected Area management in the Great Barrier Reef Region (www.gbrmpa.gov.au);
- Large-scale river basin management in the Murray-Darling catchment (www.mdbc.gov.au);
- Integrated catchment and coastal management for Moreton Bay and its catchment in southeast Queensland (www.healthywaterways.env.qld.gov.au); and
- Regional oceans planning (www.oceans.gov.au).

While these examples hold considerable hope for the future, two important caveats need to be applied. First, comprehensive examples of ecosystems approaches have generally emerged as solutions to a resource management crisis – ecosystem approaches are not yet embedded as a normal way of doing business in the culture and statutes of most resource planning and management agencies. Second, a significant gap still occurs between planning and action. Coordinated planning using an ecosystems approach is relatively easy – coordinated action across all sectors and agencies in order to achieve desired ecosystem outcomes is quite a different matter actually!

Dr David Pitts is a Director of Environment Science and Services – a small professional consulting practice that has been providing specialized resource planning and management services to Commonwealth, State and local government agencies in Australia for more than 25 years. He has been involved in a diverse range of projects including planning within the Great Barrier Reef Region, regional planning for Cape York Peninsula, the development of Australia's Oceans Policy, preparation of the Queensland Ecotourism Strategy and river basin and coastal management for the Brisbane River Catchment and Moreton Bay.

Bocking's interpretation omits one feature, however, also usually associated with the ecosystem approach. This feature is that humans are part of, not separate from, the ecosystem. One implication is that analysts and planners should not be unduly *anthropocentric* (human-centred), but during management should include the needs of non-human species with which we share the planet.

Cortner and Moote (1999: 37) argued that ecosystem management is different from traditional resource management in that the latter focused on the manipulation and harvesting of resources, with humans in a controlling role. In contrast, ecosystem management is concerned with preserving intrinsic values or natural conditions of the ecosystem, and the commodity values become secondary by-products, much as "interest" relative to "capital". The over-riding priority is to conserve ecological integrity, with levels of commodity and amenity outputs adjusted to meet that dominant goal. Given this perspective, they argued that four basic themes are associated with ecosystem management: (1) socially defined goals and objectives; (2) holistic, integrated science; (3) adaptable institutions; and (4) collaborative decision making.

Box 5.3 Problems of prediction

The structure and dynamics of all ecosystems are to a greater or lesser extent the result of stochastic processes. Indeed, most exhibit sharp shifts which are often crucial to their structure. These stochastic effects preclude deterministic management or planning policies which assume the possibility of perfect prediction. This principle emphasizes the need for ecosystem management and planning to be flexible and to make due allowance for unexpected events.

Source: Walker and Norton, 1982: 332.

5.2.1 Obstacles in developing ecological principles to use in an ecosystem approach

One of the challenges for ecologists and the ecosystem approach is to provide "sound principles" to guide resource and environmental management. However, Norton and Walker (1982) concluded that there are few unambiguous and relevant principles (Box 5.3). Several reasons account for this situation. First, many of the principles are more "normative" (moral or ethical) than "positive" (scientific). For example, the idea that we should strive to avoid foreclosure of options is a normative rather than a scientific concept. Norton and Walker concluded that mixing normative and scientific issues raises questions about the credibility of ecological principles. Normative questions have to be addressed, but ecological principles usually cannot be expected to provide answers to value-based questions.

Second, positive or scientific "principles" occur at two extremes. At one end, general statements have been produced which are informative but not readily applicable. An example would be the idea that diversity leads to stability, and therefore that diversity is a desirable condition (see comments in Box 5.2). At the other end, "principles" related to *carrying capacity* have been developed for specific situations such as range, park or lake management. Such principles are helpful for those specific conditions, but they usually are not transferable to other situations, and certainly do not and cannot provide answers to questions about the best use of a particular landscape.

Third, Norton and Walker argued that tight laws applicable in all conditions are unlikely to exist in ecology. As they noted, too many "ifs", "buts" and "maybes" exist to allow for definitive principles. Cortner and Moote (1999: 45–6) supported this view when noting that one criticism of ecosystem management has been little consensus exists regarding the meaning of new terminology. For instance, while many agree on the importance of maintaining ecological sustainability, integrity, productivity and biological diversity, huge disagreements exist regarding what these concepts mean in an operational sense and also what they imply in terms of management outcomes. This conclusion reflects the considerable complexity and uncertainty associated with ecosystems, and our limited understanding of them.

5.2.2 Major themes in ecosystem management

Notwithstanding the very real obstacles to developing ecological principles to serve as the basis for ecosystem management, attempts have been made to identify dominant themes relevant for ecosystem management. For example, Grumbine (1994: 29–30) identified ten dominant themes:

(1) **Hierarchial context.** It is not sufficient to focus upon any *one* level (genes, species, populations, ecosystems, landscapes) of the biodiversity hierarchy. Attention must be given to connections among all levels. Such an approach is often characterized as a *systems* perspective.

(2) **Ecological boundaries.** Resource and environmental management requires the attention to biophysical or ecological rather than administrative or political units. For example, migratory birds do not respect political or administrative boundaries, and any management plan for them must be based on boundaries that relate to their needs and activities. A difficulty, of course, is that an appropriate ecoregion for migratory birds may not be appropriate for managing water, and there quickly could be many, overlapping ecosystem units being used.

(3) **Ecological integrity.** Much attention has been devoted to ecological integrity, which is usually interpreted to mean protecting total natural diversity (species, populations, ecosystems) along with the patterns and processes which maintain that diversity. The emphasis normally has been upon conserving viable populations of native species, maintaining natural disturbance regimes, reintroducing native, extirpated species, and achieving representation of ecosystems across natural ranges of variation. However, as the comments in Box 5.4 illustrate, many problems must be dealt with when deciding which conditions represent "integrity".

(4) **Data collection.** To manage ecosystems, there has to be research and data collection, particularly regarding functional (what if?) rather than descriptive (what is?) questions. Data are required regarding habitat

---

Box 5.4 What represents integrity?

In essence, ecosystem management aims to restore forests to some biological condition that reflects fewer human impacts, but just *what* condition is a matter of arbitrary selection.

. . . In Europe, . . . , the distinction between forests before and after human settlement is virtually impossible to make, and, as a result, determining desired forest condition is more difficult. Should forests there be returned to their pre-Celtic condition before about 15,000 BC, to their pre-Roman condition, to their condition in the Middle Ages, or what? This question inevitably raises more fundamental questions – namely, whether less human impact is always preferable to more human impact, and, if so, why. These questions do not have scientific answers.

Source: Sedjo, 1995: 10, 19.

Box 5.5 Adaptiveness and flexibility

. . . , institutions such as organizations, laws, policies, and management practices need to be flexible, in order that they may rapidly adapt to changes in social values, ecological conditions, political pressures, available data, and knowledge. Considerable emphasis is put on the value of decentralized decision-making arrangements to avoid the rigidities of highly centralized institutional arrangements with inflexible prescriptions.

Source: Cortner and Moote, 1999: 44.

---

inventory and classification, baseline species, disturbance regime dynamics and population assessment.

(5) **Monitoring.** Managers must record the results from their decisions and actions, so that successes and failures can be measured and documented. Useful information and insight are generated by systematic monitoring. This aspect is addressed in more detail in Chapters 6 and 13.

(6) **Adaptive management.** As considered in Chapter 6, an adaptive approach assumes incomplete understanding of ecosystems, and expects both turbulence and surprise (Box 5.5). Emphasis is placed on treating management as a learning experience, and encourages management to be viewed as a series of experiments from which new knowledge leads to continuous adjustments and modifications. Monitoring is a key activity in adaptive management.

(7) **Interagency cooperation.** Whether biophysical or political boundaries are used, there will have to be sharing and cooperation among municipal, state, national and international agencies, as well as the private sector and non-government organizations. Planners and managers will have to improve their capacity to deal with conflicting legal mandates and management objectives. For example, within one government, an agricultural agency may emphasize removal of wetlands to improve crop production, while a natural resource agency may emphasize protection or restoration of wetlands to improve wildlife habitat.

(8) **Organizational change.** To implement an ecosystem approach there often must be alterations in the structures and processes used by resource and environmental management agencies. Such changes can be relatively simple (creation of an inter-agency coordinating group) to fundamental (re-allocating power and changing basic values or principles). The key point is that most agencies are not oriented or structured to use an ecosystem approach. An example of such a change is provided in Section 5.5.1.

(9) **Humans embedded in nature.** As already noted earlier in this chapter, an ecosystem approach requires people to be considered as part of, rather than as separate from, natural systems. People cannot be separated from nature. The comments in Box 5.6, referring to the USA, illustrate challenges which must be overcome in some societies.

Box 5.6 Conflicting ideas about the role of humans and nature

An ecosystem approach to land policy encounters resistance to the degree that it is inconsistent with the values, assumptions, institutions, and practices that shape the prevailing social arrangement which affect the custody and care of the land. . . . Thus, the factors involved in banking, taxation, insurance, and property law, when woven into a non-ecological matrix of public land policy, afford a very resistant, inadvertent barrier to an ecosystems approach.

. . . To conceive an ecosystem approach to public land policy, one must have first arrived at an ecological viewpoint toward the world of man and nature. But this is not the viewpoint from which pioneers, land speculators, farmers, miners, stockmen, lawyers, bankers or local government officials have commonly seen the land . . . An ecosystems approach . . . would impose constraints upon single purpose approaches to the environment and would arouse hostility among individuals whose single purpose pursuits would therefore be constrained.

Source: Caldwell, 1970: 204–5.

---

(10) **Values.** An ecosystem approach must recognize that both scientific and traditional knowledge, and human values, will be involved. Indeed, human values will have the dominant role in the setting of goals for ecosystem management. Thus, ecosystem management is not just a scientific endeavour. It must also incorporate human values.

Give the above ten themes, Grumbine (1994: 31) developed the following definition of ecosystem management: "Ecosystem management integrates scientific knowledge of ecological relationships within a complex sociopolitical and values framework toward the general goal of protecting native ecosystem integrity over the long term." If this definition is modified to include traditional as well as scientific knowledge (discussed more fully in Chapter 9), then this interpretation of ecosystem management is one that provides a good focus. The principles identified in Table 5.1 provide a further basis from which ecosystem management approaches can be designed.

5.3 Distinction between comprehensive and integrated approach

The ecosystem approach encourages analysts and planners to consider the "big picture" by emphasizing entire systems, their component parts, and the relationships among those parts. Such a perspective is important, as it reminds us that many water problems (pollution, flooding) cannot be resolved by focusing only on water. Many sources of pollution are from land-based activities, and flood damage potential is strongly influenced by land uses. Conversely, many land-based problems (dropping agricultural production, loss of biodiversity) occur from too much or too little water. Thus, it is important that we take the "big picture" into consideration, and not become unduly focused on one element or

Table 5.1 General principles and characteristics of ecosystem management contained in the US Federal Ecosystem Management Initiative

---

- Science and other disciplines are integrated into an holistic and integrated approach to managing natural resources.
- Ecosystems and biodiversity are managed in the context of natural spatial boundaries as well as temporal horizons which ecosystems constantly change.
- Ecosystem management recognizes that ecosystem components are interconnected, that they include humans, and that altering one component may have effects on others.
- For policy making, sound scientific information is used instead of subjective judgement.
- Management strategies and techniques are adapted as new information becomes available.
- Uncertainty is acknowledged in measuring and evaluating ecosystem characteristics.
- Institutions must become adaptable to new approaches and to cooperation.
- Partnerships based on resource stewardship are formed among stakeholders for collaborative democratic decision making and sharing resource costs and benefits.
- Conflict management is used to resolve differences among stakeholders.
- Ecosystem management seeks to achieve balanced socio-economic and environmental sustainability through environmental ethics and resource stewardship.

Source: Based on Malone, 2000: 10.

component of an ecosystem. However, as already noted, a danger arises in knowing how widely to cast the net, how large an ecosystem to consider, and how many components and relationships to address. If the "big picture" becomes too big, the planners and analysts may become so entangled in the complexities of multiple components and linkages that they are unable to complete their analysis in a reasonable period of time.

An ecosystem approach is synonymous with a *holistic* perspective. However, such a perspective can be interpreted in either a *comprehensive* or an *integrated* manner. It is argued here that too often analysts and planners have advocated an ecosystem or holistic approach without having clearly thought through what that implies. By default, a holistic and a comprehensive approach have been considered the same, and this has led to some problems.

By definition, *comprehensive* means all inclusive. As a result, a comprehensive interpretation of a holistic approach indicates that whatever system is defined, the analyst or planner should examine all the components and all the relationships. Such an interpretation has several implications. First, it creates expectations that if we work diligently and study everything, it will be possible to understand the ecosystem, and therefore be able to control or manage it. Second, it also almost guarantees that a significant amount of time will be required to complete the analysis and a plan. As a result, there is a high probability that "the plan" will be a historical rather than a strategic document, because by the time all of the work is completed events may have swept past the plan.

In contrast, an *integrated* approach retains most of the core ideas of being holistic, but is more focused and therefore is more practical. The key distinction is that an integrated approach does not seek to analyse all components and linkages, but concentrates upon those judged to be key components and linkages.

Eventually, if enough components and linkages are examined, the integrated approach would expand to become the same as a comprehensive approach.

The integrated interpretation results in a more limited focus being taken for a number of reasons. First, it accepts that we are unlikely to be able to understand all of the variation in a system. If analysts or planners could account for and understand the components that cause 75–80% of the variability in a system, they would usually be very satisfied. Second, usually a relatively small number of variables cause a large proportion of variation. As a result, understanding their role is usually sufficient for developing effective management strategies. All the extra effort and time needed to identify and understand the components that account for the remaining 20–25% of the variability are often all out of proportion to the benefits in achieving such understanding. Third, even if most of the variables could be identified and understood, many of them cannot be readily modified or changed by managers, so the "value added" from such insight is not high. And fourth, an integrated approach is likely to keep expectations for a plan more realistic, and also allow plans to be completed in a more reasonable time frame.

Thus, it is argued here that it is very important for analysts and planners to have a clearly thought out interpretation of what they mean by an ecosystem approach before they become advocates of it. The conceptual value of taking a big picture perspective by considering a system, its parts and their connections is very high. However, operationally, if an ecosystem that is too large or complex is defined, the product from analysis and planning is likely to have little value. If analysts and planners are not able to create useful products (strategies, plans) that help to resolve environmental and resource problems, then the credibility of the ecosystem approach will be damaged – to the extent that managers may be reluctant to use it. They may become concerned that an ecosystem approach represents a "black hole" into which management exercises may literally disappear, not to emerge until much too late to be helpful.

In Chapter 2, alternative schools or models of planning are reviewed. It is suggested here that the comprehensive approach to ecosystem management is similar to the *synoptic* or *comprehensive rational model*. What is likely to be more useful is to maintain a comprehensive perspective to scan for a broad range of issues and opportunities within an ecosystem, but then to use an integrated approach to achieve more focus for problem solving. In that manner, a blend of comprehensive and integrated approaches can be similar to the *mixed scanning model* described in Chapter 2.

Using an integrated interpretation relative to an ecosystem approach will not usually be sufficient to ensure effective application. Experience suggests that the following considerations also deserve attention (Mitchell, 1998: 39–40):

(1) **Significance of context.** It is important to understand and appreciate the context or local conditions related to a problem-solving situation, and to search for a custom-designed solution. Standardized, off-the-shelf solutions will not usually fit the conditions and needs of a particular situation.

(2) **Long-term perspective.** Since most resource and environmental problems were not usually created in a few years, it is unlikely that they will be resolved in one or two years. It is essential that participants appreciate the need for a long-term perspective. Decades often will be needed to stop degradation or to resolve scarcity problems.

(3) **Vision.** As demonstrated in Chapter 3, it is important to have a vision, or well thought out desired future condition in order that there is a clear sense of what future condition is sought.

(4) **Legitimacy.** An ecosystem approach must be given legitimacy or credibility if it is to be implemented, and that normally is best achieved through ongoing commitment from senior leaders (elected and appointed officials). Such commitment is not always easy to obtain, since elected decision makers usually are most interested in initiatives which provide tangible and short-term results. The results or outcomes from an ecosystem approach are often intangible and long term.

(5) **Leader or champion.** A key factor to introduce and implement an ecosystem approach is to have a leader or champion who will advocate the concept, and who will continue to work for and support the ecosystem approach through inevitable disappointments, setbacks and frustrations. Experience shows that a dedicated and determined leader is often *the* key factor related to success.

(6) **Redistribute power.** A willingness to share or redistribute power is usually essential if significant progress is to be made. This often requires central authorities to be willing to turn over some of their responsibility and authority to local agencies or organizations.

(7) **Collaborative approach.** A multi-stakeholder group should be established in order that the ecosystem approach can be implemented using a process which is inclusive, open, transparent and accessible (Chapter 8).

(8) **Consensus.** Decisions by a multi-stakeholder group should be based on consensus, as this is the most likely way to nurture long-term commitment from the community regarding decisions.

(9) **Sensitivity to burnout.** Volunteers from the community who participate in an ecosystem approach may become tired and "burn out" after a period of time. There should be sensitivity to this possibility, and provision made to bring new volunteers into the process.

(10) **Turbulence and surprise.** It needs to be accepted from the outset that, despite conscientious efforts, from time to time there will be surprises, and anticipated outcomes and benefits may not always materialize. Participants needs to be flexible, and prepared to learn from experience.

(11) **Communication.** There can never be enough time dedicated to sharing information, interpretations, insights and understanding during the management process. Such communication should be done in "plain language" in order to keep all participants informed and updated.

(12) **Demonstration projects.** Practical "hands-on" projects should be a component of an ecosystem approach, to create tangible evidence of

accomplishments and progress, and to allow a role for those who feel more comfortable with action-oriented rather than planning activity.

(13) **Information and education.** For long-term change and improvement to occur, information and education will be essential if basic attitudes and values are to be modified.

(14) **Implementation and monitoring.** Explicit attention should be given to the means to ensure implementation of the ecosystem approach, and to monitor progress associated with various initiatives. Only in this manner will it be possible to learn from doing, and to benefit from accumulated experience.

The above points deserve attention, as each has a role in facilitating effective use of an ecosystem approach. These points also highlight that using an ecosystem approach is not a technical exercise. Much of what is required to successfully use an ecosystem approach includes incorporation of the "human dimension" into the process. In the next section, an approach which has become known as *agroecosystem analysis* will be outlined, as in many ways it is a practical version of an integrated approach.

5.4 Agroecosystem analysis

Conway (1985; 1987) developed the concept of agroecosystem analysis, which reflects many of the attributes of an integrated interpretation of an ecosystem approach. The purpose here is to outline the basic idea of agroecosystem analysis. Conway explained that agroecosystems are ecological systems which have been modified by human activity in order to produce food, fibre or other agricultural outputs. As with all ecological systems, they are structurally and dynamically complex.

Conway's motivation for developing agroecosystem analysis was to improve our capacity to address problems which emerged from the application of new technologies in agriculture. In particular, he was interested in the environmental consequences of the agricultural revolution that had occurred in developing countries from the creation of new seeds, which, when used in combination with irrigation and agrochemicals, allowed dramatic increases in food production. However, short- and medium-term problems accompanied application of Green Revolution practices, including increasing incidence of pest, disease and weed problems, and deterioration of soil structure and fertility, along with increased indebtedness and inequity.

As the environmental and socio-economic problems from Green Revolution technology were recognized, each was addressed individually. However, as the comments in Box 5.7 highlight, in most cases the problems were inter-related. The concept of agroecosystem analysis was developed to provide a multidisciplinary and holistic approach to address such problems.

Conway explained that the departure point for his approach was the notion of *systems*, and the related idea of *system hierarchy*. Regarding hierarchies, he

---

Box 5.7 Interconnections among Green Revolution problems

. . . there has been a growing realisation that many, if not all, of the problems are essentially systematic in nature. They are linked to each other, and to the performance of the system as a whole. As a consequence problems that were initially viewed as side-effects often, it turns out, threaten directly the main objectives of development. Moreover, even where agricultural production is increased, this success may be short lived if attention is not quickly diverted to side effects which threaten other equally important development goals.

Source: Conway, 1985: 32.

---

noted that the natural world, or farming systems, can be conceived as a nested hierarchy of systems (such as organism–population–community–ecosystem–biome–biosphere, for natural systems; or plant–crop–field–cropping system–farming system–household–village–region–nation–world, for an agroecosystem). To understand the behaviour of any level in the hierarchy, it is not sufficient to consider only the levels below it. Each level has to be studied in its own right, and with regard to its connections to other systems. This approach is a strong reminder that when we are studying an ecosystem or agroecosystem, we need to give attention to the idea of layers or hierarchies of systems.

Four properties of agroecosystems were identified by Conway, which are also relevant to understanding other ecosystems. Illustrated in Figure 5.1, these are:

- **Productivity.** The output or yield of, or net income from, a valued product per unit of resource input. Productivity often is measured in terms of yield or income per hectare, or total production of goods and services per household or nation. It also can be measured as kilograms of grain, tubers, fish or meat, or it can be converted into calories, protein, vitamins or monetary units. The basic resource inputs are land, labour and capital.
- **Stability.** The constancy of productivity relative to small disturbing forces occurring from normal fluctuations and cycles in the surrounding environment. Such fluctuations may be in climate or water available in rivers or aquifers, or in the market demand for crops.
- **Sustainability.** The capacity of an agroecosystem to maintain productivity when subjected to major disturbing forces. Such disturbing forces could range from regular but relatively modest disturbances, such as soil salinity or indebtedness, to less regular, unpredictable and much larger disturbances such as floods, droughts or new pests.
- **Equitability.** The (un)evenness of the distribution of the benefits among humans from productivity. Equitability normally is measured through the distribution of benefits and costs associated with the production of goods and services from an agroecosystem.

Conway explained that the four properties can be used as either *neutral descriptors* of the behaviour of a system, or as *performance indicators*. When used as

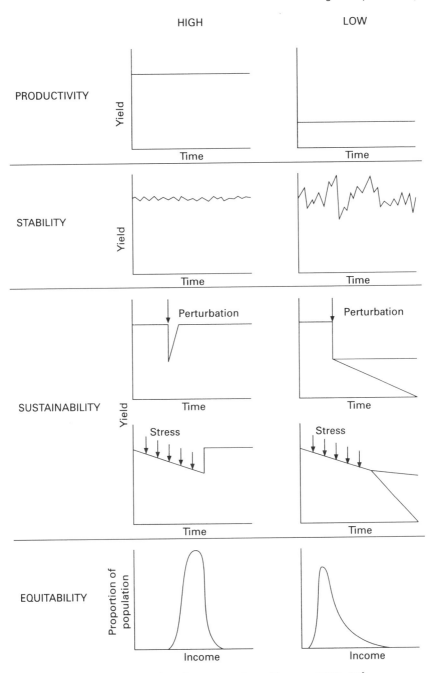

Figure 5.1 Four system properties of agroecosystems (Conway, 1985: 36)

Box 5.8 Trade-offs among agroecosystem properties

The pursuit of high productivity in both the developed and less developed countries has brought with it declines in sustainability and equitability that increasingly are being regarded as undesirable. Our present priority is for policy research, practical analytical tools and development packages aimed at increasing agricultural sustainability and rectifying undesirable inequities.

Source: Conway, 1987: 111.

indicators, they can highlight trade-offs that may occur. As an example, he indicated that a large-scale irrigation project might achieve higher overall productivity, but at the expense of equitability and sustainability. Efforts to achieve equitability could inhibit productivity (Box 5.8). Thus, choices exist, and the thrust in management of ecosystems ultimately reflects choices among sometimes mutually exclusive properties.

To illustrate, Conway noted that traditional agricultural systems such as shifting cultivation (*swidden agriculture*) usually have low productivity and stability, but achieve high equitability and sustainability (Table 5.2). Traditional cropping systems score higher in terms of productivity and stability properties, yet retain a high level of sustainability as well as some equitability. When new technologies are introduced, productivity usually goes up, with other properties normally falling. The introduction of high-yield varieties of rice during the Green Revolution fits this pattern. Overall productivity increased, but yields fluctuated widely. One consequence was development of new seeds with combined properties of high productivity and stability, but they are still rated low relative to sustainability.

This introduction to selected aspects of agroecosystem analysis illustrates several important considerations. First, what is judged to be *ecosystem integrity* is determined by the properties of an ecosystem judged to be important. Different weights can be given to productivity, stability, sustainability and equitability. No mix is "right" or "correct". Choices are available, and normally would be

Table 5.2 Types of agricultural activity as a function of agroecosystem properties

	Productivity	Stability	Sustainability	Equitability
A. Swidden cultivation	Low	Low	High	High
B. Traditional cropping system	Medium	Medium	High	Medium
C. Improved	High	Low	Low	Low
D. Improved	High	High	Low	Medium
E. ?Ideal (best land)	High	Medium	High	High
F. ?Ideal (marginal Land)	Medium	High	High	High

Source: Conway, 1985: 37.

made relative to needs and conditions for a place and a time. In that regard, this point confirms the argument of the Brundtland Commission, presented in Chapter 4, that there is no blueprint for sustainability. Custom-designed strategies must be developed. Second, the concept of *nested hierarchies* emphasizes that whatever ecosystem boundaries are chosen, there will always be other ecosystems whose properties and behaviour will be relevant to the one being examined. We must not lose sight of such connections, even though for pragmatic reasons we may chose not to examine them in the same detail as the linkages among components within the ecosystem being addressed. The examples presented in the following section show some of the opportunities, and problems, involved in transforming the ecosystem concept from idea to action.

5.5 Examples of ecosystem approaches

5.5.1 "Profound superficiality"

During the 1990s, the lead federal environmental agency (Environment Canada) in Canada was reoriented and restructured (Mitchell and Shrubsole, 1994). The purpose was to improve the capacity to achieve sustainable development through an ecosystem approach. One outcome was that the unit with federal responsibilities for water was eliminated. As part of an assessment of the reorientation and restructuring, Bruce and Mitchell (1995) concluded that this decision created a paradox.

The paradox involved the following elements. On one hand, many people believed that water was a significant management concern for the country, and would become even more significant during the 21st century in both Canada and the world. This view appeared to be confirmed by discussions at the Earth Summit during 1992 in Rio de Janeiro. So, at a time when more international recognition was being given to water, the paradox was why the federal government appeared to be withdrawing or moving out of water management. With no specific federal agency responsible for water, there was concern that it would be difficult to know who to contact in the federal government about a water problem.

The disappearance of the water agency reflected an attempt to implement an ecosystem approach and integrated resource management by Environment Canada. The dismantling of the water agency was a conscious and deliberate attempt to break down what was viewed as an overly sectoral approach. Instead of having people concentrated in a water agency, the intent was to relocate water specialists into a variety of Environment Canada divisions and branches, to ensure that the connection of water to other environmental components was made. In this manner, the decision could be viewed as responding to the challenge issued by the Brundtland Commission's *Our Common Future*. The Commission had concluded that too many national and international organizations had been created on the basis of "narrow preoccupations and compartmentalized concerns". It challenged countries to move toward a more integrated approach to environmental and development problems. The Commission had argued that the real world of interconnected economic and ecological systems was unlikely to change, and therefore the policies and institutions would have to. In its words, such a modification was "one of the chief institutional challenges of the 1990s and beyond" (World Commission on Environment and Development, 1987: 10).

The down side of this initiative was that in-depth capacity regarding any component of the ecosystem, such as water, could be weakened significantly by dispersing people with expertise throughout an organization. If an integrated approach is to occur, there must be substantive knowledge and understanding to integrate. A major concern was that, regarding water, this substantive capability might not be maintained. It was this situation that led one commentator to conclude that what was occurring "smacks of profound superficiality". The restructuring was "profound" in that it recognized the need to create the capability to use an ecosystem approach in which the entire environmental system would be addressed. However, it had the danger of being "superficial" in that lack of attention to some of the basic components of the environment, such as water, could lead to poor science and analysis.

Two implications can be identified from this experience. First, every country and organization has the challenge to find an appropriate balance between breadth (ecosystem approach) and depth (sectoral approach). Each offers advantages and disadvantages, and the task is to try and incorporate as many of the strengths from each approach, while minimizing the disadvantages. Second, a distinction should be made between conceptual and operational decisions. In this example, while eliminating the water resources agency made conceptual sense, it created operational problems since a farmer or industrialist with a water problem would find it much easier to find a water agency, than to find one that might be labelled as an "aquatic systems branch". Thus, while conceptually we need to strive to build an ecosystem approach more explicitly into planning and management, it may be that organizational structures should be maintained along sectoral lines, as those are the ones most easily recognized by the public.

5.5.2 The Baltic Sea ecosystem

The previous example from Canada illustrated some of the problems that can be encountered when applying the ecosystem approach. The problems increase when the ecosystem spans a number of countries, as is the case in the Baltic Sea, which is the largest brackish body of water in the world. As recently as 1950, the Baltic Sea was judged to be environmentally "healthy". However, by the 1980s it had severe water pollution and environmental degradation problems (Kindler and Lintner, 1993; Jansson and Dahlberg, 1999).

The catchment area for the Baltic Sea includes 14 countries and more than 85 million people (Figure 5.2). Nine countries share the coastline of the Baltic Sea: Sweden, Finland, Russia, Estonia, Latvia, Lithuania, Poland, Germany and Denmark. Portions of five other countries (Belarus, Norway, Ukraine, Slovak Republic, Czech Republic) are included in the catchment area because the

SOURCE: Helsinki Commission, The Baltic Sea Joint Comprehensive Environmental Action Programme, Helsinki, 1993.

Figure 5.2   The Baltic Sea ecosystem (Kindler and Lintner, 1993: 9)

headwaters of rivers which drain into the Baltic begin there. The Baltic Sea is significant for several reasons. The coastline has been a popular recreational area, leading to establishment of facilities catering to tourists. The coastal areas also provide spawning, nursery and feeding grounds for both freshwater and marine fish, and fishing is an important economic activity in the region.

With a total surface area of only 415,000 km^2, the Baltic Sea is vulnerable to pollution because of its confined nature and specific hydrography. The Baltic Sea is connected to the North Sea by several narrow channels between Sweden and Denmark. The shallowest depth in these channels is only 18 m. A consequence is that the Baltic Sea is primarily dominated by inputs of fresh water, and is not "flushed" by tides coming from the North Sea. Furthermore, the loss of wetlands in the nineteenth century to accommodate expansion of agriculture production, and more recently to support urban and industrial development, reduced the natural buffering capacity of the system. Pollution inputs come from various sources. Non-point sources include airborne emissions and agricultural runoff. Point sources concern sewage treatment plants in cities, including untreated sewage from more than 30 million people, and industrial factories, especially pulp and paper mills. The importance of addressing the environmental degradation problems is highlighted by the fact that the principal sources of pollution – municipalities, industries, agriculture – are not only located on the coastline but are also found in the headwaters of the rivers draining into the Baltic Sea. As a result, resolving the water pollution and degradation problems will not be successful if attention focuses only on the sea itself. The entire ecosystem needs to be considered.

The role of agriculture as a source of water pollution illustrates the need to consider land-based activities. Significant agricultural areas are found in Russia, Estonia, Latvia, Lithuania and Poland, with Poland by itself accounting for about 40% of the total arable land in the catchment. Agriculture is also important, and intensive, in Denmark and southern Sweden, with fertilizers being heavily used. For example, farming activity in the Danish portion of the Belt Sea catchment, a subsystem of the Baltic Sea, is only some 12,400 km^2 but discharges about 30,000 tons of nitrogen into the sea annually. In contrast, farming in the much larger Vistula River catchment, which covers 166,000 km^2 and includes two-thirds of Poland and small parts of Belarus, the Slovak Republic and the Ukraine, annually discharges about 50,000 tons. The main discharges are nitrogen and phosphorus from agrochemicals, and these contribute to the eutrophication of the Baltic Sea.

With fourteen countries sharing the Baltic Sea ecoregion, unilateral action by any one country is unlikely to make a significant impact. The Baltic countries had recognized the problem for several decades, and in 1974 signed the Baltic Marine Environmental Protection Convention, known more popularly as the Helsinki Convention. The countries informally began to implement its provisions, and then in May 1980 its provisions formally went into effect. The Helsinki Commission (HELCOM) was established as the coordinating organization for the convention. However, until the end of the Cold War, efforts were concentrated on the open sea, with the coastal zone and inland areas being neglected.

As a result of changing political conditions, a meeting was held in Ronneby, Sweden, in September 1990. The outcome was the first Baltic Sea Declaration. The Declaration stated that the countries agreed to begin specific initiatives to achieve ecological restoration of the Baltic Sea, and to create the possibility of self-restoration of the marine environment as well as preservation of its ecological balance. The necessary activities are occurring through the Baltic Sea Joint Comprehensive Environmental Action Programme, a joint initiative of the Baltic Sea nations, inland countries contained within the Baltic Sea catchment, the European Union, and four international financial organizations. The Programme was formally approved in April 1992 in Helsinki by the Ministers of the Environment from the involved countries.

The strategy. Given experience in Canada, reported in Section 5.5.1 above, it is interesting to note that the main thrust of the Baltic Sea Joint Comprehensive Environmental Action Programme is to strengthen the ability of each participating country to implement an ecosystem approach. In the words of Kindler and Lintner (1993: 12), the emphasis is upon

> . . . actions by each concerned government to carry out needed policy and regulatory reforms; to build capacity; and to invest in controlling pollution from point and nonpoint sources, safely disposing of or reducing the generation of waste, and conserving ecologically sensitive and economically valuable areas. To complement these activities, the program also includes elements to support applied research, environmental awareness, and environmental education.

The programme. Six components have been designed to realize the overall ecological restoration of the Baltic Sea. The components are:

- Policy, legal and regulatory changes to establish a long-term environmental management framework in each country.
- Institutional strengthening and human resource development to create the capacity to plan, design and implement environmental management systems, to manage efficiently, and to enforce regulations.
- A programme for infrastructure investment in specific activities to control point and non-point sources of pollution, and to minimize and dispose of wastes – with priority to municipalities, industries, and agriculture.
- A programme to help the management of coastal lagoons and wetlands.
- Support for applied research to create the knowledge base required to develop solutions, transfer technology and widen appreciation about the critical problems.
- Improve public awareness and environmental education to develop a widely based and sustainable foundation of support for implementation of the other five components.

The "strategy" and "programme" components illustrate that the Baltic Sea Joint Comprehensive Environmental Action Programme reflects the essential components of an ecosystem approach. It has defined a system, has identified key components of that system, and recognizes that linkages among the components must be considered. It also recognized that people are part of the problem, and that therefore human resources have to be improved to facilitate effective management, and that education and information have to be provided if people are to understand the issues and support the implementation measures.

Such a programme will not be inexpensive, nor will it be achieved in a few years. Kindler and Lintner (1993: 15) stated that the total cost of the 20-year programme for all the countries included in the Baltic Sea catchment had been estimated to be about 18 billion ECU (European currency unit, or at an exchange rate of 1 ECU to US$1.20, about US$25.6 billion). To achieve documentable results, the programme focuses on remedial action for 132 "hot spots", 98 of which are located in Russia, Estonia, Latvia, Lithuania, Belarus, Ukraine, Poland, the Czech Republic and the Slovak Republic.

Anticipated benefits. Numerous benefits are expected. The quality of water in the rivers draining into the Baltic Sea should improve notably. As these rivers are the principal sources of water supply for domestic needs of 85 million people as well as industry and agriculture, this improvement is expected to improve the health and well being of the local people. Some of the fastest improvements are expected to occur in the coastal waters. Such changes would allow several contaminated beaches to be re-opened, which would contribute to re-establishment of tourism and local recreation. Reduction in nutrients should reduce algae blooms, lower eutrophication and improve oxygen levels, thereby having a positive impact on the fishery resource. It is expected that the open sea will improve at a slower rate, due to the difficulty for the Baltic Sea countries of controlling the impact of long-range atmospheric transportation of various pollutants from countries outside of the basin. This issue is a reminder that no matter how broad the net is cast in defining the ecosystem, the ecosystem selected usually is a subcomponent of some larger ecosystem which affects the unit chosen for analysis and action.

Finally, the Baltic Sea programme is a reminder that an ecosystem approach normally has to be conceived and designed with the long term in mind. This reality is often a challenge, as the time horizon of elected officials too often is only until the next election, which in most countries is four or five years (Box 5.9).

By 1999, Jansson and Dahlberg (1999) reported on progress in the Baltic Sea. They observed that because the Joint Comprehensive Environmental

---

Box 5.9 Need for political commitment as well as technical understanding

We now have adequate knowledge of how the Baltic Sea ecosystem functions, and of what is needed to restore the Baltic environment. This requires large societal changes especially in agriculture, transportation and industry. . . . But success will be delayed, as long as political issues are given higher priority than environmental action.

Source: Jansson and Dahlberg, 1999: 312.

Action Programme and the subsequent Baltic Agenda 21 did not have concrete environmental goals, cost-effective implementation of the strategy was "impossible". At the same time, they noted that the present load of nitrogen to the Baltic Sea was about three-times higher than in the 1940s, and the load of phosphorus was about five-times greater, primarily due to the increased use of chemical fertilizers and intensified industrialization after World War II. To reduce the nutrient load to levels in the 1940s would require a reduction of 65% for nitrogen and 80% for phosphorus. By the late 1990s, significant reductions had been achieved from point sources, such as industrial and municipal sources, and these were very significant for the coastal environment but had little impact on conditions in the open sea. In contrast, progress on reducing emissions from diffuse sources had been much poorer. For Jansson and Dahlberg, a worrisome point was that much of the improvements were due to an economic recession in the Baltic countries during the early and mid 1990s, but by the end of the decade their economies were growing. In their view, "the present environmental gains in the Baltic States, Poland, and Russia will be lost if western-style, intensive agricultural production is copied, wetlands are drained, traffic volumes grow, and the use of chemicals increases" (p. 319). In their view, the restoration of a healthy ecosystem in the Baltic region will only be achieved once agricultural and forest land are designed to retain nutrients, nutrients are recycled, and emissions and discharges of hazardous pollutants are stopped. For such changes to occur, they concluded that there would have to be strong political and private commitment to the goal of restoring the Baltic Sea.

5.5.3 The Himalayas

The Himalayan ecosystem in India involves two subsystems. The mountainous area covers 523,000 km^2, or 16% of the total area of the country. The lowland area, fed by Himalayan rivers, and including the Indo-Gangetic and Brahmaputran plains, includes 726,250 km^2. Combined, these two areas cover 38% of India, and are home to more than 300 million people (Figure 5.3). Forests cover about 35% of the total area. On the plains, altitudes begin at about 300 m above sea level, with the highest mountain peaks reaching above 8,000 m. Degradation of mountain ecosystems occurs around the world, but the Himalayan one is under great stress.

Box 5.10 Mountain ecosystems

Mountain ecosystems are sensitive to quite small disturbances and the consequences of disturbance are often irreversible. This is especially true of tropical high mountains where relief and steep slopes are combined with huge contrasts in climate, soil and vegetation cover.

Source: Ahmad, 1993: 4.

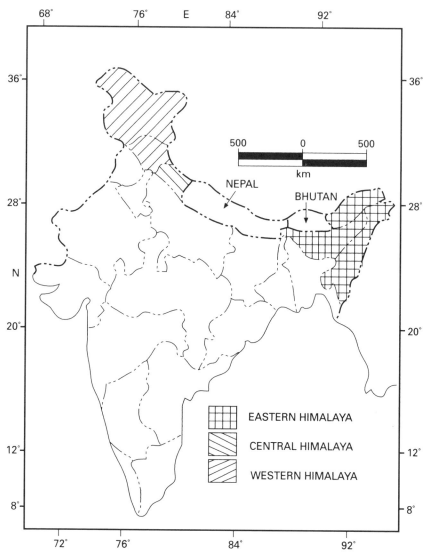

Figure 5.3   The Himalayas, India (Ahmad, 1993: 4)

Causes.   Many causes contribute to the environmental degradation, including: (1) unplanned land use; (2) cultivation on steep slopes; (3) overgrazing of naturally grassed areas, (4) major structural projects (roads, mines, dams, irrigation systems); (5) overexploitation of village or community forests; (6) harvesting of broad-leaved plant species; and (7) shifting cultivation, especially in northeastern India (Ahmad, 1993).

Consequences.   Degradation of the physical environment is reflected in many ways, such as through soil erosion and impoverishment, water impairment,

Figure 5.4 The Kullu Valley, Himachal Pradesh, India. Located in the Western Himalayas, on the far side of the Beas River, the village of Old Manali is on a hillside which has been denuded of forest cover, contributing to erosion, sedimentation in the river and increased likelihood of landslides (Alan Diduck)

atmospheric changes, avalanches, landslides and land-use practices (Figure 5.4). Soil erosion and impoverishment are one of the most serious problems, and illustrate the need for an ecosystem approach to deal with the issues.

Ahmad (1993: 5) suggested that about one-third of the total Himalayan land area has become "derelict" due to inappropriate land use, such as agricultural production on very steep slopes without appropriate actions to counter erosion. Other contributing variables have been shifting cultivation based on short cycles, monoculture in mixed forests, destruction of native plant species, and overgrazing. In his words, "soils are nutrient deficient and desertification is widespread".

The degradation of soil conditions has negative consequences for vegetation. Mountain ecosystems in general and the Himalayan system in particular normally have high biological diversity, but are fragile due to low resilience to changing conditions. Creation of large-scale plantations to grow chirpine (*Pinus roxburghii*) has changed physical and chemical properties of soils. Clear cutting and overgrazing have contributed to poor recycling of nutrients in deep soil horizons. A partial reason for clear cutting and overgrazing has been population growth, and the need to obtain fuel wood, fodder, construction timber and food (Figure 5.5). As a result, forest ecosystems have failed to recover and that has contributed to desertification.

Solutions.   The solution to the multi-faceted causes and problems was viewed by Ahmad to involve two key aspects: (1) an ecosystem approach in which natural and cultural landscapes were viewed in an integrated manner, and (2) locally designed and appropriate strategies. Both aspects were to be used to achieve "ecologically sustainable development". A selection of the strategies is presented below, to illustrate the importance of dealing with various components of the Himalayan ecosystem, as well as the connections among them.

Watersheds and water resources.   Detailed understanding about soil characteristics and land-use patterns is essential for development of watershed management programmes. Initiatives to counter erosion should include water- and land-based actions, including deep ploughing, mulching and well-constructed terraces on steep slopes, as well as constructing check dams and subsurface drains. Agroforestry programmes could stabilize soil conditions, and contribute food, fuel and fodder for the growing populations. A people's committee should be created from the villages in a watershed for which a programme is to be developed, and in that way local knowledge could be drawn upon. The key point here is that many of the initiatives to deal with water resources involve land-based activities, which is consistent with an ecosystem approach.

Land.   The capability of the land should be determined through a classification and inventory system. Steeply sloping lands (at a gradient of plus 45%) should not be used for agriculture, but should be used for revegetation. Lands with slopes of 30% could be used for agriculture, with appropriate actions to counter potential erosion. Numerous measures could be used, such as deep ploughing, contour cultivation, wing and strip cultivation, crop rotation, mulching, changes in the relief by contour furrows, ditches and dams, and well-prepared terraces.

Forests.   Large-scale afforestation and reforestation programmes are needed. Marginal and common lands should be included in a plantation programme, with priority to broad-leaved tree species which provide good green cover. Villagers,

Figure 5.5 The Kullu Valley, Himachal Pradesh, India. A woman carrying forest fodder, near Manali (John Sinclair)

including school children, should become involved in plantation activities, through contributing labour for planting. Villagers should have a right to use forest resources that grow as a result of their planting efforts to meet their needs for fodder, fuel and timber.

Wildlife.   Assessment of wildlife habitat is needed, especially regarding the carrying capacity. Exotic species (weeds and shrubs) should be controlled by the

121

villagers. Alternative habitats should be created for rare and endangered species. Special consideration should be given to creating "biosphere reserves" or protected areas. Tribal cooperatives could be established to allow local people to make wildlife-based products through cottage industries.

Health. Primary health centres should be established, and should provide basic health education, immunization programmes and family-planning services. Local health guidance and pharmacies should be provided in each village, so that villagers do not have to make long treks to obtain basic medical advice or supplies.

Energy. Large-scale hydro-electric projects should be replaced by small-scale projects. Wind and solar energy should be used, as well as wood from fast-growing and better coppicing tree species. These practices are suitable for both urban and rural areas. More bio-gas plants should be used to take advantage of cattle dung, instead of burning the dung in traditional stoves which result in atmospheric pollution and health risks.

Women. Women should be consulted in all planning and development decisions. Water supplies for both irrigation and domestic needs should be improved, along with supplies of fuel wood and fodder, to reduce the work load for women. Opportunities for alternative work should be provided, such as training for mushroom cultivation, handicrafts, bee-keeping/honey, fruit preservation, cultivation of herbs, sericulture (silkworm breeding) and small-animal (such as rabbits) breeding. See Chapter 10 for further discussion about the role of women.

The examples noted above, for initiatives regarding water and watersheds, land, forests, wildlife, health, energy and women, indicate that in an ecosystem approach it is necessary to consider a mix of activities, each of which addresses a number of ecosystem components. They also highlight that it is not sufficient to focus only on the biophysical or natural system. The initiatives planned regarding health and women indicate that people are part of the ecosystem, and thus can be both part of the problem and the solutions. And, as with the Baltic Sea programme, the time horizon has to be a long one. Many of the necessary changes will take years, if not decades, to show results.

5.6 Implications

The experiences in Canada, the Baltic Sea countries and the Himalayas indicate that resource and environmental managers in many countries are explicitly using an ecosystem approach. This initiative is a positive one when it forces attention of planners and managers on the interconnections among components of systems. However, it also can be problematic if the planners and managers seek to understand the totality of ecosystems. Our knowledge is usually not adequate for such detailed understanding. Furthermore, from a practical perspective, it is usually not necessary to understand the entire system. The key surely is to be able to understand those variables and their interactions that: (1) cause the greatest variation in system behaviour, and (2) are amenable to modification through management intervention. For these reasons, an *integrated* approach appears to be the best interpretation regarding an ecosystem perspective. If an ecosystem approach does not produce insights and strategies that have practical value for managers, its credibility will be damaged, and people may move away from it. As a result, resource and environmental managers have a responsibility to have a well thought out interpretation of what they mean by an ecosystem approach, and how it can be applied to resolve real world problems.

Ecosystem theory is still poorly developed, primarily because of the difficulty in general statements being applicable to conditions in a specific place and time. Ecosystem theory is also discredited when it becomes a mix of moral and scientifically oriented statements. Ecosystem understanding by itself cannot, and should not be expected to, provide answers to moral questions. As a result, it appears that it will be some time before ecologists will be able to provide scientifically based guidelines to guide decisions. This situation should not be viewed as a fundamental weakness, however, since most disciplines and professions experience this problem. Indeed, it is much more common for people, regardless of profession or discipline, to have to draw on experience and judgement in reaching decisions about what is the most appropriate decision or course of action.

References and further reading

Ahmad A 1993 Environmental impact assessment in the Himalayas: an ecosystem approach. *Ambio* 22: 4–9

Alario M 2000 Urban and ecological planning in Chicago: science, policy and dissent. *Journal of Environmental Planning and Management* 43: 489–504

Banks A 1994 Environmental cooperation in the Baltic Sea region. *GeoJournal* 33: 37–43

Barrett B 1994 Integrated environmental management. *Journal of Environmental Management* 40: 17–32

Barth H-J 1999 Desert ecosystems in Eastern Province of Saudi Arabia – a realistic way to improve their economic value. *Arab World Geographer* 2: 265–74

Bartlett J G, D M Mageean and R J O'Connor 2000 Residential expansion as a continental threat to US coastal ecosystems. *Population and Environment* 21: 429–68

Beatley T 2000 Preserving biodiversity: challenges for planners. *Journal of the American Planning Association* 66: 5–20

Bellamy J A and A K L Johnson 2000 Integrated resource management: moving from rhetoric to practice in Australian agriculture. *Environmental Management* 25: 265–80

Berkes F and C Folke (eds) 1998 *Linking Social and Ecological Systems: Management Practices and Social Mechanisms for Building Resilience*. Cambridge, Cambridge University Press

Berry J, G D Brewer, J C Gordon and D R Patton 1998 Closing the gap between ecosystem management and ecosystem research. *Policy Sciences* 31: 55–80

Bocking S 1994 Visions of nature and society: a history of the ecosystem concept. *Alternatives* 20(3): 12–18

Born S M and W C Sonzogni 1995 Integrated environmental management: strengthening the conceptualization. *Environmental Management* 19(2): 167–81

Bruce J and B Mitchell 1995 *Broadening Perspectives on Water Issues*. Canadian Global Change Program Incidental Report Series No. IR95-1, Ottawa, The Royal Society of Canada, August

Burroughs R H and T W Clark 1995 Ecosystem management: a comparison of Greater Yellowstone and Georges Bank. *Environmental Management* 19: 649–63

Cairns J 1994 Ecosystem health through ecological restoration: barriers and opportunities. *Journal of Aquatic Ecosystem Health* 3: 5–14

Cairns J and T V Crawford (eds) 1991 *Integrated Environmental Management*. Chelsea, Michigan, Lewis Publishers

Caldwell L K 1970 The ecosystem as criterion for public land policy. *Natural Resources Journal* 10: 203–21

Callicott J B 2000 Harmony between men and land – Aldo Leopold and the foundations of ecosystem management. *Journal of Forestry* 98(5): 4–13

Castillo A and V M Toledo 2000 Applying ecology in the Third World: the case of Mexico. *BioScience* 50: 66–76

Cawley R M and J Freemuth 1993 Tree farms, mother earth, and other dilemmas: the politics of ecosystem management in Greater Yellowstone. *Society and Natural Resources* 6: 41–53

Cocklin C and P Doorman 1994 Ecosystem protection and management in New Zealand: a private land perspective. *Applied Geography* 14: 264–81

Conway G R 1985 Agroecosystem analysis. *Agricultural Administration* 20: 31–55

Conway G R 1987 The properties of agroecosystems. *Agricultural Systems* 24: 95–117

Cooperrider A Y 1996 Science as a model for ecosystem management – panacea or problem? *Ecological Applications* 6: 736–7

Cortner H J and M A Moote 1999 *The Politics of Ecosystem Management*. Washington, DC, Island Press

Courtney C A and A T White 2000 Integrated coastal management in the Philippines: testing new paradigms. *Coastal Management* 25: 183–204

Czech B 1995 Ecosystem management is no paradigm shift: let's try conservation. *Journal of Forestry* 93(12): 17–23

Daniels S E and G B Walker 1996 Collaborative learning: improving public deliberation in ecosystem-based management. *Environmental Impact Assessment Review* 16: 71–102

de Nooij F 1999 The Dutch coast: battlefield and zone of opportunity. *Tijdschrift voor Economische en Sociale Geografie* 90: 432–7

Dedrick J P, T E Hall, R B Hull and J E Johnson 2000 The Forest Bank™: an experiment in managing fragmented forests. *Journal of Forestry* 98: 22–5

Duane T P 1997 Community participation in ecosystem management. *Ecology Law Quarterly* 24: 771–97

Ecological Society of America 1995 *The Report of the Ecological Society of America Committee on the Scientific Basis for Ecosystem Management*. Washington, DC, Ecological Society of America

Egan A F, K Waldron, J Raschka and J Bender 1999 Ecosystem management in the Northeast: a forestry paradigm shift? *Journal of Forestry* 97(10): 24–30

Falkenmark M 1997 Society's interaction with the water cycle: a conceptual framework for a more holistic approach. *Hydrological Sciences Journal* 42: 451–66

Fedkiw J 1997 The Forest Service's pathway toward ecosystem management. *Journal of Forestry* 95: 30–4

Forbes G, S Woodley and B Freedman 1999 Making ecosystem-based science into guidelines for ecosystem-based management: the Greater Fundy ecosystem experience. *Environments* 27: 15–23

Francis G 1993 Ecosystem management. *Natural Resources Journal* 33: 315–45

Freemuth J 1996 The emergence of ecosystem management: reinterpreting the gospel? *Society and Natural Resources* 9: 411–17

Galliano S J and G M Loeffler 1999 *Place Assessment: How People Define Ecosystems*. General Technical Report PNW-GTR-462, Portland, Oregon, US Department of Agriculture, Forest Service, Pacific Northwest Research Station, September

Garcia S M and M Hayashi 2000 Division of the oceans and ecosystem management: a contrastive spatial evolution of marine fisheries and governance. *Ocean and Coastal Management* 43: 445–74

Gerlach L P and D N Bengston 1994 If ecosystem management is the solution, what is the problem?: eleven challenges to ecosystem management. *Journal of Forestry* 92: 18–21

Grigg N S 1999 Integrated water resources management: who should lead, who should pay? *Journal of the American Water Resources Association* 35(3): 527–34

Grumbine R E 1994 What is ecosystem management? *Conservation Biology* 8: 27–38

Grumbine R E 1997 Reflections on "What is ecosystem management?" *Conservation Biology* 11: 41–7

Gurtner-Zimmermann A 1998 The effectiveness of the Rhine Action Program: methodology and results of an evaluation of the impacts of international cooperation. *International Environmental Affairs* 10: 241–66

Hartig J H and D M Dolan 1995 Successful application of an ecosystem approach – the restoration of Collingwood Harbor. *Journal of Great Lakes Research* 21: 201–11

Hartig J H, M A Zarull, T M Heidtke and H Shah 1998 Implementing ecosystem-based management: lessons from the Great Lakes. *Journal of Environmental Planning and Management* 41: 45–75

Harwell M A 1997 Ecosystem management of South Florida. *BioScience* 47: 499–512

Haynes R, R Graham and T Quigley 1996 *A Framework for Ecosystem Management in the Interior Columbia River Basin*. Portland, Oregon, USDA Forest Service

Holling C S 1987 Simplifying the complex: the paradigms of ecological function and structure. *European Journal of Operational Research* 30: 139–46

Holling C S 1998 Editorial: two cultures of ecology. *Conservation Ecology* 2: 1–5, online at http://www.consecol.org/vol2.iss2.art4

Hooper B P, G T McDonald and B Mitchell 1999 Facilitating integrated resource and environmental management: Australian and Canadian perspectives. *Journal of Environmental Planning and Management* 42: 747–66

Hrenegaard G T and P Dearden 1998 Linking ecotourism and biodiversity conservation: a case study of Doi Inthanan National Park, Thailand. *Singapore Journal of Tropical Geography* 19: 183–211

Imperial M T 1999a Institutional analysis and ecosystem-based management: the institutional analysis and development framework. *Environmental Management* 24: 449–65

Imperial M T 1999b Analyzing institutional arrangements for ecosystem-based management: lessons from the Rhode Island Ponds SAM Plan. *Coastal Management* 27: 31–56

Interagency Ecosystem Management Task Force 1995 *The Ecosystem Approach: Healthy Ecosystems and Sustainable Economies*. Washington, DC, The White House

Ireland L 1994 Getting from here to there: implementing ecosystem management on the ground. *Journal of Forestry* 92(8): 12–17

Jacobs J 2000 *The Nature of Economies*. Toronto, Random House

Jansson B-O and K Dahlberg 1999 The environmental status of the Baltic Sea in the 1940's, today, and in the future. *Ambio* 28: 312–19

Kay J J and E Schneider 1994 Embracing complexity: the challenge of the ecosystem approach. *Alternatives* 20: 32–9

Kay J J, H A Regier, M Boyle and G Francis 1999 An ecosystem approach for sustainability: addressing the challenge of complexity. *Futures* 31: 721–42

Keiter R 1996 Toward legitimizing ecosystem management on the public domain. *Ecological Applications* 6: 727–30

Keiter R 1997 Ecological policy and the courts: of rights, processes, and the judicial role. *Human Ecology Review* 4: 2–8

Kidd S and D Shaw 2000 The Mersey Basin and its river valley initiatives: an appropriate model for the management of rivers? *Local Environment* 5: 191–209

Kindler J and S F Lintner 1993 An action plan to clean up the Baltic. *Environment* 35(8): 6–15; 28–31

Levins R 1995 Preparing for uncertainty. *Ecosystem Health* 1: 47–57

MacKenzie S H 1996 *Integrated Resource Planning and Management: An Ecosystem Approach in the Great Lakes Basin*. Washington, DC, Island Press

Malone C R 2000 State governments, ecosystem management, and the enlibra doctrine in the US. *Ecological Economics* 34: 9–17

Margerum R D 1999a Integrated environmental management: lessons from the Trinity Inlet Management Program. *Land Use Policy* 16: 179–90

Margerum R D 1999b Integrated environmental management: the foundations of successful practice. *Environmental Management* 24: 151–66

Margerum R D and S M Born 1995 Integrated environmental management: moving from theory to practice. *Journal of Environmental Planning and Management* 38: 371–91

Margerum R D and S M Born 2000 A co-ordination diagnostic for improving integrated environmental management. *Journal of Environmental Planning and Management* 43: 5–21

McDonald B and J Cook 1998 The Malpai Borderlands Group: ecosystem management in action. *Environments* 26(1): 48–55

McIntosh R P 1976 Ecology since 1900. In B T Taylor and T J White (eds) *Issues and Ideas in America*. Norman, University of Oklahoma Press, pp. 353–72

Melious J O and R D Thornton 1999 Contractual ecosystem management under the Endangered Species Act: can federal agencies make enforceable commitments? *Ecological Law Quarterly* 26: 489–542

Meyer J L and L T Swank 1996 Ecosystem management challenges ecologists. *Ecological Applications* 6: 738–40

Mitchell B (ed.) 1990 *Integrated Water Management: International Experiences and Perspectives*. London, Belhaven Press

Mitchell B 1998 *Sustainability: Search for Balance*. Waterloo, Ontario, University of Waterloo, Faculty of Environmental Studies, First Annual Research Lecture

Mitchell B and D Shrubsole 1994 *Canadian Water Management: Visions for Sustainability*. Cambridge, Ontario, Canadian Water Resources Association

Naveh Z 2000 The total human ecosystem: integrating ecology and economics. *BioScience* 50: 357–61

Nelson G, P Lawrence and H Black 2000 Assessing ecosystem conservation plans for Canadian national parks. *Natural Areas Journal* 20: 280–7

Norton G A and G H Walker 1982 Applied ecology: towards a positive approach I. The context of applied ecology. *Journal of Environmental Management* 14: 309–24

Nunn P D 2000 Coastal changes over the past 200 years around Ovalau and Moturiki Islands, Fiji: implications for coastal zone management. *Australian Geographer* 31: 21–39

Omernik J M and R G Bailey 1997 Distinguishing between watersheds and ecoregions. *Journal of the American Water Resources Association* 33: 935–49

Pastor J 1995 Ecosystem management, ecological risk, and public policy. *BioScience* 45: 286

Peck J 2000 Seeing the forest through the eyes of a hawk: an evaluation of recent efforts to protect Northern Goshawk populations in Southwestern forests. *Natural Resources Journal* 40: 125–56

Quigley T M, R W Haynes and W J Hann 1998 Using an ecosystem's assessment for integrated policy analysis. *Journal of Forestry* 96(10): 33–8

Quigley T M, R T Graham and R W Haynes 1999 Interior Columbia River Basin Ecosystem Management Project: a case study. In K N Johnson, F Swanson, M Herring and S Greene (eds) *Bioregional Assessments: Science at the Crossroads of Management and Policy*. Washington, DC, Island Press, pp. 270–87

Rapport D J 1995 Ecosystem health: exploring the territory. *Ecosystem Health* 1: 1–13

Robinson G M and M Lind 1999 Set-aside and environment: a case study in Southern England. *Tisdschrift voor Economische en Sociale Geografie* 90: 296–311

Ruhl J B 1999 The (political) science of watershed management in the ecosystem age. *Journal of the American Water Resources Association* 35: 519–26, plus comments in 36, 2000: 229–32

Samson F B and F L Knopf 1996 Putting "ecosystem" into natural resource management. *Journal of Soil and Water Conservation* 51: 288–92

Sedjo R A 1995 Ecosystem management: an uncharted path for public forests. *Resources* 121: 10, 18–21

Skjæreth J B 1993 The "effectiveness" of the Mediterranean Action Plan. *International Environmental Affairs* 5: 313–34

Slocombe D S 1993a Environmental planning, ecosystem science, and ecosystem approaches for integrating environment and development. *Environmental Management* 17: 289–303

Slocombe D S 1993b Implementing ecosystem-based management: development of theory, practice, and research for planning and managing a region. *BioScience* 43: 612–22

Slocombe D S 1998a Defining goals and criteria for ecosystem-based management. *Environmental Management* 22: 483–93

Slocombe D S 1998b Lessons from experience with ecosystem-based management. *Landscape and Urban Planning* 40: 31–9

Smith P D, M H McDonough and M T Mang 1999 Ecosystem management and public participation: lessons from the field. *Journal of Forestry* 97(10): 32–8

Sproule-Jones M 1999 Restoring the Great Lakes: institutional analysis and design. *Coastal Management* 27: 291–316

Stanley T R 1995 Ecosystem management and the arrogance of humanism. *Conservation Biology* 9: 255–62

Suman D O 1997 The Florida Keys National Marine Sanctuary: a case study of an innovative federal–state partnership in marine resource management. *Coastal Management* 25: 293–324

Thomas J W 1996 Forest Service perspectives on ecosystem management. *Ecological Applications* 6: 703–5

United States Department of Agriculture, Forest Service, Northwest Research Station and United States Department of the Interior, Bureau of Land Management 1999

The Interior Columbia Basin Ecosystem Management Project: Scientific Assessment. CD-ROM, Portland, Oregon, Pacific Northwest Research Station

Walker B H and G A Norton 1982 Applied ecology: towards a positive approach. II. Applied ecological analysis. *Journal of Environmental Management* 14: 325–42

Woodley S, J J Kay and G Francis (eds) 1993 *Ecological Integrity and the Management of Ecosystems.* Delray, Florida, St Lucie Press

World Commission on Environment and Development 1987 *Our Common Future.* Oxford and New York, Oxford University Press

Chapter 6

Learning Organizations and Adaptive Environmental Management

6.1 Introduction

As outlined in Chapter 2, complexity and uncertainty present challenges for resource and environmental analysts, planners and decision makers. In this chapter, two inter-related responses to deal with change, complexity and uncertainty are examined: learning organizations and adaptive environmental management. In Section 6.2, the characteristics of learning organizations are outlined. Then, in Section 6.3, basic ideas associated with adaptive environmental management are considered, and Section 6.4 provides an example of the application of adaptive environmental management. A final section highlights the main conclusions about adaptive environmental management, as well as the implications of this concept.

6.2 Learning organizations

Based on his analysis of resource and environmental management in various regions of the world, Holling (1995) concluded that understanding of ecosystems is imperfect and surprise is inevitable. In his view, the ecosystems to be managed are moving targets, evolving due to the impact of management initiatives and the influence of human activity. As a result, he observed that ". . . the path of learning is not easy, partly because the new class of complex issues is sufficiently novel that the science is incomplete and the future is unpredictable" (p. 11).

Holling (1995: 9) suggested that crisis, conflict and gridlock arise when the following characteristics are present during resource and environmental management: (1) a single target and piecemeal policy; (2) a single scale of focus, typically on the short term and the local level; (3) no appreciation that all policies are experimental; and (4) inflexible management, with no priority to designing interventions to test hypotheses underlying policies. In situations with such characteristics, the usual management response is a demand for more data or more precision in data, as well as for more certainty and more control of information and of individuals.

In contrast, he believed that the following characteristics often allow the problems of crisis, conflict and gridlock to be overcome: (1) integrated rather than piecemeal policies; (2) flexible and adaptive policies instead of rigid ones; (3) management and planning for learning, rather than a focus on economic or social products; (4) monitoring intended to be part of active interventions to achieve understanding and to help identify responses, not just for the sake of monitoring; (5) investments in many kinds of science, not only in controlled science; and (6) citizen involvement to build active partnerships rather than reliance on public information to inform passively. Key desirable attributes highlighted above are the capacity for learning, and to be flexible and adaptive. In the remaining part of this section, attention turns to the desired characteristics of "learning organizations", drawing on the ideas of Peter Senge (1994).

Senge (1994: xiv) has argued that "there is growing awareness that the present trends of unsustainable resource consumption, pollution, social disintegration and ungovernability pose unprecedented threats to our future. Many now recognize the needs for organizationwide learning capabilities not possessed by traditional authoritarian, hierarchical organizations". To achieve, or enhance, learning organizations, he argued that profound changes are required. In his view, the fundamental shift is to move away from a belief that the world is created of separate, unrelated forces, and instead, to see the whole altogether. As Senge observed, at least in Western societies, from childhood onward people are taught to break problems into parts. The rationale is that such an approach makes complex problems and tasks manageable. However, an enormous cost occurs when people lose ability to see the connections of the parts to each other and to a larger whole, and thus do not always see or appreciate the consequences of actions for the larger system. Or, if they try to reassemble the parts of the problem, they often overlook that the whole is greater than the sum of all the parts. In Senge's view, once a person or an organization discards the illusion that the world is created from separate and unrelated forces, the foundation has been established for a learning organization (Box 6.1).

Senge suggested that five basic elements or disciplines are required for a learning organization. Each is essential, and are discussed in turn below.

(1) Systems thinking

Systems thinking, consistent with the *ecosystem approach* reviewed in Chapter 5, is the core foundation on which learning organizations are built. Processes and structures of systems – whether biophysical, economic, social, technological,

---

Box 6.1 Definition of a learning organization

. . . an organization that is continually expanding its capacity to create its future. For such an organization, it is not enough merely to survive. "Survival learning" or what is more often termed "adaptive learning" is important – indeed it is necessary. But for a learning organization, "adaptive learning" must be joined by "generative learning", learning that enhances our capacity to create.

Source: Senge, 1994: 14.

administrative, political – are linked and interconnected. Each normally has an influence on the others, but the influence often is hidden or subtle. For example, after a rainfall, some water ends up in rivers and streams, but other water infiltrates into the ground to feed into groundwater systems and provides a source of water for springs and creeks many kilometres away. To understand the hydrology of a surface stream, it often is necessary to also understand the structure and processes in aquifers which may be so far removed from the stream that the connection is not self-evident.

Senge argued that systems thinking is the "fifth discipline" because it is the foundation beneath all of the learning disciplines. All learning disciplines involve the need for a shift in mind set from seeing parts to seeing wholes, from seeing individuals as powerless to being engaged and capable of shaping their reality, and from being reactive to the present to being able to create the future.

Systems thinking also directs participants to think about *dynamic complexity*, meaning those situations in which cause and effect are subtle, and for which the results of initiatives are not obvious in the short or medium term, or in a local area and in another region. The outcome is a shift which focuses on identifying inter-relationships and networks rather than linear cause-and-effect chains, and on visualizing processes rather than snapshots of change. Another outcome is the ability to recognize that every influence is both a cause and an effect, since nothing is influenced only in one direction.

(2) Personal mastery

Personal mastery has two key aspects. First, people with high personal mastery are able to systematically and continually recognize what is important. Second, they are always clarifying their understanding of current reality. The benefits of personal mastery are numerous: clarifying and deepening a sense of vision, focusing energy, developing patience and seeing reality "objectively". When a vision (what is wanted) and a clear picture of the present reality (where we are now relative to where we want to be) are considered simultaneously, the result is *creative tension*. When this occurs, learning does not take the form of collecting more information, but instead of enhancing the capacity to achieve desired results.

As Senge argued, people with high personal mastery are always expanding their capacity to create the results that they truly want and seek. Such people maintain a sense of purpose, as they are able to "see" the present situation in a realistic light, and seek to use forces of change to help them, rather than trying to resist such forces. In addition, those with high personal mastery are always in a learning mode. The journey is never completed, as it is a lifelong process to reach an improved future state.

(3) Mental models

Mental models mean the ingrained assumptions, generalizations and images that influence how individuals see the world and what action is taken. To be able to learn, it is essential first to understand what mental model(s) are being used to filter the reality of the current world, as well as to imagine a different future. To appreciate mental models, individuals must look deep into themselves to identify their underlying internal images of the world and then to subject those images to critical scrutiny through dialogue with others. While advocating a particular mental model is fine, it is also essential to be open to other mental models, which might lead to modification or rejection of existing ones.

It is important to understand prevailing mental models, since often new ideas fail to get implemented because they conflict with internal images of reality, or what is possible. By recognizing, testing and, where necessary, modifying mental models, it is possible to understand the present reality in a different light, as well as to conceive previously unimaginable futures. When this occurs, a major step forward is made in creating capacity for a learning organization.

Mental models are powerful because they strongly influence how we behave. For example, if a partnership process is being considered (see Chapters 8 and 9), and if we believe groups are predisposed to cooperate and collaborate, rather than to compete, we are likely to consider different types of partnership arrangements. If trust and respect are preconditions for alternative dispute resolution, and those attributes are not thought to be present, the resource and environmental manager may not consider an alternative dispute resolution approach, but instead turn to a more legalistic one (see Chapter 11).

The issue is not whether mental models are correct or incorrect. A more basic concern is that the models are too often implicit or not recognized, and therefore tend to limit the range of options considered. If people and organizations are to be learning oriented, they need to be willing and able to "think outside the box", and not confine themselves only to continuing with business as it has always been done. Explicit recognition and critical assessment of mental models can help to overcome institutional inertia, and foster willingness to innovate and experiment with unfamiliar but promising practices.

The alternative, no examination of dominant mental models, can lead to what Senge has called *profound incompetence* – a characteristic of many adults who protect themselves from pain and discomfort by never moving outside the box, but at the cost of not learning new things or how to create the results really wanted.

(4) Shared vision

A shared vision can help to bind people together in an organization with regard to a common identity and sense of destiny (Box 6.2). When there is a vision to which people are committed, then Senge argued that people both excel and learn, because they want to learn rather than because they are told to learn.

Box 6.2 The impact of a shared vision

One is hard pressed to think of any organization that has sustained some measure of greatness in the absence of goals, values, and missions that become deeply shared throughout the organization.

Source: Senge, 1994: 9.

To create a shared vision, there must be the ability to identify shared pictures of a desirable future, and then to craft a vision that contains common elements which generate commitment and enthusiasm. It also is important to distinguish between a purpose and a vision. The former represents a direction or general outcome, and is usually abstract, such as "being the best I can be" or "advancing humankind's ability to explore space". The latter is a specific destination, or a specific desired future, such as "breaking the four minute mile" or "placing a person on the moon by the end of the 1960s".

Senge argued that it is not possible to have a learning organization without a shared vision, such as the National Conservation Strategy for Pakistan discussed in Chapter 4. Without a shared vision, powerful forces supporting the status quo are often difficult to overcome. A shared vision can also become a first step to help people who previously have not trusted each other to begin to work together, as the shared vision fosters a common cause and identity. A shared vision also provides a common bond to help keep people motivated when the almost inevitable stresses or obstacles are encountered. Finally, a shared vision encourages team members to take risks and experiment in order to achieve the desired end state.

(5) Team learning

Team learning requires development of capacity in the team to create and achieve the results identified in the vision. Team learning also requires members to become aware of the skills and abilities of other members, so that the team can take advantage of each member's strengths, and minimize the weaknesses.

A starting point for team learning is *dialogue*, which requires members to enter into a process of "thinking together" while suspending individual assumptions and positions. The key to dialogue is willingness to participate in a free-flowing exploration of goals, assumptions and actions. Dialogue contrasts with "discussion" in which emphasis is on the identification and promotion of ideas from which someone "wins". If dialogue is to be successful, it is essential to recognize barriers that will undermine team learning. One of the basic barriers is that individuals can become defensive about past decisions or behaviour, and therefore do not allow themselves to become or stay open to new ideas or approaches.

An innovative and effective team is not automatically one which experiences no conflict. Indeed, often a sign of an outstanding team is the continuous conflict of ideas being raised. However, for truly exceptional teams, the members are able to channel conflict in a constructive and productive way, to develop new and better ideas. In contrast, mediocre teams have often one of two attributes: (1) no appearance of conflict on the surface, as people believe that they should suppress conflicting views in order to maintain harmony on the team, or (2) polarization of competing ideas, in which people speak out strongly about their views, the different views are entrenched, and there is unwillingness to consider the merits of competing views. In either situation, the likelihood of creating a learning organization becomes low.

Box 6.3 The essentials of a learning organization

At the heart of a learning organization is a shift of mind – from seeing ourselves as separate from the world to connected to the world, from seeing problems as being caused by someone or something "out there" to seeing how our own actions create the problems we experience. A learning organization is a place where people are continually discovering how they create their reality. And how they can change it.

Source: Senge, 1994: 12–13.

Team learning therefore is an important component for a learning organization, since it is generally believed that teams rather than individuals are the basic learning units in organizations. If teams cannot learn, the organization will not learn.

(6) The collective value of the five dimensions

Senge called the above five elements, the basic disciplines of a learning organization (Box 6.3). What makes them distinctive is that they are "personal". In other words, each relates to how individuals think, what they want, and how they interact and learn with others. Furthermore, he argued that the five disciplines must develop together. In his view, this need for the disciplines to be integrated is the reason why *systems thinking* is the "fifth" discipline, since it integrates all of the others, shaping them into a coherent whole.

To be more specific about the importance of integrating the disciplines, Senge remarked that a *vision* without *systems thinking* can lead to imaginative and interesting pictures of the future, but without the necessary understanding of the forces that must be harnessed to move from the present to the desired future state. At the same time, *systems thinking* needs a *shared vision*, *mental models*, *team learning* and *personal mastery*. Creating a shared vision facilitates the essential commitment to the long term. Mental models generate openness, essential to identify weaknesses of the present condition, and new ways to view the world. Team learning facilitates the capacity of a group to search for the big picture transcending individual perspectives. And, personal mastery encourages the motivation for individuals to continue to learn about how decisions and actions affect the world.

6.3 Adaptive environmental management

Holling (1978) has been a major contributor to developing the concept of adaptive environmental management (Box 6.4). He edited a book entitled *Adaptive Environmental Assessment and Management*, whose purpose was to develop an alternative approach to environmental impact assessment and management for policy makers and managers who were dissatisfied with traditional principles and methods. The central message was that a new process was needed to deal

---

Box 6.4 Rationale for an adaptive approach

. . . But however intensively and extensively data are collected, however much we know of how the system functions, the domain of our knowledge of specific ecological and social systems is small when compared to that of our ignorance.

Thus, one key issue for design and evaluation of policies is how to cope with the uncertain, the unexpected, and the unknown.

Source: Holling, 1978: 7.

---

with a fundamental challenge: ". . . to cope with the uncertain and the unexpected. How, in short, to plan in the face of the unknown." (Holling, 1978: 7). The following comments summarize some of the main arguments presented by Holling and his colleagues in support of an adaptive approach to environmental management.

Holling concluded that people have always lived in an unknown world, and yet generally have prospered. The traditional way of dealing with the unknown has been through *trial-and-error methods*. What is known becomes the departure point for a trial. Errors provide new information and understanding, and become the basis from which new experiments are designed. "Failures" are accepted as necessary to gain understanding about previously unknown conditions, and to improve our capability to deal with them. With experience, new understanding is achieved, and progress is realized.

However, three minimum conditions must be met if the trial-and-error method is to work. First, the experiment cannot destroy the experimenter. Or, at least someone has to be able to learn from the experience. Second, the experiment should not create irreversible changes in the environment. If that did occur, then it would be difficult, perhaps impossible, for the experimenter to gain from the new knowledge. Third, the experimenter must be willing to start again, having learned from failures. Holling concluded that for resource and environmental management it was increasingly difficult to meet these three minimum conditions. "Trials", such as release of greenhouse gases into the atmosphere, were becoming capable of producing "mistakes" with greater costs than societies could bear.

Other concerns existed. Errors might theoretically be reversible, but the commitment of resources and prestige could make it extremely unlikely that decisions would be reversed. Humans and their organizations seem to have great difficulty in acknowledging failure or mistakes, or in learning from such experiences. Egos may be bruised, reputations tarnished or face lost. As a result, rather than admitting mistakes, cutting losses and starting anew, it is more common for people to try and eliminate or "fix" problems. The outcome usually is further investment of resources and reputation, growing costs associated with maintenance and repair, and eventual loss of options. In addition, in some instances, such as the construction of a major dam and reservoir, the scale and implications of trying to modify decisions make them effectively irreversible.

Notwithstanding the many real difficulties and challenges, Holling argued that the need for innovative solutions required trial-and-error approaches, and that it was not sensible to try and *eliminate* the uncertain and the unknown. The proper direction, in his view, was to design approaches that allowed trial-and-error procedures to work. While we should strive to *reduce* uncertainty, Holling (1978: 9) also believed that:

> But if not accompanied by an equal effort to design for uncertainty and to obtain benefits from the unexpected, the best of predictive methods will only lead to larger problems arising more quickly and more often. This view is the heart of adaptive environmental management – an interactive process using techniques that not only reduce uncertainty but also benefit from it. The goal is to develop more resilient policies.

By *resilience*, Holling meant the ability of a system, natural or human-made, to absorb and use (ideally even to benefit from) change.

The concept of adaptive management has been addressed by others, two of whose ideas are considered here (Box 6.5). Lee (1993) considered adaptive environmental management relative to achieving an environmentally sustainable economy. He argued that as part of earning a living, we use the resources of the world, even if we do not understand natural systems enough to know how to stay within environmental limits or thresholds. Adaptive managers explicitly consider uncertainty and lack of understanding, by treating human intervention in natural systems as *experimental probes*. More specifically, adaptive managers take particular care about information. In particular, regarding information, adaptive managers:

- are explicit about their objectives and what they expect as outcomes, so that they can design methods and techniques to monitor and measure what happens;

Box 6.5 Adaptive management

Adaptive management is an approach to natural resource policy that embodies a simple imperative: policies are experiments; *learn from them*. . . . Linking science and human purpose, adaptive management serves as a compass for us to use in searching for a sustainable future.

Source: Lee, 1993: 9.

Adaptive management . . . is the posing of *management-as-experiment* in complex, dynamic situations where controls and strict replication are not possible. The experiments become learning and iterative processes of review and revision. At the core is the structuring of management so that hypotheses can be posed and tested, combining the rigours of the scientific method and the realities of management. Policies and management become "experimental probes" designed to learn more about the system, not confident prescriptions.

Source: Dovers and Mobbs, 1997: 41.

- collect and assess information so that outcomes and impacts can be compared with expectations; and
- take their new understanding and learn from it by correcting errors, and changing both plans and actions.

Given the above, what is the core or essence of adaptive environmental management? According to Lee, an adaptive approach is designed from the outset to test clearly expressed ideas or hypotheses about the behaviour of an ecosystem being changed through human use. The ideas or hypotheses usually represent predictions regarding how one or more components of the ecosystem will respond or behave as a result of implementation of a policy. When the policy is successful, the hypothesis is validated. However, when the policy fails, the adaptive approach is designed so that learning occurs, adjustments can be made, and future initiatives can be based on the new understanding. As Lee explained, experiments often produce surprises, but if resource and environmental management is accepted to have inherent uncertainty, then surprises become viewed as opportunities to learn from, rather than as failures to predict or avoid.

An adaptive approach is most appropriate when uncertainty is high. Nevertheless, it is not free from problems. The costs of collecting the necessary information from which to learn can be very high. In addition, the political risks of clearly documenting failures can also be very high. Furthermore, such an approach assumes there is capability and willingness to learn from errors. Organizations can vary markedly in their capacity for learning, and hence this aspect can represent a significant obstacle for effective application of an adaptive approach.

Challenges also exist in our ability to understand or analyse large ecosystems. Lee suggested that three obstacles create serious difficulties for analysts:

- *Sparse data*. Measurements of the natural and human world are inexact and imperfect. Time series data often are necessary, but it can take many years to assemble a credible data set. For example, data on the number of spawning fish in a run can only be determined once each year.
- *Limited theory*. Poor understanding makes it difficult to extrapolate very far from our experience (the rationale for the *incrementalist* approach described in Chapter 2). Human impacts on natural systems can also be both so large and unprecedented that it is difficult to determine which of several alternative theories is most relevant.
- *Unexceptional surprises*. Predictions frequently are incorrect, and expectations are not fulfilled. Uncertainty ensures that errors and surprises will be inevitable.

To manage adaptively, Lee therefore argued that several aspects needed careful attention. These include the following three key points:

- The focus of adaptive management is *ecosystematic* rather than jurisdictional. In other words, the adaptive approach uses ecosystem rather than political or administrative boundaries. One outcome is that almost inevitably the adaptive approach ends up using a spatial unit which spills over or straddles one or more human-made boundaries and management functions.

- The focus of adaptive management is upon a *population* or *ecosystem*, not individual organisms or projects. Failures at the individual level have to be accepted or tolerated in order to gain understanding about the population or ecosystem. Risk taking relative to individuals is accepted in order to enhance the population.

- The time scale of adaptive management is a *biological generation* rather than the business cycle, electoral term of office, or budget period.

Consideration of the above three points could lead to the conclusion that an adaptive approach would be ponderous and slow. However, as Lee (1993: 63) commented, "Just the opposite should be true. The adaptive approach favors action, since experience is the key to learning." And, consistent with Holling, Lee (1993: 63) argued that:

> . . . the adaptive approach does not aim for a fixed end point. . . . The goal, instead, is resilience in the face of surprize. Surprize can be counted on. Resilience comes from the constant testing, that is, from change and stress, from survival of the fittest in a turbulent environment.

Table 6.1 highlights those contextual conditions that favour the successful application of adaptive environmental management. In Section 6.4.1, we will consider in more detail the role of these aspects in the application of the adaptive approach in the Columbia River basin, USA.

Rondinelli (1993a, b) shares Holling's and Lee's enthusiasm about an adaptive approach. He remarked that there has been a growing agreement among organizations in the public and private sectors that they operate in increasingly complex environments, as well as under conditions of rapid change, reduced resources and greater uncertainty. These aspects have to be addressed if organizations are to be effective and efficient in realizing goals and objectives (Rondinelli, 1993a: 1).

Rondinelli's interest is particularly with development issues in Third World countries. He noted that, in the 1970s and 1980s, donor agencies emphasized what he called *blueprint approaches* in which systems analysis techniques were used to maintain control, and to minimize variation from specified objectives. Such an approach continues to be used for development projects for which it is relatively straightforward to define objectives and purposes, and for which general agreement exists about methods. Thus, this approach is usually well suited for projects emphasizing physical infrastructure and construction of facilities (roads, sewage treatment plants).

However, as donor agencies and recipient countries moved away from emphasis on infrastructure and construction, and toward social and human development projects (rural development, poverty alleviation, health, education), they soon realized that conventional blueprint approaches were less appropriate. Political, economic and social conditions also were evolving in Third World countries, creating further difficulties to be able to predict what would be most appropriate.

Table 6.1   Contextual conditions affecting adaptive management

There is a mandate to take action in the face of uncertainty. *But experimentation and learning are at most secondary objectives in large ecosystems. Experimentation that conflicts with primary objectives will often be pushed aside or not proposed.*

Decision makers are aware that they are experimenting anyway. *But experimentation is an open admission that there may be no positive return. More generally, specifying hypotheses to be tested raises the risk of perceived failure.*

Decision makers care about improving outcomes over biological time scales. *But the costs of monitoring, controls, and replication are substantial, and they will appear especially high at the outset when compared with the costs of unmonitored trial and error. Individual decision makers rarely stay in office over times of biological significance.*

Preservation of pristine environments is no longer an option, and human intervention cannot produce desired outcomes predictably. *And remedial action crosses jurisdictional boundaries and requires coordinated implementation over long periods.*

Resources are sufficient to measure ecosystem-scale behaviour. *But data collection is vulnerable to external disruptions, such as budget cutbacks, changes in policy, and controversy. After changes in the leadership, decision makers may not be familiar with the purposes and value of an experimental approach.*

Theory, models, and field methods are available to estimate and infer ecosystem-scale behaviour. *But interim results may create panic or a realization that the experimental design was faulty. More generally, experimental findings will suggest changes in policy; controversial changes have the potential to disrupt the experimental program.*

Hypotheses can be formulated. *And accumulating knowledge may shift perceptions of what is worth examining via large-scale experimentation. For this reason, both policy actors and experimenters must adjust the trade-offs among experimental and other policy objectives during the implementation process.*

Organizational culture encourages learning from experience. *But the advocates of adaptive management are likely to be staff who have professional incentives to appreciate a complex process and a career situation in which long-term learning can be beneficial. Where there is tension between staff and policy leadership, experimentation can become the focus of an internal struggle for control.*

There is sufficient stability to measure long-term outcomes; institutional patience is essential. *But stability is usually dependent on factors outside the control of experimenters and managers.*

Source: Lee, 1993: 85.

As a result, Rondinelli (1993a: 3) commented that due to "high levels of uncertainty, complexity, and risk in development activities, and the diminishing control . . . over factors affecting their success", various new challenges were emerging, such as:

- precise goals and objectives were more difficult to state as a result of development problems being less easy to understand or define, solutions not always being obvious or transferable from country to country, impacts from activities being less easily predictable, and expectations and objectives of many participants being inconsistent, and often in conflict;
- difficulties in estimating the feasibility of possible interventions, due to inadequate knowledge about the nature of the problem and the most appropriate interventions;

- problems in being able to pre-design projects in detail – technical experts were no longer accepted as being the only experts, and local participants increasingly expected to be involved in definition of the problem and in formulation of solutions;
- difficulties in separating the design and implementation of projects, since activities needed to be capable of modification based on experience during implementation; and
- challenges in using standard criteria to judge the effectiveness or success of projects.

Given the above, Rondinelli (1993a: 4) concluded that:

> The complexity, uncertainty, lack of control, inability to predict behavior, inability to predetermine outcomes, and inadequate knowledge about the most appropriate ways of promoting economic and social development, in reality, made all development projects and programs "experiments."

The obvious implication, in his view, was an extension of the last word in the above quote. Development initiatives had to be considered as experiments in problem solving. Management strategies needed to encourage and reward experimentation, innovation and adaptation. As a result, he concluded that "the lessons of experience from more than 40 years of development assistance has led many observers to call for a more 'adaptive' approach to planning and management that is more strategic, iterative and responsive" (Rondinelli, 1993a: 4). He then offered Table 6.2 as a tool to determine when an adaptive approach would be most appropriate, and when a blueprint approach would be most effective. The information in Table 6.2 reflects the differences between an *incremental* (right-hand column) and *synoptic* or *comprehensive rational* schools of planning. In the guest statement by Jim McIver from the USA, further insight is provided about the benefits from adaptive management, but also about some of the obstacles that hinder its use.

In Section 6.4, we will examine one example of an adaptive approach, and consider the insight which the ideas of Holling, Lee, McIver and Rondinelli provide.

6.4 Adaptive management in the Columbia River basin, USA

In this section, one example is considered from the US Pacific Northwest. In different ways, it reflects many of the ideas contained in Box 6.6.

6.4.1 Columbia River basin

Holling (1989/90: 81) has stated that one of the best examples of the application of an adaptive approach has been the experiment with sustainable development in the Columbia River basin, USA (Box 6.7). It is this experience that we will now consider.

Table 6.2   Management strategies

	Management strategy	
Characteristics	Mechanistic	Adaptive
Environment	Certain	Uncertain
Tasks	Routine	Innovative
Management processes		
Planning	Comprehensive	Incremental
Decision making	Centralized	Decentralized
Authority	Hierarchical	Collegial
Leadership style	Command	Participatory
Communications	Vertical, formal	Interactive, formal and informal
Coordination	Control	Facilitation
Monitoring	Conformance to plan	Adjust strategy and plan
Controls	Ex-ante	Ex-post
Use of formal rules and regulations	High	Low
Basis of staffing	Functions	Objectives
Structures	Hierarchical	Organic
Staff values	Low tolerance for ambiguity	High tolerance for ambiguity

Source: Rondinelli, 1993a: 5a.

Box 6.6 Uncertainty and the adaptive approach

When policies are defined, management begins and the same process of design and analysis occurs, but now in an environment where action has to be taken, however uncertain the outcome. That is where active adaptive management can play a central role, because its premise is that knowledge of the system we deal with is always incomplete. Not only is the science incomplete, the system itself is a moving target, evolving because of the impacts of management and the progressive expansion of the scale of human influences on the planet. Hence, the actions needed by management must be ones that achieve ever-changing understanding as well as the social goals desired. That is the heart of active experimentation at the scales appropriate to the questions. Otherwise, the pathologies of management are inevitable – increasingly fragile systems, myopic management, and social dependencies leading to crises.

Source: Walters and Holling, 1990: 2067.

Background.   The Columbia River is the fourth largest river in North America. Flowing over 1,930 km, the river basin drains parts of two Canadian provinces and seven US states (Figure 6.1). The Columbia River was once a major salmon spawning river, and provided an important source of food for native Americans.

Guest Statement

Adaptive management in US national forests

Jim McIver, USA

While adaptive management makes great sense in improving the care of complex systems like forests, its application challenges how managers and scientists normally do business. Traditionally, forest managers are trained to apply existing knowledge to build the best prescriptions possible, given what they see on the land. They are also encouraged to apply their prescriptions to as large a land base as possible, especially when environmental conditions indicate that a change is needed to achieve a desired forest condition.

In contrast to the traditional approach, adaptive management works best when managers compare the prescription they think is best with alternative prescriptions, including untreated controls. But managers are often reluctant to try different approaches, especially under the public eye, because this could be perceived to indicate uncertainty and indecisiveness. Furthermore, the concept of alternative prescriptions is difficult to accept, especially if it means that acres in need of management are left untreated. Applied scientists are trained to use the tools of replication and control to answer questions relevant to their clients, which include land managers. They are particularly rewarded when their results can be applied broadly, and this requires experiments in which treatments are randomly assigned to a replicated set of units. Further, the effects of an imposed treatment are best isolated when some units are left untreated – these controls can then be used as a comparative baseline for what the system would have done if left alone.

Traditionally, scientists have been reluctant to apply a rigorous scientific model to complex systems like forests, especially when it means working with management that occurs on a large scale, and on projects that are subject to a variety of operational constraints. As a result, treatments can be imprecise, replication is typically minimal, and experiments sometimes lack control or the opportunity to collect pre-treatment data. While managers are typically sensitive to the needs of scientists, they are bound by a much broader set of operating conditions, constrained by contractual and public obligations. Whether or not scientists take the time to understand these constraints, their response to these conditions is often to remain in the world of tightly controlled experiments that produce results with more narrow application.

Future adaptive management would be applied more commonly on US national forest lands if managers and scientists were introduced to it during school, and rewarded for it on the job. Forestry schools in North America have a long tradition of producing highly trained and qualified people to manage our forests. Yet while training is clearly based on scientific knowledge, it would be helpful if more emphasis were placed on teaching the scientific methods that produced the knowledge in the first place.

Compared with traditional management, adaptive management amounts to a change in methods, from learning in sequence, to learning in parallel. Throw in an unmanipulated control, and you've got a management experiment, or informed tinkering using the most robust method of scientific learning. Similarly, scientists learn the value of replication and control in school, but not about the realities of how these tools can be applied to address natural resource issues in a management context.

Future managers and scientists would benefit from seeing examples of adaptive management applied to forest ecosystems, and would better appreciate its value in later years. Adaptive management would also flourish if managers and scientists were rewarded for it during their course of work. For the manager, rewards would emphasize creativity in prescription writing, careful monitoring and communication of results. Scientists would be rewarded for participating in interdisciplinary projects, working with managers and other scientists, and for communicating results in management-oriented venues. These changes would increase the extent to which adaptive management is applied in dynamic forest ecosystems.

James McIver is a Research Ecologist at the Forestry and Range Sciences Lab (US Forest Service, PNW Research Station), La Grande, Oregon. He received his PhD in Entomology at Oregon State University in 1983, following undergraduate studies at the University of Colorado and graduate work at Idaho State University and the University of Cincinnati. His post-doctoral training included a year as visiting professor of biology at Idaho State, and seven years as a research associate at Oregon State and the University of California (Berkeley). Since 1991, McIver has worked with the US Forest Service in La Grande, on education and research in applied forestry, with a focus on interdisciplinary studies and the application of adaptive management.

Box 6.7 Adaptive management in the Columbia River basin

It is an explicit effort to apply the principles of adaptive management in a region the size of France, with different political jurisdictions, conflicting agencies, and the full range of resource and environmental conflicts. It is a mix of science, negotiation, planning and politics, from which remarkable lessons are emerging.

Source: Holling, 1989/90: 81.

Then, in the early nineteenth century, people from Europe began to arrive. The native Americans were decimated by diseases brought by the newcomers, and logging, mining and farming significantly altered the landscape. The new economic activities had an impact on the capacity of the river to support spawning of salmon. No activity had a more dramatic impact than the construction of a series of dams and reservoirs under the strong influence of the Bonneville Power Administration, a federal organization with a mandate to market electricity from the federal dams.

Beginning in the 1930s, 19 major dams were constructed, and, in combination with more than 60 smaller hydro projects, became the world's largest system for hydro-electricity generation. Low-priced electricity became a magnet for

Figure 6.1   The Columbia River basin (Lee, 1993: 18)

industry, attracting companies which were large consumers of energy. As a result of the development of the hydro-electricity capacity of the Columbia River, by the late 1970s the annual salmon runs had fallen to 2.5 million fish, in comparison with runs ranging from 10 to 16 million prior to industrialization. Furthermore, the priorities for use of the river had been established to be: (1) power, (2) urban and industrial use, (3) agriculture, (4) flood control, (5) navigation, (6) recreation and (7) fish and wildlife. The operating guidelines for the dams emphasized maximizing electricity generation, even when that adversely affected fish and wildlife.

Northwest Power Act 1980.   Until 1970, electric power capacity had steadily been expanded in the Columbia River basin. With demand growing, the utilities pressed for approval to construct new facilities. However, Indian people and recreationists began to lobby for greater emphasis upon energy conservation,

---

Box 6.8 Fish and Wildlife Program

Adaptive management is learning by doing: by treating measures in the Fish and Wildlife Program as experiments, the implementation of the program becomes a set of opportunities to test and improve the scientific basis for action. Those opportunities, in turn, structure a systemwide planning regime that makes uses of information produced by implementation of the program.

. . . Adaptive management should be distinguished, however, from incremental policymaking generally. The emphasis in adaptive management on clear specification of outcomes *before* action is undertaken contrasts sharply to the incrementalist assumption that programs must be adjusted in light of political reaction. Although "disjointed incrementalism" does rely upon action to stimulate information, there is no expectation that the information will improve understanding of the system being affected by government action.

Source: Lee and Lawrence, 1986: 442.

---

arguing that such an approach made sense both economically and environmentally. The outcome was that Congress tried to accommodate both the hydro-power and fish interests through the Northwest Power Act. From the perspective of adaptive management, the interesting aspect is that the Northwest Power Planning Council, formed in 1981 under the legislation, created a Fish and Wildlife Program whose purpose was to elevate the position of fish and wildlife, especially salmon and steelhead, among the multiple uses of the Columbia River. As Lee and Lawrence (1986: 433) explained, the Fish and Wildlife Program became "the most ambitious and costly effort at biological restoration on the planet". When initiated, the estimated input costs were about US$100 million annually, an amount which included forgoing 1% (US$54 to US$74 million) each year in federal wholesale power revenues. From the outset, the Fish and Wildlife Program explicitly used an adaptive approach (Box 6.8).

Adaptive management in the Columbia River basin was based on five principles. These included:

1. Protection and restoration of fish and wildlife is a common objective. Nevertheless, short-run interests often overshadow long-term needs of the natural system. The focus of the Program is the shared, long-term interest in protecting and restoring fish stocks.

2. Projects have to be considered as experiments. Given incomplete understanding of the fish and their habitat, the outcome of any initiative cannot be known in advance. Some will do better than anticipated; others will fail.

3. Action is overdue and required. Action cannot be deferred until "enough" knowledge is gained. New knowledge is best obtained if we seek it by expecting to encounter surprises.

4. Information offers value in two ways: as a basis for action and as a result of action. To obtain such value, it is essential to pose questions that address

management needs, and to use experimental designs that will provide answers to such questions. The adaptive approach seeks information not just from conducting an experiment, but to guide action.

5. Enhancement initiatives may be constrained to a specific time period. However, management is ongoing and forever. In that manner, it is possible to learn from and benefit from both successes and mistakes. One purpose of adaptive management is to ensure that managers in the future will have a better understanding on which to base their decisions.

Adaptive and consensus approaches. Lee and Lawrence (1986) explained that the experience in the Columbia River basin provides insight as to how adaptive management is different from a *consensus approach* (Table 6.3). If a problem has the characteristics identified in the left-hand column of Table 6.3 then a consensus approach is likely to be appropriate. Attention focuses upon achieving agreement about the best solution for a specific problem. In contrast, the adaptive approach emphasizes exploring a mix of solutions, each of which has different strengths and weaknesses. A choice then can be made as to which alternative is likely to be most resilient and robust in the context of the many uncertainties that will prevail.

The consensus approach can be effective if agreement exists about goals and objectives. However, when there is disagreement about them, as was the case regarding fisheries in the Columbia River basin, the consensus approach quickly hits a major problem: "consensus management is vulnerable to value differences clothed as scientific dispute" (Lee and Lawrence, 1986: 450). No consensus among experts can result in inaction – exactly the situation that the *precautionary principle*, discussed in Chapter 2, was designed to overcome. A basic belief in consensus management is that inaction is better than taking action regarding a matter for which disagreement exists. In contrast, the adaptive approach does not expect consensus about the most appropriate short-term answer. Instead, it

Table 6.3 Consensus and adaptive management

Characteristics	Consensus management	Adaptive management
Process	Answer oriented	Question oriented
Design strategy	Optimal solution to problem at hand	Multiple solutions (resilient mix)
Burden of proof	Bias towards study (e.g. acid rain)	Bias towards action plus monitoring (e.g. water budget)
Purpose of monitoring	Compliance and crediting	Learning and adjusting
Range of utility	Problem curable	Continuing management
	Project not repeatable	Project repeatable
	Experiments too risky (e.g. to individuals)	Experiments acceptable (e.g. populations more important than individuals)

Source: Lee and Lawrence, 1986: 448–9.

seeks to identify key questions to facilitate choices among options that provide for flexibility.

Notwithstanding the advantages of an adaptive approach, Lee and Lawrence (1986) also noted that it is not always appropriate. Specifically, they commented that four aspects can create problems for its application. These occur when:

- *The problem is curable rather than chronic.* The need to manage over an extended period of time makes learning from an adaptive approach more valuable than if a one-time remedy is feasible. Thus, restoration of fish and wildlife necessitates continuing management of living populations which can adjust to changing situations. Methods for ongoing restoration thus are different from those needed to clean up a polluted landfill site, a one-time solution.
- *The remedy is unique to the issue.* If lessons are not transferable, then the lessons from the adaptive approach may be less useful. In the Columbia River basin, five dams in the middle stretch of the river are uniquely different in design, meaning that lessons from improving fish passage at any one of them would unlikely be transferable to the other four. Other dams were more similar, indicating that for them the time and cost of using an adaptive approach could be worthwhile.
- *Experiments are too risky.* In some situations, the benefits to the population may not justify the risk to individuals. This requires a *harm–benefit* calculation, one of the standard concepts associated with ethical matters in research (Mitchell and Draper, 1982: 5–7).
- *Project failure is a reasonable basis for holding managers accountable.* If uncertainty is low, management should be blamed for negative results. However, when uncertainty is high, failures in a project may contain valuable lessons that can help the entire system. Failure is not always from bad luck and/or lack of skill. Both success and failure can be instructive, if there is willingness to learn from them.

Redefining success. A final note needs to be made about adaptive management, based on the Columbia River experience. As Lee and Lawrence (1986: 431) remarked, "perhaps the most difficult part of adaptive management is the need to redefine success" (Box 6.9). Adaptive management stipulates that failure should be expected and anticipated, so that evaluation procedures can be designed to determine why expectations were not met. However, planning for, or accepting, failure can be politically hazardous for decision makers. Even conceding the existence of uncertainty can be interpreted as a sign of weakness in an adversarial environment in which different interests are in conflict. Furthermore, sponsoring or funding agencies are more interested in supporting initiatives that they believe will succeed, even if the measures of success are short-term ones.

Thus, the concept that some failures are inevitable, and indeed learning from them can help to achieve longer-term successes, may not be accepted enthusiastically in a turbulent setting in which interests and values are in conflict. It is for these reasons that the Columbia River initiative is even more important, and deserves close attention in the years to come.

Box 6.9 Failure or success?

In the long run, the greatest hurdle may turn out to be the problem of adverse results. . . . Adaptive management requires clear specification, in advance, of anticipated outcomes. Given biological uncertainty, one must expect even perfectly implemented measures to fail sometimes, either because of natural fluctuations or because the underlying concept is flawed. Thus, adaptive management increases the likelihood of both visible success and visible failure. Resource managers whose professional reputations hang in the balance will regard this prospect with mixed feelings.

Source: Lee and Lawrence, 1986: 457.

6.5 Implications

Two key concepts have been examined in this chapter: learning organizations and adaptive environmental management. It has been argued that learning organizations are more likely to be able to be effective in dealing with resource and environmental management problems characterized by change, complexity, uncertainty and conflict. Such learning organizations are based on what Senge has called five disciplines: systems thinking, personal mastery, mental models, shared vision and team learning.

The position in this book is that learning organizations are highly desirable, and are required if adaptive environmental management approaches are to be implemented successfully. The concept of adaptive environmental management matches well the conditions outlined in Chapter 2 (complexity, uncertainty, turbulence). The appropriate management strategy related to an adaptive approach would often challenge the conventional wisdom associated with the comprehensive rational planning model, also outlined in Chapter 2. The comprehensive rational planning model implies that by careful problem definition and diligent research it is possible to gain understanding of resource and environmental systems, and then to control or manage them. In contrast, the adaptive approach explicitly accepts that resource and environmental systems will contain surprises, and that often even the most carefully crafted policies and actions will turn out to be inappropriate. When that occurs, whether we have followed hedging or flexing strategies, change and adjustment will be necessary. Box 6.10 summarizes some of the key findings related to experience with the adaptive approach.

The adaptive approach encourages planners and managers to approach their work and make their decisions with the expectation that they may well be wrong, but that the experience and lessons gained from mistakes can allow them to improve policies and practices. One of the largest obstacles against more widespread adoption of an adaptive approach is for planners, managers and decision makers to be able to acknowledge mistakes, and to make appropriate adjustments.

Box 6.10 Key findings and management implications related to adaptive environmental management

Findings:

1. Concepts of adaptive management are confused. Existing definitions range from simply completing the planning–doing–monitoring–evaluating cycle, local public participation, and "fiddling" with new approaches in an unstructured way. The focus should be on accelerating learning, adapting through new partnerships, and changing management and research institutions.

2. New citizen–manager–scientist partnerships are essential to learning to achieve sustainable ecosystems. Society no longer accepts expert-based learning and decision making, or separating learning by scientists from doing by managers.

3. Learning needs to be balanced with other resource objectives. By making learning central to the mission of management, research, regulatory agencies and the public, we can move away from short-term, reactive management.

4. The purpose of adaptive management is to expand the range of alternatives available to managers and society in their efforts to meet the needs of both societal values and ecological capacity.

5. Conscientious adaptive management provides rapid learning opportunities for scientists, managers and the public.

Implications:

1. The focus on learning needs to be expanded. People learn and adapt in many ways, and the process of learning and adapting also must evolve over time. Particularly important is adapting to changes in understanding of society's needs and wants, and of ecological capacity.

2. Assume that various pathways can meet a given objective. Recognize that the pathways may represent conflicting worldviews. Designing and testing a wide range of pathways to achieve the current generation's objectives will provide future generations with better choices.

3. Some important questions can only be addressed at large scales. Many environmental, social, and organizational dynamics cannot be measured at the site scale. Assessments and . . . plan revisions and amendments can add learning objectives and approaches to begin effective learning at this scale.

4. No-treatment areas and comparative treatments are crucial to enhancing the adaptive management process, and to transferring usable management information into practice more rapidly.

5. Conscientious adaptive management requires a tolerance for uncertainty, a flexibility in approach, and a commitment to both teaching the public and learning on the fly.

Source: Duncan, 1998: 2–3; Duncan, 1999: 2–3.

Box 6.11 Adaptive management not a panacea

Adaptive management is neither a panacea nor a means to escape difficult management decisions. In fact, by requiring constant, timely interchange among researchers, resource managers and other stakeholders, adaptive management has the potential to increase the number of difficult management decisions by revealing new problems and new possibilities more rapidly. However, the same mechanisms can also permit rapid resolution of problems and increased protection of natural resources.

Source: Meretsky, Wegner and Stevens, 2000: 584.

References and further reading

Agardy M T 1994 Advances in marine conservation: the role of protected areas. *Trends in Ecology and Evolution* 9: 267–70

Baker J A 2000 Landscape ecology and adaptive management. In A H Perera, D L Euler and I D Thompson (eds) *Ecology of a Managed Terrestrial Landscape: Patterns and Processes of Forest Landscapes in Ontario*. Vancouver, UBC Press, pp. 310–22

Bennett J W 1969 *Northern Plainsmen: Adaptive Strategies and Agrarian Life*. Chicago, Aldine

Blaikie P, T Cannon, I Davis and B Wisner 1994 *At Risk: Natural Hazards, People's Vulnerability, and Disasters*. London, Routledge

Bormann B T, P G Cunningham, M H Brookes, V W Manning and M W Collopy 1994 *Adaptive Ecosystem Management in the Pacific Northwest*. General Technical Report PNW-GTR–341, Portland, Oregon, US Department of Agriculture, Forest Service, Pacific Norwest Research Station

Bormann B T, J R Martin, F H Wagner, G W Wood, J Alegria, P G Cunningham, M H Brookes, P Fiesema, J Berg and J R Hensaw 1999 Adaptive management. In W T Sexton, A J Maik, R C Szaro and N C Johnson (eds) *Ecological Stewardship: A Common Reference for Ecosystem Management*. Volume III, New York, Elsevier, pp. 505–34

Brinkerhoff D W and M D Ingle 1989 Integrating blueprint and process: a structured flexibility approach to development management. *Public Administration and Development* 9: 487–503

Bunch M J 2000 *The Cooum River Environmental Management Research Program: Exploration of an Adaptive Ecosystem Approach for Rehabilitation and Management of the Cooum River, Chennai*, Unpublished PhD dissertation, Waterloo, Ontario, University of Waterloo, Department of Geography

Commonwealth Scientific and Industrial Research Organization 2000 *Sustainable Land Management for the Murray Darling Basin – An Integrated Approach*. Online 10 January 2000 at http://www.csiro.au

Daniels S E and G B Walker 1996 Collaborative learning: improving public deliberation in ecosystem-based management. *Environmental Impact Assessment Review* 16: 71–102

Dovers S and C Mobbs 1997 An alluring prospect? Ecology and the requirements of adaptive management. In N Klomp and I Lunt (eds) *Frontiers in Ecology: Building the Links*. London, Elsevier, pp. 39–52

Duncan S 1998 Adaptive management: good business or good buzzwords? *Science Findings*. Portland, Oregon, US Department of Agriculture, Pacific Northwest Research Station, Issue 7

Duncan S 1999 Confronting illusions of knowledge: how should we learn? *Science Findings*. Portland, Oregon, US Department of Agriculture, Pacific Northwest Research Station, Issue 11

Fratkin E 1986 Stability and resilience in East African pastoralism: the Rendille and the Ariaal of northern Kenya. *Human Ecology* 14: 269–86

Gibbs J P, H L Snell and C E Causton 1998 Effective monitoring for adaptive wildlife management: lessons from the Galápagos Islands. *Journal of Wildlife Management* 63: 1055–65

Gilmour A, G Walkerton and J Scandol 1999 Adaptive management of the water cycle in the urban fringe: three Australian case studies. *Conservation Ecology* online at http://www.consecol.org/vol3/iss1/art1

Grayson R B, J M Doolan and T Blake 1994 Application of AEAM (Adaptive Environmental Assessment and Management) to water quality in the Latrobe River catchment. *Journal of Environmental Management* 41: 245–58

Gunderson L H, C S Holling and S S Light (eds) 1995 *Barriers and Bridges to the Renewal of Ecosystems and Institutions*. New York, Columbia University Press

Haney A and R B Power 1996 Adaptive management for sound ecosystem management. *Environmental Management* 20: 879–86

Havens K E and N G Aumen 2000 Hypothesis-driven experimental research is necessary for natural resource management. *Environmental Management* 25: 1–7

Hennessey T M 1995 Governance and adaptive management for estuarine ecosystems: the case of Chesapeake Bay. *Coastal Management* 22: 119–45

Hewitt K (ed.) 1983 *Interpretations of Calamity: From the Viewpoint of Human Ecology*. London, Allen and Unwin

Hewitt K 1996 *Regions of Risk*. Harlow, Longman

Hilborn R and J Sibert 1988 Adaptive management of developing fisheries. *Marine Policy* 12: 112–22

Holling C S (ed.) 1978 *Adaptive Environmental Assessment and Management*. Chichester, Wiley

Holling C S 1989/90 Integrating science for sustainable development. *Journal of Business Administration* 19: 73–83

Holling C S 1995 What barriers? What bridges? In L H Gunderson, C S Holling and S S Light (eds) *Barriers and Bridges to the Renewal of Ecosystems and Institutions*. New York, Columbia University Press, pp. 3–34

Holling C S and G K Meffe 1996 Command and control and the pathology of natural resources management. *Conservation Biology* 10: 328–37

Horsefield R S 1998 *Benchmarks for Assessing Environmental Management Practices against an "Ideal" Adaptive Environmental Assessment and Management Protocol*. [online]. Available on the Internet: URL: http://www.gse.mq.edu.au/Research/adaptive/report1/html

Hunter C 1997 Sustainable tourism as an adaptive paradigm. *Annals of Tourism Research* 24: 850–67

Iles A 1996 Adaptive management: making environmental laws and policy more dynamic, experimentalist and learning. *Environmental and Planning Law Journal* 12: 331–54

Imperial M T 1993 The evolution of adaptive management for estuarine ecosystems: the National Estuary Program and its precursors. *Ocean and Coastal Management* 20: 147–80

Jachowski R 1998 *Adaptive Management and the Assessment of Habitat Changes on Migratory Birds*. United States Geological Survey, Patuent Wildlife Research Center, Maryland. [online]. Available from the Internet: URL: http://www.pwrc.nbs.gov/research/sis98/jachow1s.htm.

Kromm D E and S E White 1984 Adjustment preferences to groundwater depletion in the American High Plains. *Geoforum* 15: 271–84

Lancia R A, C E Braun, M W Collopy, R D Duser, J G Kie, C J Martinka, J D Nichols, T D Nudds, W R Porath and N G Tilghman 1996 ARM! for the future: adaptive resource management in the wildlife profession. *Wildlife Society Bulletin* 24: 436–42

Lee K N 1993 *Compass and Gyroscope: Integrating Science and Politics for the Environment*. Washington, DC, Island Press

Lee K 1997 Implementing adaptive management: a conversation with Kai N. Lee (interviewed by Cindy Halbert). *Northwest Environmental Journal* 7: 136–50

Lee K N and J Lawrence 1986 Restoration under the Northwest Power Act: adaptive management: learning from the Columbia River Basin Fish and Wildlife Program. *Environmental Law* 16: 431–60

Lessard G 1998 An adaptive approach to planning and decision-making. *Landscape and Urban Planning* 40(1–3): 81–7

Lindenmayer D B, R B Cunningham and M L Pope 1999 A large-scale "experiment" to examine the effects of landscape context and habitat fragmentation on mammals. *Biological Conservation* 88: 387–403

Ludwig D, R Hilborn and C Walters 1993 Uncertainty, resource exploitation, and conservation: lessons from history. *Science* 260 (2 April): 17 and 36

Manaaki Whenua Landcare Research 1999 *Community-based Research, Monitoring and Adaptive Management*. [online]. Available on the Internet: URL: *http://www.landcare.cri.nz/sal/index.shtml/prograt* (14 November 2000)

McAllister M K and R M Peterman 1992 Experimental design in the management of fisheries. *North American Journal of Fisheries Management* 12: 1–18

McCabe J T 1987 Drought and recovery: livestock dynamics among the Ngisonyoka Turkana of Kenya. *Human Ecology* 15: 371–89

McDaniels T, M Healey and R Paisley 1994 Cooperative fisheries management involving First Nations in British Columbia: an adaptive approach to strategy design. *Canadian Journal of Fisheries and Aquatic Science* 51: 2115–25

McGinnis M V 1995 On the verge of collapse: the Columbia River system, wild salmon and the Northwest Power Planning Council. *Natural Resources Journal* 35: 63–92

McLain R J and R G Lee 1996 Adaptive management: promises and pitfalls. *Environmental Management* 20: 437–48

Meretsky V J, D L Wegner and L E Stevens 2000 Balancing endangered species and ecosystems: a case study of adaptive management in the Grand Canyon. *Environmental Management* 25: 579–86

Mitchell B and D Draper 1982 *Relevance and Ethics in Geography*. Harlow, Longman

Mulvihill P R and R F Keith 1989 Institutional requirements for adaptive EIA: the Kativik Environmental Quality Commission. *Environmental Impact Assessment Review* 9: 399–412

Norton B G 1999 Pragmatism, adaptive management, and sustainability. *Environmental Values* 8: 451–66

Nudds T D 1999 Adaptive management and the conservation of biodiversity. In R K Baydack, H Campa and J B Haufler (eds) *Practical Approaches to the Conservation of Biodiversity*. Washington, DC, Island Press, pp. 179–93

Palm R 1990 *Natural Hazards: An Interactive Framework for Research and Planning.* Baltimore, Johns Hopkins University Press

Perevolotsky A, A Perevolotsky and I Noy-Meir 1989 Environmental adaptation and economic change in a pastoral mountain society: the case of the Jabaliyah Bedouin of the Mt Sinai region. *Mountain Research and Development* 9: 153–64

Reed M G 1999 Collaborative tourism planning as adaptive experiments in emergent tourism settings. *Journal of Sustainable Tourism* 7: 331–55

Rogers K 1998 Managing science/management partnerships: a challenge of adaptive management. *Conservation Ecology* [online] 2(2): R1. http://www.consecol.org/vol2/iss2/repl

Rondinelli D A 1993a *Strategic and Results-based Management: Reflections on the Process.* Ottawa, Canadian International Development Agency, June

Rondinelli D A 1993b *Development Projects as Policy Experiments: An Adaptive Approach to Development Administration.* London, Routledge

Schindler B and K A Cheek 1999 Integrating citizens in adaptive management: a propositional analysis. *Conservation Ecology* 39 [online]. Available from the Internet: http://www.consecol.org.vol3/iss1/art9

Schmieglelow F K A and S J Hannon 1993 Adaptive management, adaptive science and the effects of forest fragmentation on boreal birds in northern Alberta. *Transactions of the 58th North American Wildlife and Natural Resource Conference*, pp. 584–98

Sendzimir J, S Light and K Szymanowska 1999 Adaptively understanding and managing for floods. *Environments* 27: 115–36

Senge P M 1994 *The Fifth Discipline: The Art and Practice of the Learning Organization.* New York, Doubleday

Shindler B, B Steel and P List 1996 Public judgments of adaptive management: a response from forest communities. *Journal of Forestry* 94(6): 4–12

Shretha M and R Angliss 1997 *Adaptive management*, University of Minnesota [online]. Available from the Internet: http://www.consbio.umn.edu/Cbclass97/_free/00000005.htm

Smith C L, J Gilden, B S Steel and K Mrakovcich 1998 Sailing the shoals of adaptive management: the case of salmon in the Pacific Northwest. *Environmental Management* 22: 671–81

Smith K 1996 *Environmental Hazards: Assessing Risk and Reducing Disaster.* Second edition. London, Routledge

Stankey G H and B Schindler 1997 *Adaptive Management Areas: Achieving the Promise, Avoiding the Peril.* General Technical Report PNW-GTR–394, Portland, Oregon, US Department of Agriculture, Forest Service, Pacific Northwest Research Station

Stromgaard P 1989 Adaptive strategies in the breakdown of shifting cultivation: the case of Manbwe, Lamba, and Lala of northern Zambia. *Human Ecology* 17: 427–44

Taylor B, L Kremsater and R Ellis 1997 *Adaptive Management of Forests in British Columbia.* Victoria, BC, British Columbia Ministry of Forests, Forest Practices Branch

Thomas D H L and W M Adams 1999 Adapting to dams: agrarian change downstream of the Tiga dam, Northern Nigeria. *World Development* 27: 919–35

Torell E 2000 Adaptive learning in coastal management: the experience of five East African initiatives. *Coastal Management* 28: 353–63

Volkman J M and W E McConnaha 1993 Through a glass, darkly: Columbia River salmon, the Endangered Species Act, and adaptive management. *Environmental Law* 23: 1249–72

Walters C J 1986 *Adaptive Management of Renewable Resources.* New York, McGraw-Hill

Walters C 1997 Challenges in adaptive management of riparian and coastal ecosystems. *Conservation Ecology* [online] 1(2): 1. Available from the Internet: http://www.consecol.org/vol1/iss2/art1

Walters C J and R Green 1997 Valuation of experimental management for ecological systems. *Journal of Wildlife Management* 61: 987–1006

Walters C J and C S Holling 1990 Large-scale management experiments and learning by doing. *Ecology* 71: 2060–8

Weiringa M J and A G Morton 1996 Hydropower, adaptive management, and biodiversity. *Environmental Management* 20: 831–40

Williams B K and F A Johnson 1995 Adaptive management and the regulation of waterfowl harvests. *Wildlife Society Bulletin* 23: 430–6

Williams B K, F A Johnson and K Wilkins 1996 Uncertainty and the adaptive management of waterfowl harvests. *Journal of Wildlife Management.* 60: 223–32

Wilson M V and L E Lantz 2000 Issues and framework for building successful science-management teams for natural areas management. *Natural Areas Journal* 20: 381–5

Chapter 7

Assessing Alternatives

7.1 Introduction

In previous chapters, attention focused upon change; surprise, turbulence and conflict; looking to the future; sustainability; the ecosystem approach; and learning organizations as well as adaptive management. For all of these matters, alternatives exist and choices must be made. In resource and environmental management, people have relied heavily on *environmental impact assessment* as a tool to help in assessing alternatives and in making choices. More recently, *life cycle analysis* has emerged as another tool or method. In this chapter, each of these is examined. However, before doing that, Section 7.2 indicates the importance of being able to recognize different perspectives, and understanding that regardless of the assessment method applied, different conclusions can be drawn.

7.2 Differing perspectives can affect assessment

Various conditions, needs, values, assumptions and criteria can lead to different ideas about what alternatives are appropriate to consider, as well as which one should be preferred. In Chapter 4, we discovered that people from developed and developing countries have divergent interpretations about *sustainability* due to their different needs, expectations and interests. We also know that if *efficiency* is the criterion to judge a policy, programme or project, an assessment might be quite different than if the criteria were *equity* or *effectiveness*. We also appreciate that different evaluations are likely if priority is given to economic development over environmental protection (or vice versa). To emphasize the important role of different *perspectives*, this section summarizes an analysis by Russell (1994), a civil engineer, regarding the Three Gorges project in China.

Three Gorges project. This project, proposed for the Yangtze River in China, is one of the largest water resource projects in the world, and involves the construction of a multi-billion dollar dam across the river. Multiple benefits are anticipated by project designers, including downstream flood control, hydro-electricity generation and improved navigation. The impetus for the project is a desire to "control" or at least minimize devastating floods which for centuries have caused loss of life and major economic disruption, and which today threaten about 10 million people. Concerns about the Three Gorges project focused on the need to relocate nearly one million people from the reservoir area, and the flooding of a gorge with high amenity value.

During the design of the project, a feasibility study by a consortium of engineering companies concluded that the project was viable, with a benefit–cost ratio of 1.5:1. In contrast, an environmental non-government organization noted many flaws in the design, and argued that the project could be disastrous for the Chinese people. Russell asked how two apparently well-qualified groups could have such different perspectives. His conclusion was that ". . . it gradually became clear that they came from two quite different 'cultures': the large dam heavy civil engineering culture, and what might be called the new green culture" (Russell, 1994: 541).

Heavy engineering culture.   Russell suggested that for civil engineers responsible for building large dams, one concern is dominant: the safety of the project. Such massive projects cannot be pre-tested, yet any failure could have catastrophic results. As a result, "it is not surprising that civil engineers are conservative and always think safety first" (Russell, 1994: 543). Furthermore, since there are never adequate resources for such projects, civil engineers establish priorities, and marginal or peripheral concerns are quickly set aside. Attention is reserved for those aspects judged to be the most important. As the comment in Box 7.1 shows, environmental concerns often do not fall into the list of most important concerns.

Green culture.   Russell suggested that the *green culture* includes people who usually are well educated, articulate and well intentioned. Many are based in universities. They normally work as individuals or as part of loosely connected coalitions. Thus, they are relatively independent, and often do not have to accommodate interests other than the ones in which they believe (Box 7.2). They have relatively few resources.

Green culture people are often highly idealistic, such as when they suggested that China does not need additional power and should concentrate on achieving energy savings through more efficient industrial processes. In this regard, they

---

Box 7.1 The large dam, civil engineering culture

Although engineers have an increasing sensitivity to environmental issues, such concern is not "built in" to the typical engineer to nearly the same extent, as is the case with safety issues. In summary, present-day civil engineers (the ones in control) have a deep personal and professional commitment to the safety of the projects for which they are responsible, a somewhat lesser commitment to economic viability of their projects (despite the old adage about engineers being able to do for $1, what anyone else can do for $2), and a still lesser commitment to the newer environmental issues.

Source: Russell, 1994: 544.

Box 7.2 Green culture

They do not have the same need to compromise as do team workers and, not being part of a professional group, can feel free to challenge any point, however important or inconsequential. . . . the writing is usually very good and very convincing and can raise important points that may not have been given due weight. But it can also lack balance and be quite unfair at times.

Source: Russell, 1994: 544.

---

argued for energy and environmental standards for a country in the early stages of development which are often not met in much more developed and industrialized nations. At the same time, such observers often identify aspects that the more focused engineers overlook or consider to be lower priority. Thus, Russell agreed that valid criticisms were made about underestimation of costs for the project, especially those linked to resettlement, insufficient consideration of many environmental problems, and use of too low interest rates in the benefit–cost analysis.

The existence of two "cultures" during the assessment of this project is a timely reminder that "truth" is rarely absolute, and can be influenced by the lenses through which the world is viewed. This point should be kept in mind when considering the following methods which are often used for assessing alternatives.

7.3 Impact assessment

7.3.1 Origins, evolution and key attributes

Environmental impact assessment (EIA) was formally introduced through the National Environmental Policy Act of 1969 in the USA, which required federal agencies to consider explicitly the environmental implications of proposed development. Over time, other countries have incorporated impact assessment (IA) into resource and environmental management, either through law or policy. Initially, the focus in EIA studies was on the impacts on the biophysical environment. However, it was not long before criticism arose that the social impacts were being overlooked. Such criticism led to the development of what became known as social impact assessment (SIA). Today, assessments normally include both "environmental" and "social" dimensions, and hence the use of the abbreviated term, "impact assessment". It is usually assumed that economic dimensions are addressed through a parallel benefit–cost analysis.

As with sustainable development, many definitions and interpretations exist for impact assessment which focus on the environmental and social implications of development. However, most people would agree that an impact statement should: (1) identify the overall goals and objectives of the project; (2) describe

157

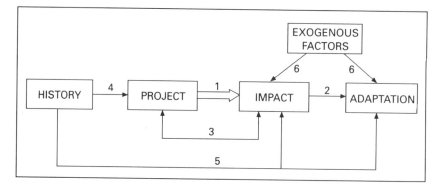

Figure 7.1   Temporal context for environmental impact assessment (Mitchell, 1989: 201)

alternative actions which could achieve the goals and objectives; (3) describe what environmental and social changes might occur without the development; (4) describe the nature, magnitude and duration of impacts from alternative actions; (5) assess the significance of the impacts; (6) identify remedial action to mitigate or eliminate negative impacts, as well as actions to enhance positive impacts; (7) identify a preferred action; and (8) outline necessary action for implementation and monitoring.

In preparing impact statements, it is important for the analyst or planner to recognize that all impacts have a *temporal dimension* (Figure 7.1). In other words, direct impacts (1) occur from changes triggered by the project on initial conditions in the natural or social environment. However, it is not adequate only to examine direct impacts. Changed conditions lead to adjustment and adaptation (2) (discussed more in Chapter 6). Furthermore, during the planning or implementation phase, the development may be modified (3) in response to public reaction or experience. Any development also has a connection with earlier decisions and developments (4), and such "history" may condition willingness to accept a proposal or to consider various adaptation or mitigation measures. Finally, other factors external to the project (such as inflation, economic stagnation, taxation policies) may influence the project (6). Thus, we must be aware of the spatial and temporal boundaries which are important for examining the impacts of a proposed development.

7.3.2 Principles for impact assessment

Gibson (1993) has suggested that a set of principles can be identified for the design of impact assessments. While no one design can be perfect for all circumstances, he believed that consideration of eight principles should help to improve designs. He stressed that the principles are interdependent and form a package. In his view, all have to be used together. The eight principles are considered below.

(1)   *An integrated approach.* By an integrated approach, Gibson meant that societies must consider the impacts of their activities at local, national and

global scales; ensure that basic needs are met and poverty is alleviated; analyse patterns of consumption to determine implications from extraction of resources and from returning wastes to the environment; and focus not only on environmental implications but also on social, cultural and economic aspects and their inter-relationships. It also is necessary to examine implications in the short, medium and long term. An integrated approach would also consider cumulative effects, which is best done by relating impact assessment to regional land-use planning.

(2) *All decision making should be environmentally responsible.* Impact assessment should apply as broadly as possible, including public and private sector initiatives for new projects or for expansion, modification, decommissioning or abandonment of existing ones. This principle also requires that proponents must know from the outset what their impact assessment obligations are. This principle means that policies and programmes as well as projects should be subject to IAs.

(3) *Impact assessment should focus on identifying best options rather than acceptable options.* This principle requires that the purpose and relative merits of alternative means are critically examined. This could result in questioning of the objectives, but more frequently would focus on the rationale for a preferred set of actions to realize objectives.

(4) *Impact assessments should be based in law, and should be specific, mandatory and enforceable.* This principle makes it clear that IA is an attack on the status quo, and is intended to lead to change in planning and decision making. Given the emphasis on change, voluntary adoption is inappropriate. The assessment expectations must be clearly understood, key tenets must be based in law, and compliance needs to be legally enforceable (Box 7.3).

(5) *Assessment processes and related decision making must be open, participative and fair.* This principle reflects the concepts of equity, empowerment and justice embodied in sustainable development (Chapter 4), and the concept of a participatory approach (Chapter 8). The rationale is that impact assessment is as much value-laden as scientific, and thus "broad

---

Box 7.3 Weaknesses in laws for IA

In practice, . . . , environmental assessment laws and processes have not been automatically effective. Their purpose is difficult and delicate: They are intended to force open and careful consideration of a new and generally ill-understood set of concerns, and they are directed at decision makers who are generally hostile to greater openness and to additional, imposed duties. Failure is easy. Moreover, laws and processes that are weak, unclear or simply difficult to administer with consistency and efficiency do not just fail to foster greater environmental sensitivity in planning and decision making; they tend also to undermine the general credibility of government efforts to encourage environmental responsibility.

Source: Gibson, 1993: 12.

participation and scrutiny is the best means of combating narrow biases and encouraging careful attention to matters of public concern" (Gibson, 1993: 19). Openness and participation also should contribute to an even-handed approach to all parties and interests. These views are elaborated upon in the guest statement by Tina Artini, reflecting on her experience in Indonesia.

(6) *Terms and conditions of approvals must be enforceable, and capacity must exist to monitor effects and to enforce compliance during implementation.* Approvals after a systematic review will have little value if there is not the capacity or commitment to track post-approval activity and effects, and to ensure compliance. While such provisions are common sense, many IA processes do not provide for what might be termed "enforceable approvals", or for monitoring. The role of *auditing* as a technique to facilitate compliance is considered in Chapter 13.

(7) *Efficient implementation should occur.* While efficiency, hopefully, is a central concern in all regulatory processes, Gibson argued that it is particularly needed in environmental assessment because inefficiency will breed hostility and antagonism, which together become a "formidable enemy". The long-term goal of IA is to change proponents into people who automatically think, plan and act with regard to environmental and social consequences. Hostility and antagonism can become a major obstacle for achieving such a long-term goal.

(8) *Provisions must be made to connect impact assessment to higher-level decision making.* This final principle is tied closely to the first principle, which advocates an integrated approach. Thus, it is important that the results of IA be fed back into more general policy and programme deliberations, and be used to help shape and develop criteria to be used for judging environmental significance.

Trade-offs to be considered. The eight criteria collectively set a high standard, and to a considerable extent reflect the *programmed approach* to be explained in Chapter 12. Thus, they minimize discretion, and seek to ensure that IA is automatically included as part of the process to approve development initiatives. After considering the arguments in Chapter 12 regarding *programmed* and *adaptive* approaches, you may wish to re-examine these criteria, to decide whether you believe any modifications should be made to them.

7.3.3 Strategic issues in impact assessment

Predicting effects

Given our imperfect understanding of ecological and social systems, it is often difficult to anticipate or predict what the effects of proposed development might be. In many instances, *baseline information* is either missing or incomplete. As a result, the state of the existing system is not well understood. It is

Guest Statement

Partnerships and participation in environmental impact assessment

Tina Artini, Indonesia

In Chapter 8, the importance of partnerships and participation is discussed. Here, I highlight ideas about the role of participation regarding environmental impact assessment (EIA), a concept addressed in this chapter. The new EIA regulation No. 27 of 1999 in Indonesia emphasizes the importance of public participation, and requires the provision of information from developers or the government to the public.

The rationale for public participation in EIA is to: (1) protect the public interest, particularly the interests directly affected by decisions regarding the exploitation of natural resources; (2) identify problems more precisely; and (3) achieve transparency. Timing for public input also is very important, and participation should be sought from the outset of the development process.

Public participation should empower the public in the decision-making process, and create an equitable partnership among the public, developers and regulatory authorities. In the ideal EIA process, the creation and implementation of partnerships should be based on mutual respect and trust. In this context, government should view its role as a facilitator to allow all stakeholders to participate effectively. I agree with the statement in Chapter 8 that "there is not one best model for partnerships. The kind of partnership and the nature of participation have to be determined by the various people or groups involved". Furthermore, various mechanisms exist to obtain effective input, suggestions or advice. Each can be assessed by how it enhances the degree of contact, and the ability to handle the interests of the participants (Canter, 1991). Communication processes also are extremely important. For instance, public displays and public meetings can be used to provide information to the public. However, the best techniques are those which facilitate two-way communication, especially when they help to provide information about project activities, assist in identifying issues, and generate feedback. The government agencies and private sector need to be creative to determine the best participatory approach based on the context of a community or region, instead of following formal guidelines in a mechanical fashion. Furthermore, the ability of a facilitator to obtain the views of EIA stakeholders can be of critical importance in achieving an effective partnership

Public participation is a complex process because it is influenced by social and political conditions in a community, including the degree of awareness and understanding about the benefits of participation. In Indonesia, public participation has traditionally been very limited. Generally, it has involved not much more than the public attending a meeting or becoming an object in a study. Thus, one problem with EIA in Indonesia is the lack of power by the public relative to government agencies. The need to empower the public is the reason why the Indonesian government created a regulation requiring public participation and transparent (continuous, accurate) information in the EIA process.

In this book, Mitchell states that "drawing upon many people and groups should help to achieve a balanced perspective relative to an issue". However, when many stakeholders are involved, public participation can become challenging. Experience with EIA for the pulp and paper industry in North Sumatra is one example. Many non-governmental organizations sought to represent the local community and other stakeholders. One result was that regulatory agencies had difficulty in determining who were the proper representatives. The criteria to select appropriate representatives of a community or the public for a partnership or participation programme are very important, if the stakeholders are to be viewed as credible by the community.

Reference

Canter L W 1991 *Environmental Impact Assessment*. New York, McGraw-Hill

Tina Artini obtained a MA from the Faculty of Environmental Studies at the University of Waterloo, Canada. She is the Head of the Subdirectorate for EIA Monitoring and Evaluation in the Environmental Impact Management Agency (BAPEDAL) in Indonesia. She is responsible for dealing with environmental impacts, especially social impacts, from projects at local and national scales, and for designing and applying policies, regulations and guidelines related to EIA. She has participated in several national and international conferences, workshops and training courses on topics such as environmental management, urban planning, gender analysis, community development and public participation.

for this reason that state of the environment and other types of monitoring are being advocated or initiated, as discussed in Chapter 13.

Our *theories* or *concepts* about ecological and human systems also may be incomplete, inconsistent or contradictory. For example, it is not universally accepted that diversity in a natural system always leads to stability. There is also questioning about whether "stability" is a state that should be expected or desired. Many argue that "resilience" is a more desirable end state for a system (see Chapter 5). We also often do not understand the conditions that may make a system "flip" or transform in some catastrophic manner. These aspects highlight the complexity and uncertainty with which analysts must deal, and our frequent ignorance or incomplete understanding.

Synergistic effects also can make predictions difficult. In other words, it may be possible to predict the outcome of a development on a particular component of a system, if it is assumed that the component is isolated from all other components of that system. Rarely if ever is the assumption of "all other things being constant or equal" satisfied. When changes in one component of a system interact with changes in other components of that system, the final changes may be totally different than the changes that might be expected for any one component. It is this aspect which makes estimating *cumulative effects* so challenging, and this is discussed in more detail following the next section on intangible effects.

Intangible effects

Problems are encountered during impact assessment in valuing environmental or social components which are not readily or easily measured in monetary terms. Biodiversity, ecological integrity, human health and cultural integrity are examples of aspects not easily quantified in monetary terms, or readily comparable. Yet, one of the basic issues in impact assessment is determining the relative merits of alternative actions, just as the relative economic efficiency of alternatives is estimated in benefit–cost analysis. Ideally, having established the relative merits, the analyst or planner can then begin to explore trade-offs, to design a project that will enhance positive environmental and social impacts, and minimize negative ones. However, when it is difficult to measure some impacts, considering trade-offs becomes challenging.

Cumulative effects

In order to be practical, IA procedures usually have a cut-off threshold, determined by variables such as capital costs or number of employees, below which assessments will not be conducted. This practice has evolved to avoid unnecessary regulations or restrictions on relatively small and innocuous activities. The danger, of course, is that the total sum of many small developments may be greater than the sum of the individual parts (Box 7.4). This dilemma is characterized as one of *cumulative effects*.

Cumulative effects can occur for many reasons. The most obvious is the *additive effect* of many small activities. One home owner burning coal to heat a house may not create a serious air pollution problem. But if every house in a major metropolitan area is burning coal, then smog can become a serious problem. *Time crowding* can create cumulative effects, such as the congestion experienced on major highways in cities during rush hour. For most of the day, the capacity of the highways and roads is adequate, but at the beginning and end of the regular working day there can often be delays caused from the congestion of too many users relative to the capacity. *Space crowding* also can occur, in combination with or separate from time crowding. Habitat fragmentation in forests or estuaries illustrates space crowding. *Compounding effects* occur when there are delayed consequences from continuous emissions, such as gaseous emissions into the atmosphere. At some point, the accumulation of emissions results in the air quality crossing a threshold from one which is acceptable to one which is dangerous for living beings. *Time lags* can also occur, such as when small

---

Box 7.4 Cumulative effects

[the] process of predicting and minimizing the consequences of a single action has not adequately considered the accumulative nature of some effects, the nonlinear responses of some natural systems, nor the linkages between a single action and other related activities.

Source: Constant and Wiggins, 1991: 298.

---

amounts of carcinogenic materials have no obvious impact until after the carcinogenic trigger has accumulated sufficiently, which may take years or decades.

All of the above pathways of or triggers for cumulative effects represent major challenges for analysts and planners. It is difficult or impossible for us to isolate the impact of individual variables from others. When we would like to identify the cumulative effects of several variables, usually our knowledge is not adequate to predict all the various ways in which multiple variables might interact. From a planning point of view, if a time-lag effect could be decades, it is difficult to persuade people to take action today if that means forgoing benefits in the short term. And yet, time-lag effects represent one dimension of *intergenerational equity* related to sustainable development, which was discussed in Chapter 4.

Compensation

Even with systematic and careful IA, not all negative impacts can be removed. These undesirable outcomes particularly occur with what are called *noxious facilities*, the type of facility we all require but no-one wants close by (see Chapter 14). Thus, we collectively create a demand for sand and gravel for construction material for roads and foundations for houses. But few people are enthusiastic about having a working sand and gravel pit as a nearby neighbour. The noise and dust from heavy trucks coming and going from the pit, the danger to children walking along the road from increased levels of traffic, and the noise and dust from the operating pit can all be negative effects for nearby land owners. Even if a rehabilitation plan is developed and implemented for the site, if it were to operate for 15 or 20 years, that could be a major part of the adult lifetime of a nearby resident.

When society makes decisions to allow developments such as *noxious facilities* which serve general societal needs, but cause inconvenience to a small number of people, equity suggests that the larger community should be prepared to compensate those who suffer negative effects. Compensation could vary from monthly or annual payments during the lifetime of the operation, to acquisition of the property or exchange of properties for a new location which would not be impacted by the activity, along with funds to relocate. As noted earlier in this chapter, however, compensation can become almost overwhelming for major projects such as the Three Gorges in which one million people have to be relocated. Generally, societies have not treated very adequately the minority who are affected negatively by development. There is scope for much work here, especially since often the effects have an intangible component, so determining a fair compensation package is not always easy.

Strategic impact assessment

While environmental impact assessment normally focuses on specific projects, strategic impact assessment (SEA) addresses the environmental implications of a proposed policy, plan or programme. As a result, SEA considers the broader context within which specific projects will be positioned (Noble, 2000). SEA is generally considered to have been introduced through the *California Environmental Quality Act*. Subsequently, countries such as Canada, Japan, New Zealand and the Netherlands have formally incorporated SEA into planning and management processes.

SEAs are normally focused on specific policy, plan or programme fields (such as transportation, agriculture, energy). Once the focus is determined, then SEA usually contains the same components associated with EIAs: needs justification, scoping, identification of alternatives, prediction of impacts, assessment of significance of impacts, evaluation, public participation, implementation, mitigation and monitoring.

What are the benefits from SEAs? The following can be expected:

- Proposed policies, plans and programmes are considered with regard to their environmental implications, and final selection of a policy, plan or programme occurs with explicit understanding about environmental merits and shortcomings of all options. This is often achieved by developing a set of environmental planning and management objectives, and using them to judge all options.
- When negative environmental consequences are part of the chosen option, attention then is directed toward their mitigation, and to establishment of monitoring procedures to track the outcomes.
- SEAs help to identify cumulative effects, something which often does not occur with EIAs, which deal with the project level.
- SEAs help to identify priority environmental concerns, helping to ensure that subsequent project level EIAs address a manageable number of priority concerns.
- SEAs allow fundamental, normative questions (should nuclear power be one source of energy for a region?) to be raised, rather than focusing on operational questions (which of three sites is most suitable for a nuclear power plant?).

SEAs do not resolve all problems. Indeed, as Novakowski and Wellar (1997: 242) have observed, the playing field regarding environmental and economic issues is not level. As a result, once SEA results begin to be incorporated into a planning process, they may encounter resistance or opposition, both from participants within key public agencies involved in the planning and approvals process, as well as from decision makers both internal and external to the planning agencies. From their comments, it can be concluded that land-use planning, and resource and environmental management, are ultimately political processes in which competing interests interact regarding choices to be made.

Critical learning

Diduck and Sinclair (1997) have suggested that general consensus exists for education being a key element for public participation in resource and environmental management, and many methods being available to facilitate education

Table 7.1 Education techniques identified in the EA literature, adopted in the legislation of selected jurisdictions or identified in literature from other environmental contexts, categorized according to format

---

audio/visual/electronic

Slide presentations	Film presentations
Computerized participation	Videotape
Knowledge-based systems	
Electronic publishing	
Information retrieval systems	
Interactive computer software	

Traditional publishing (printed) (verbal)

publications	Brochures
newspaper inserts	Notices
feature articles	Position papers
reports	Newsletters
information kits	Central depositories
decisions and reasons	Translation
plain language legislation	Posters
photonovel	Manuals

direct/individualized

direct mail	Phone lines
field offices	Technical assistance
direct e-mail	

media

public service announcements	News releases
news conferences	Advertising
call-in television	Talk radio
coverage of hearings	Interviews

public presentations/events

workshops	Conferences
panels	Open houses
exibits/displays	Contests
simulation exercises	Song contests
meetings	Town hall meetings
dialogues/coffee klatches	Brainstorming
speakers bureau	Special event days
discussion group conferencing	

formal education

integration into existing curricula	
discussion in literacy programmes	

Source: Diduck and Sinclair, 1997.

(Table 7.1). At the same time, however, they concluded that education is often restricted to information dissemination and communication practices.

They argue that education should be designed to facilitate *critical learning*, which includes both "education about EA" and "education through EA". In their view, education about EIA is needed to prepare participants before they become engaged in involvement processes. Education about EIA ensures that participants are informed regarding both process and substantive issues, covering matters as diverse as engineering aspects of a project, ecological and economic analyses of project options, how communities and ecosystems work, how status quo decision-making processes and project decisions can be challenged, and how members of the public can collaborate to define and pursue common goals.

In contrast, education through EIA happens when, through participating, members of the public have their critical awareness increased or improved. For example, they learn how to make more effective presentations, how to lobby and file an appeal, and how to interact effectively with "experts". Diduck and Sinclair (1997: 299) argue that "education through EA provides the means of providing informed, critical social activists capable of engaging their communities in meaningful and critical dialogue and of mounting efforts for real social change". In their view, the combination of education about EA and through EA creates two important outcomes: empowerment and social action.

They argue that some educational techniques are more suited for a critical learning approach than others. The most well-suited techniques are those which emphasize interactive learning, and are people centred, With regard to the methods shown in Table 7.1, *workshops* are desirable because they are participatory, dialogical, research oriented, activist and reliant on group learning. Methods not suited for critical learning are ones which focus on presentation of "facts" and information dissemination, with little or no interaction with the affected public. Examples of such non-interactive techniques are advertising, direct mail (either e-mail or snail mail), posters and central depositories.

More discussion about participatory methods will be found in Chapter 8. However, for impact assessment, this discussion highlights the opportunities for making the EIA process accessible and available to the public who will be affected by development decisions.

Cultural aspects

Boyle (1998) has noted that in many fundamental ways EIA is a cultural issue. What impacts deserve attention and how their relative importance is determined require answers which are shaped by the cultural characteristics of the people in a society who are grappling with them. Furthermore, the design of an EIA process must fit the cultural values and norms of a given society.

Boyle argued that one of the problems of EIA in developing countries has been that most of them have based their process on one designed for application in the USA, an industrialized democracy. Boyle argued that the EIA process for the USA reflects the following democratic principles: (1) politicians and governments are accountable to the public; (2) political and business elites do not have an unfettered right to do what they want; (3) government bureaucratic and decision-making processes should be open, accessible and transparent; (4) natural resources such as air, water, forests, wildlife and landscape beauty are a common heritage and cannot be unilaterally appropriated for private purposes; and (5) individuals affected by development projects have a right to information,

---

Box 7.5 Importance of cultural characteristics in Thailand, Indonesia and Malaysia for EIA

The reliance on paternalistic authority, hierarchy, and status as principles of social organization; the dependence on patron–client relationships for ensuring loyalty and advancement; and the desire to avoid conflict and maintain face in personal relations are all cultural characteristics that tended to isolate decision-makers from the concerns of people and communities affected by major projects, and reinforce their power to act in their own or the "national" interest. These factors effectively circumscribed the ability of individuals, communities and public-interest groups to participate constructively in the process of project planning and decision-making. Moreover, they resulted in government bureaucracies that were strongly hierarchical with little opportunity for the interagency communication, cooperation, and coordination needed for integrated environmental and natural resource management in general and effective EIA in particular.

Source: Boyle, 1998: 114.

---

to question the need for and design of projects, and to participate in the planning process.

In contrast, many developing nations have markedly different cultural and sociopolitical heritages and practices. Boyle reviewed experience in Thailand, Indonesia and Malaysia to examine differences in culture and to consider their implications for EIA. While it is difficult to generalize about cultures in three countries, he did suggest that certain aspects of Southeast Asian cultures were significantly different from that in the USA (Box 7.5). For example, he noted in Southeast Asia a strong desire for paternalistic authority and for dependency and loyalty to a group. People generally are very aware of their relative position in a hierarchy and of their status, and deference is commonly granted to people of higher status. Another characteristic is the value placed on self-control, avoidance of conflict and sensitivity to saving of reputation or "face". This second characteristic directs individuals to contain inner feelings, to avoid explicit criticism of others, and to conduct interpersonal relations in a smooth and unthreatening manner. Self-control is a fundamental virtue. When these cultural characteristics are contrasted to the principles identified by Boyle for Western democracies, it is hardly a surprise that an EIA process designed for the USA may not always work well in a nation with such significantly different cultures. The EIA process in the USA is based on the assumption that information will be available to the public, and that the public will continuously press to hold accountable the elected and appointed individuals responsible for decisions about development projects. In contrast, in societies in which there is strong awareness of respective hierarchy and status, and in which overt criticism is avoided, it is unlikely that a process assuming accessibility to information and mechanisms for public participation is likely to be effective.

Thus, while technical issues always will require attention in the design and implementation of EIA processes, it also is important that there be awareness of cultural factors when decisions are being taken regarding how to create or modify an impact assessment process.

Summary. Impact assessment was introduced to ensure that environmental and then social considerations were given due attention relative to economic ones. In that manner, impact assessment is a useful tool for examining options that are consistent with sustainable development. Used in conjunction with benefit–cost analysis, impact assessment helps to integrate economic and environmental considerations.

7.4 Life cycle assessment

7.4.1 Definition and evolution

Life cycle assessment (LCA) was formalized during the early 1990s. It emerged from a conviction that it was important to conduct "cradle-to-grave" assessments of products, packages, processes and activities (Box 7.6). The Ecobalance Study approach developed in Europe, emphasizing waste reduction to curtail water and air pollution, and Resource and Environmental Profile Analysis work in the USA during the 1980s, were the forerunners of LCA. However, it was at a workshop organized by the Society of Environmental Toxicology and Chemistry (SETAC) in the USA during the early 1990s that the basic approach for LCA was developed (SETAC 1991; 1993; 1994).

7.4.2 Stages in life cycle analysis

As indicated in Figure 7.2, LCA can be divided into four stages or phases. Each is briefly considered below.

Box 7.6 Life cycle assessment

A concept and a method to evaluate the environmental effects of a product or activity holistically, by analysing its entire life cycle. This includes identifying and quantifying energy and materials used and wastes released to the environment, assessing their environmental impact, and evaluating opportunities for improvement.

Life cycle stages

The set of major sequential stages that a product or service passes through over the course of its existence from cradle to grave. For all products, four generic stages apply: raw materials and energy acquisition, manufacturing (including materials manufacture, product fabrication, and filling/packaging/distribution steps), use/reuse/maintenance, and recycle waste management.

Source: Canadian Standards Association, 1994a: 6.

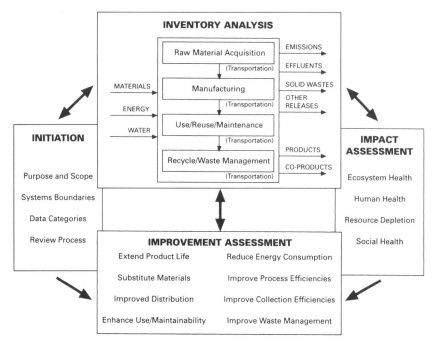

Figure 7.2 The four phases of life cycle assessment (with the permission of the Canadian Standards Association, 1994a: 13 material is reproduced from the CSA Standard: Z760–94 (Life Cycle Assessment), which is copyrighted by CSA, 178 Rexdale Blvd., Etobicoke, Ontario, M9W 1R3.)

Initiation. This stage is very similar to scoping in impact assessment. The main purposes are to establish the objectives for and detail of an assessment, to define the "system" to be assessed, and to identify types of data needed.

Inventory. Primary attention is allocated to collecting data about raw materials needed for inputs, including energy and water, and about wastes produced as outputs during the process and at the end of the useful life of the product, process, package or related activity.

Impact. This component is similar to the comparable phase in impact assessment. The focus is upon identifying the effects, and reaching a judgement about the significance of such effects. Effects on the environment, economy and health or well being are usually included.

Improvement. This can be characterized as the *normative* or *prescriptive* stage in which attention is directed to possible actions to reduce or mitigate any negative impacts identified in the previous stage.

The four phases or components identified above and in Figure 7.2 are intended to assist resource and environmental managers in encouraging better product design, more effective processes regarding raw material inputs or waste outputs, improved transportation methods, more careful consumer use and better waste disposal practices. More specifically, in the context of resource and environmental management, LCA is intended to lead to decisions which result in greater conservation of resources and the environment, increased energy conservation and decreased waste generation, improved industrial processes related to providing resource-based products, and fewer problems in final disposal. Thus, regarding toxic wastes, for example, life cycle analysis would consider whether a toxic chemical needed to be used at all; and, if it were to be used, what should be done to ensure its proper production, transportation, storage, use and disposal.

Some key aspects of each of the four components are considered in more detail below.

Initiation

Defining the *system boundaries* for the life cycle assessment is a critical aspect during initiation. For example, the scope of a LCA for a box-making plant in Germany should not only include the raw materials and energy required at the plant to manufacture each box, and the wastes that must be disposed of. It also should examine the raw material and energy inputs for the other firms that provide component parts for the production of the boxes, even when those suppliers are located outside Germany. Furthermore, a decision must be taken as to whether the analysis only focuses on the primary manufacturers, or whether the assessment should include the forestry operations that provide the paperboard for the box, the natural gas industries which provide the resin and polyethylene, and the aluminum industry that produces the aluminum foil, and so on. Figure 7.3 outlines in schematic form the various levels that could be involved in conducting a life cycle assessment of the process to produce laminated cartons. The choice of *system boundaries* is thus not always obvious or automatic. Furthermore, the definition of such boundaries can greatly complicate and extend the LCA, and add to the time and cost to complete it.

Decisions also have to be taken regarding who will complete the life cycle assessment. There are at least three choices. Employees of the firm could do the assessment. They likely are in the best position to be able to identify all of the inputs and outputs. However, the credibility of a self-evaluation can be low. The alternative is to have outside experts conduct the assessment. If such an external team does the work with reference to some generally accepted standards, then the credibility of the assessment usually is greater. The third option is to have a team involving both employees and outside experts conduct the task. For practical reasons, it often is some form of the third alternative which is used, since a firm often does not have all of the expertise to conduct a LCA, yet the outside experts need the input and advice of people within the firm.

Inventory

As Figure 7.2 indicates, inventory is required for raw material acquisition, manufacture, use/reuse/maintenance and recycling/waste management. Particular attention is given to the material, energy and water inputs, and to the emission, effluent, solid waste and other release outputs, as well as to products and

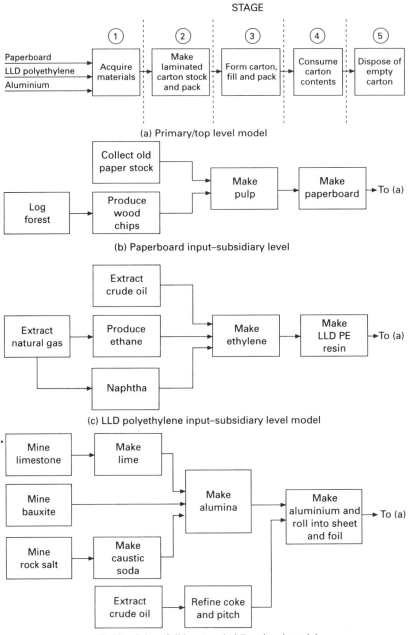

STAGE

(a) Primary/top level model

(b) Paperboard input–subsidiary level

(c) LLD polyethylene input–subsidiary level model

(d) Aluminium foil input–subsidiary level model

Figure 7.3 Life cycle assessment for production of laminated cartons (with the permission of the Canadian Standards Association, 1994b: 7, 8 material is reproduced from the CSA Publication: PLUS 1107 (User's Guide to the Life Cycle Assessment: Conceptual LCA in Practise, copyrighted by CSA, 178 Rexdale Blvd., Etobicoke, Ontario, M9W 1R3.)

Box 7.7 Inventory: four key components

1. **Raw materials acquisition.** All activities necessary to collect a raw material or energy source from the earth. This includes transportation of the raw material to the place of material manufacture.

2. **Manufacturing.** (1) *Materials manufacture*: This includes the activities needed to process raw material into a form useful for fabricating a product or package. The production of many intermediate materials or chemicals often is included in this subcomponent. Transport of intermediate materials also is included. (2) *Product fabrication*: This includes the activities which use raw or manufactured materials to fabricate a product to the point where it can be filled or packaged. The output here could be a product that may be distributed for retail sale, or that could be distributed for use by other industries. (3) *Filling/packaging/distribution*: Preparation of the final products for shipment to users, as well as transportation of products to the retail or other recipient.

3. **Use/reuse/maintenance.** This starts following the distribution of products or materials for use, and includes any activity which reconditions, maintains or services the product or package.

4. **Recycle/waste management.** This begins after the product, material or package has served its intended function, and either enters a new system through recycling or enters the environment through a waste management process.

Source: After Canadian Standards Association, 1994a: 14.

---

co-products resulting from recycling or waste management. A thread common to many of the activities is transportation, as it is necessary to determine which inputs and outputs are needed to move material from one stage to another. As Box 7.7 indicates, inventory involves four major activities.

Impact assessment

The activities here are similar to those described in the previous section dealing with environmental impact assessment. Particular concern usually focuses upon the implications of resource consumption associated with the creation of the product, material or packaging, and the releases into the environment during each of the life cycle phases. Ideally, the information here is in the form of *impact indicators* which indicate whether critical thresholds have been passed. The challenge, of course, is that for many outputs we do not yet know what the critical thresholds are, and this aspect becomes even more problematic when the synergistic effects of two or more waste outputs are involved. One of the criticisms of LCA is that it is a very detailed, and potentially very expensive, procedure, but at the conclusion of the impact assessment stage there still are not enough generally agreed upon environmental standards to allow effective use of all the information which has been collected.

Improvement

The improvement component is often done in parallel with the impact component. An over-riding purpose is to identify opportunities to reduce raw material inputs and waste outputs. Other benefits could be to streamline the production process so as to reduce costs to the firm and therefore make it more profitable. The alternatives normally considered include changes in product design to extend its life; to allow material or other input substitutions; to improve production, process or collection effectiveness; to improve distribution or transportation systems; to enhance consumer use or maintenance procedures; and to improve waste management.

7.4.3 Motivation for doing life cycle assessment

What would motivate a firm to conduct LCAs? Several incentives could provide the motivation. First, a firm may believe that with growing environmental awareness in many countries, completing a LCA may give it a marketing advantage by being able to promote its products as "clean". Second, with the introduction of *environmental labelling*, firms may decide that LCAs will help them become eligible to have their products certified as "green" or "environmentally friendly", and again provide a marketing advantage over competitors (CSA, 1993). Third, the International Standards Organization (ISO) guidelines, such as the ISO 14000 series, will be used to certify firms that have met environmental management and other requirements (Figure 7.4). While such

Figure 7.4 ISO 14001 certification. ISO certification is used by companies to promote themselves and their products. This picture is of the sign on the property of the Toyota car manufacturing plant in Cambridge, Ontario. The ISO 14001 certification acknowledges that the firm has a recognized environmental management system explicitly incorporated into its operations (Bruce Mitchell)

standards are presently voluntary for industries, it is conceivable that in the near future countries such as those in the European Union may prohibit the import of products or materials unless they are certified as having been manufactured by ISO 14000 designated firms (Box 7.8). Such an initiative would provide a "level playing field" for environmental standards among countries, and would definitely provide a major incentive for firms to have their products so identified.

Box 7.8 Environmental management system (EMS) principles specified in ISO 14004

Principle 1: Commitment and Policy. An organization should define its environmental policy and ensure commitment to its environmental management system. Key elements to achieve this are (1) top management commitment and leadership, (2) initial environmental review, and (3) establishment of environmental policy.

Principle 2: Planning. An organization should formulate a plan to fulfill its environmental policy. Key elements are (1) identification of environmental aspects and evaluation of associated environmental impacts, (2) identification of legal requirements, (3) development and adoption of internal performance criteria, (4) specification of environmental objectives and targets, and (5) establishment of environmental management plans and programs.

Principle 3: Implementation. (a) ensure capability – To implement the EMS, an organization should develop the capabilities and support mechanisms needed to achieve its environmental policy, objectives and targets, and key elements are (1) dedicated human, financial and physical resources, (2) aligning and integrating EMS with other management systems, (3) establishing accountability and responsibility, (4) developing environmental awareness and motivation, and (5) improving knowledge, skills and training; and **(b) support action** – Develop the capabilities and support mechanisms needed to achieve environmental policies, objectives and targets, and key elements are (1) communication and reporting, (2) creation of EMS documentation, (3) operational controls, and (4) emergency preparedness and response capacity.

Principle 4: Measurement and evaluation. An organization should measure, monitor and evaluate its environmental performance. Key elements are (1) measuring and monitoring of ongoing performance, (2) ensuring corrective and preventive action, (3) establishing EMS records and information management procedures, and (4) providing for audits of EMS (see Chapter 13).

Principle 5: Review and improvement. An organization should review and continually improve its EMS, with the objective of improving its overall environmental performance. Key elements are (1) reviews at regular intervals, and (2) continual improvement by identifying opportunities.

Source: Based on Auditor General of Canada in cooperation with the Federal Committee on Environmental Management Systems, no date; and Auditor General of Canada, 1995.

7.4.4 Availability of LCA studies

There is not yet a readily accessible literature on LCA. Most assessments have been completed by companies, or by consulting firms for clients. In either case, the reports have become the property of the company for which the LCA was completed. And, for competitive reasons, many companies are hesitant to make the reports generally available, since in most cases the assessment was completed to help them become more competitive in the market place. Nevertheless, LCA can be anticipated to be a "growth" industry for those interested in resource and environmental management, and people hoping to work in this broad field should be familiar with LCA.

7.5 Implications

Two methods or techniques – impact assessment, life cycle assessment – have been examined in this chapter. Each provides an entry point for the task of assessing alternatives. Environmental impact assessment appeared in the early 1970s in response to concern that benefit–cost analysis was not giving sufficient attention to environmental matters. In turn, environmental impact assessment became criticized for focusing unduly upon biophysical issues, and by the end of the 1980s procedures to deal with social impacts had appeared. In the 1990s, life cycle assessment emerged, reflecting concern about the importance of tracking inputs and outputs in production processes from "cradle-to-grave". Such LCAs are receiving impetus from the development of the ISO guidelines which certify firms as following "good" management practices. While ISO has been a voluntary procedure, it can be anticipated that national governments will eventually require companies to conform to it, suggesting that LCA will become a growth industry in its own right.

However, it is important to emphasize that none of the methods helps resource and environmental managers to know which vision, goals or objectives are the most desirable. These methods most usually have been designed, and applied, to determine the best *means* to achieve *ends* which already have been chosen. The one exception is the relatively limited application of impact assessment to policies and programmes, as opposed only to projects. Nevertheless, resource and environmental planners ideally should have a vision or direction established before they begin to make choices about which one or more methods they will use to compare alternative means. Once that vision is in place, then planners and managers should be encouraged to use a combination of these methods to make choices. Any one method by itself has enough weaknesses that it probably is unwise to rely only on one.

References and further reading

Annandale D 2000 Mining company approaches to environmental approvals regulation: a survey of senior environment managers in Canadian firms. *Resources Policy* 26: 51–9

Auditor General of Canada 1995 *Environmental Management Systems: A Principle-based Approach*. Ottawa, Minister of Supply and Services Canada

Auditor General of Canada, with the Federal Committee on Environmental Management Systems (no date) *EMS Self-assessment Guide*. Ottawa, Minister of Supply and Services Canada

Bailey I 2000 Principles, policies and practice: evaluating the environmental sustainability of Britain's packaging regulations. *Sustainable Development* 8: 51–64

Banks G 1998 Compensation for communities affected by mining and oil developments in Melanesia. *Malaysian Journal of Tropical Geography* 29: 53–67

Boyle J 1998 Cultural influences on implementing environmental impact assessment: insights from Thailand, Indonesia and Malaysia. *Environmental Impact Assessment Review* 18: 95–116

Brown A L and R Thérivel 2000 Principles to guide the development of a strategic environmental assessment methodology. *Impact Assessment and Project Appraisal* 18: 183–9

Buckley R 2000 Strategic environmental assessment of policies and plans: legislation and implementation. *Impact Assessment and Project Appraisal* 18: 209–15

Burgham M C J, S H Brassat, W A Ross and D A R Thompson 1998 Extending environmental management systems principles to environmental assessment: the use of environmental audits. *Impact Assessment and Project Appraisal* 16: 155–61

Burris R K and L W Canter 1997 A practitioner survey of cumulative impact assessment. *Impact Assessment* 15: 181–94

Buström F 2000 Environmental management systems and co-operation in municipalities. *Local Environment* 5: 271–84

Cada G F and C T Hunsaker 1990 Cumulative impacts of hydropower developments. *Environmental Professional* 12: 2–9

Canadian Standards Association 1993 *Guideline on Environmental Labelling*. Publication Z761-93, Rexdale (Toronto), Canadian Standards Association

Canadian Standards Association 1994a *Life Cycle Assessment*. Publication Z760-94, Rexdale (Toronto), Canadian Standards Association

Canadian Standards Association 1994b *User's Guide to Life Cycle Assessment: Conceptual LCA in Practice*. Publication PLUS 1107, Rexdale (Toronto), Canadian Standards Association

Carroll M S, A K Findley, K A Blatner, S R Mendez, S E Daniels and G B Walker 2000 *Social Assessment for the Wenatchee National Forest Wildfires of 1994: Targeted Analysis for the Levenworth, Entiat, and Chelan Ranger Districts*. General Technical Report PNW-GTR-479, Portland, Oregon, US Department of Agriculture, Forest Service, Pacific Northwest Research Station, January

Chin C L M, S A Moore, T J Wallington and R K Dowling 2000 Ecotourism in Bako National Park, Borneo: visitors' perspectives on environmental impacts and their management. *Journal of Sustainable Tourism* 8: 20–35

Ciccantell P S 1999 Making aluminum in the rainforest: the socio-economic impact of globalization in the Brazilian Amazon. *Journal of Developing Areas* 33: 175–97

Cocklin C, S Parker and J Hay 1992 Notes on cumulative environmental change I: concepts and issues. *Journal of Environmental Management* 35: 31–49

Constant C K and L L Wiggins 1991 Defining and analyzing cumulative environmental impacts. *Environmental Impact Assessment Review* 11: 297–309

Cooper T A and L W Canter 1997 Substantive issues in cumulative impact assessment: a state-of-practice survey. *Impact Assessment* 15: 15–31

de Roo G and D Miller 1997 Transitions in Dutch environmental planning: new solutions for integrating spatial and environmental policies. *Environment and Planning B* 24: 427–36

de Snoo G R and G W J van de Ven 1999 Environmental themes in ecolabels. *Landscape and Urban Planning* 46: 179–84

Devuyst D 2000 Linking impact assessment and sustainable development at the local level: the introduction of sustainability assessment systems. *Sustainable Development* 8: 67–78

Diduck A P and A J Sinclair 1997 The concept of critical environmental assessment (EA) education. *Canadian Geographer* 41: 294–307

Dixon J and B E Montz 1995 From concept to practice: implementing cumulative impact assessment in New Zealand. *Environmental Management* 19: 445–56

Dowie W A, D M McCartney and J A Tam 1998 A case study of an institutional solid waste environmental management system. *Journal of Environmental Management* 53: 137–46

Duinker P N and G L Baskerville 1986 A systematic approach to forecasting in environmental impact assessment. *Journal of Environmental Management* 23: 271–90

Ebisemiju F S 1993 Environmental impact assessment: making it work in developing countries. *Journal of Environmental Management* 38: 247–73

Elling B 2000 Integration of strategic environmental assessment into regional spatial planning. *Impact Assessment and Project Appraisal* 18: 233–43

Fearnside P M 1999 Social impacts of Brazil's Tacuruí dam. *Environmental Management* 24: 483–95

Fortin M-J and C Gagnon 1999 An assessment of social impacts of national parks on communities in Quebec, Canada. *Environmental Conservation* 26: 200–11

Gagnon C, P Hirsch and R Howitt 1993 Can DIS empower communities? *Environmental Impact Assessment Review* 13: 229–53

Gallagher T J and W S Jacobson 1993 The typography of environmental impact statements: criteria, evaluation, and public participation. *Environmental Management* 17: 99–109

Gibson R B 1993 Environmental assessment design: lessons from the Canadian experience. *Environmental Professional* 15: 12–24

Gilpin A 1995 *Environmental Impact Assessment: Cutting Edge for the Twenty-First Century.* Cambridge, Cambridge University Press

Glasson J 1995 Regional planning and the environment: time for a SEA change. *Urban Studies* 32: 713–31

Goodenough R A and S J Page 1994 Evaluating the environmental impact of a major transport infrastructure project: the Channel Tunnel high-speed rail link. *Applied Geography* 14: 26–50

Harrison K 1999 Racing to the top or the bottom? Industry resistance to eco-labelling of paper products in three jurisdictions. *Environmental Politics* 8: 110–36

Heijungs R and J B Guinée 1995 On the usefulness of life cycle assessment of packaging. *Environmental Management* 19: 665–8

Hendriks C, R Obernosterer, D Muller, S Kytzia, P Caccini and P H Brunner 2000 Material flow analysis: a tool to support environmental policy decision making: case-studies on the city of Vienna and the Swiss lowlands. *Local Environment* 5: 311–28

Hollick M 1993 *An Introduction to Project Evaluation.* Melbourne, Longman Cheshire

Hunt R G, J D Sellars and W E Franklin 1992 Resource and environmental profile analysis: a life cycle environmental assessment for products and procedures. *Environmental Impact Assessment Review* 12: 245–69

Island Resources Foundation 1998 *Guidance for Best Management Practices for Caribbean Coastal Tourism.* On line at http://irf.org/ir_bmp.html

Jolliet O, K Cotting, C Drexler and S Farago 1994 Life-cycle analysis of biodegradable packing materials compared with polystyrene chips: the case of popcorn. *Agriculture, Ecosystems and Environment* 49: 253–66

Jones T 1993 The role of environmental impact assessment in coal production and utilization. *Natural Resources Forum* 17: 170–9

Kakonge J O 2000 A review of refugee environmental-oriented projects in Africa: a case for environmental impact assessment. *Impact Assessment and Project Appraisal* 18: 23–32

Keoleian G A and D Memerey 1993 *Life Cycle Design Guidance Manual: Environmental Requirements and the Product System*. Ann Arbor, University of Michigan, National Pollution Prevention Center, prepared for the USA Environmental Protection Agency

Krut R and H Gleckman 1998 *ISO 14001: A Missed Opportunity for Sustainable Global Industrial Development*. London, Earthscan

Lathrop K W and T J Canter 1998 Eco-labeling and ISO 14000: an analysis of US regulatory systems and issues concerning adoption of Type II standards. *Environmental Management* 22: 163–72

Lawrence D P 1997 Integrating sustainability and environmental impact assessment. *Environmental Management* 21: 23–42

Lee N and R Colley 1991 Reviewing the quality of environmental statements: review methods and findings. *Town Planning Review* 62: 239–48

Lee N and F Walsh 1992 Strategic environmental assessment: an overview. *Project Appraisal* 7: 126–36

Little A D 1990 *Life Cycle Assessment of Disposable versus Cloth Diaper Systems*. Report commissioned and sponsored by Procter and Gamble, San Diego, California

McCold L and J Holman 1995 Cumulative impacts in environmental assessments: how well are they considered? *Environmental Professional* 17: 2–8

McLeod H 2000 Compensation for landowners affected by mineral development: the Fijian experience. *Resources Policy* 26: 115–25

Miner R A and A A Lucier 1994 Considerations in performing life-cycle assessments on forest products. *Environmental Toxicology and Chemistry* 13: 1375–80

Mitchell B 1989 *Geography and Resource Analysis*. Harlow, Longman

Mohammed M 2000 The ISO 14001 EMS implementation process and its implications: a case study of central Japan. *Environmental Management* 25: 177–88

Nilsson J, C Bjuggren and B Frostell 1998 Greening of a campus restaurant at Stockholm University: sustainable development audits by means of the SDR methodology. *Journal of Environmental Management* 52: 307–15

Noble B F 2000 Strategic environmental assessment: what is it? & what makes it strategic? *Journal of Environmental Assessment Policy and Management* 2: 203–24

Novakowski E N and B Wellar 1997 Strategic environmental assessment and regional planning: practical experience from the National Capital Commission. In A J Sinclair (ed.) *Canadian Environmental Assessment in Transition*. Department of Geography Publication Series No. 49, Waterloo, Ontario, University of Waterloo, pp. 221–50

Odum W E 1982 Environmental degradation and the tyranny of small decisions. *BioScience* 32: 728–9

Orians G H 1995 Thought for the morrow: cumulative threats to the environment. *Environment* 6–14, 33–6

Owens J W 1997 Life-cycle assessment in relation to risk assessment: an evolving perspective. *Risk Analysis* 17: 359–65

Palerm J R 1999 Public participation in environmental impact assessment in Spain: three case studies evaluating national, Catalan and Balearic legislation. *Impact Assessment and Project Appraisal* 17: 259–71

Partidário M 1996 Strategic environmental assessment: key issues emerging from recent practice. *Environmental Impact Assessment Review* 16: 31–55

Pendall R 1998 Problems and prospects in local environmental assessment: lessons from the United States. *Journal of Environmental Planning and Management* 41: 5–23

Prost R A M, A J Kolhoff and B J A M Velthuyse 1998 Towards integration of assessments, with reference to integrated water management projects in third world countries. *Impact Assessment and Project Appraisal* 16: 49–53

Radchenko V N and M Y Aleyev 2000 Environmental and social impacts of management approaches in Sevastopol Bay in a historic retrospective: a case from the Black Sea. *Ocean and Coastal Management* 43: 793–817

Richardson N 1994 Moving toward planning for sustainability: integrating environmental assessment and land use planning in Ontario. *Plan Canada* March, 18–23

Rosenthal I and D F Theiler 1998 Use of an ISO 14000 option in implementing EPA's rule on risk management programs for chemical accidental release prevention. *Risk Analysis* 18: 199–203

Ross W A 1999 Cumulative effects assessment: learning from Canadian case studies. *Impact Assessment and Project Appraisal* 16: 267–76

Russell S O D 1994 Insights from the Three Gorges Study. *Canadian Journal of Civil Engineering* 21: 541–6

Schindler D W 1998 A dim future for boreal waters and landscapes. *BioScience* 48: 157–64

SETAC 1991 *A Technical Framework for Life-cycle Assessment.* Washington, DC, United States Society of Environmental Toxicology and Chemistry, and the SETAC Foundation for Environmental Education, Inc.

SETAC 1993 *A Conceptual Framework for Life-cycle Assessment.* Washington, DC, United States Society of Environmental Toxicology and Chemistry, and the SETAC Foundation for Environmental Education, Inc.

SETAC 1994 *Guidelines for Life-cycle Assessment: A Code of Practice.* Washington, DC, United States Society of Environmental Toxicology and Chemistry

Shoemaker D J 1994 *Cumulative Environmental Assessment.* Department of Geography Publication Series Number 42, Waterloo, Ontario, University of Waterloo

Sinclair A J and A P Diduck 2000 Public involvement in environmental impact assessment: a case study of hydro development in Kullu District, Himachal Pradesh, India. *Impact Assessment and Project Appraisal* 18: 63–75

Smith L G 1993 *Impact Assessment and Sustainable Resource Management.* Harlow, Longman

Spaling H and B Smit 1993 Cumulative environmental change. *Environmental Management* 17: 587–600

Spaling H and B Smit 1995 A conceptual model of cumulative environmental effects of agricultural land drainage. *Agriculture, Ecosystems, and Environment* 53: 99–108

Street P and B Barker 1995 Promoting good environmental management: lessons from BS 5750. *Journal of Environmental Planning and Management* 38: 484–503

Tarrant M A, A D Bright and H K Cordell 1999 A world-wide-web human dimensions framework and database for wildlife and forest planning. *Human Dimensions of Wildlife* 4: 18–43

Taylor C N, C Goodrich and C H Bryan 1995 Issues-oriented approach to social assessment and project appraisal. *Project Appraisal* 10: 142–54

Teisl M F and B Roe 2000 Environmental certification: informing consumers about forest products. *Journal of Forestry* 98: 36–42

Therivel R 1993 Systems of strategic environmental assessment. *Environmental Impact Assessment Review* 13: 145–68

Therivel R, E Wilson, S Thompson, D Heaney and D Pritchard 1993 *Strategic Environmental Assessment*. London, Earthscan Publications

Thomas I G 1998 *Environmental Impact Assessment in Australia: Theory and Practice*. Sydney, Federation Press

Tortajada C 2000 Environmental impact assessment of water projects in Mexico. *International Journal of Water Resources Development* 16: 73–87

Tzilivakis J, C Broom, K A Lewis, P Tucker, E Drummond and R Cook 1999 A strategic environmental assessment method for agricultural policy in the UK. *Land Use Policy* 16: 223–34

United States Environmental Protection Agency 1992 *Life-cycle Assessment: Inventory Guidelines and Principles*. EPA/600/R-92/036, Washington, DC, Office of Research and Development, Risk Reduction Engineering Laboratory

Verheem R A A and J A M W Tonk 2000 Strategic environmental assessment: one concept, multiple forms. *Impact Assessment and Project Appraisal* 18: 177–82

von Seht H 1999 Requirements of a comprehensive strategic environmental assessment system. *Landscape and Urban Planning* 45: 1–14

Walker D H and D Lowes 1997 Natural resource management: opportunities and challenges in the application of decision support systems. *AI Applications* 11: 41–51

Walker J L, B Mitchell and S Wismer 2000 Impacts during project anticipation in Molas, Indonesia: implications for social impact assessment. *Environmental Impact Assessment Review* 20: 513–35

Wherrett J R 1999 Issues in using the Internet as a medium for landscape preference research. *Landscape and Urban Planning* 45: 209–17

Wood C 1995 *Environmental Impact Assessment: A Comparative Review*. Harlow, Longman

Wood C 1999 Pastiche or postiche? Environmental impact assessment in South Africa. *South African Geographical Journal* 81: 52–9

Chapter 8

Partnerships and Stakeholders

8.1 Introduction

Key aspects of sustainability include empowerment of local people, self-reliance and social justice (Chapter 4). One way to achieve those aspects is to move away from traditional forms of environmental and resource management which are dominated by professional experts in the government and private sector, and toward approaches which combine the experience, knowledge and understanding of various groups and people (Box 8.1). The words *partnerships* and *stakeholders* are often used to characterize an approach to include both organized interest groups and the general public into resource and environmental planning (Box 8.2).

In this chapter, discussion focuses initially upon characteristics of public participation. Attention then turns to examining some experiences with partnerships.

Box 8.1

The changes in human attitudes that we call for depend on a vast campaign of education, debate, and public participation.

. . . , environmental and economic problems are linked to many social and political factors. . . . It could be argued that the distribution of power and influence within society lies at the heart of most environment and development challenges. Hence new approaches must involve . . . local participation in decision making.

Source: World Commission on Environment and Development, 1987: 23, 38.

Box 8.2 Partnership

A *partnership* is a mutually agreed arrangement between two or more public, private or non-governmental organizations to achieve a jointly determined goal or objective, or to implement a jointly determined activity, for the benefit of the environment and society.

8.2 Fundamental aspects regarding participation and partnerships

8.2.1 Rationale for participation

Many reasons can be given for involving the public in resource and environmental management. By consulting with people living in a region who will be affected by a policy, programme or project, it is possible to: (1) define the problems more effectively, (2) access information and understanding that fall outside the scientific realm, (3) identify alternative solutions that will be socially acceptable, and (4) create a sense of ownership for the plan or solution which facilitates implementation. While a participatory approach may extend the time needed during the initial stages of analysis and planning, such an investment normally is "returned" later in the process by avoiding or minimizing conflict. While some elected and technical officials may feel challenged or threatened by a participatory approach, believing that it is their job to define the problem and develop solutions, in democratic countries most now appreciate that the complexity of problems means that it is sensible to draw on all possible sources of knowledge and understanding (Box 8.3).

Given the above considerations, partnerships can be helpful for both idealistic and pragmatic reasons. With growing complexity, interdependence and uncertainty of issues, and the rapid rate at which conditions change, drawing upon many people and groups should help to achieve a balanced perspective relative to an issue. Furthermore, there is growing public expectation and demand for greater involvement, and less willingness to accept that "experts" know what is best. Members of the public also are increasingly willing to accept responsibilities and risks which accompany re-allocation of power or authority to them when they become partners with government agencies which have legal mandates and responsibilities. And, when economic conditions become difficult, and less public funding is available for resource and environmental initiatives, partners outside of government often can contribute, in money or in kind, to expedite activity which otherwise would be difficult to support. In this manner, partnerships can help to maintain or to

---

Box 8.3 Creating volunteer partnerships

Tens of thousands of volunteers across the country are increasingly becoming involved in activities – such as planting trees, hauling mulch, and assisting with public outreach – that contribute to the health of the urban forest. Enhancing and developing partnerships with volunteers can further strengthen the growing field of urban forestry. But urban forestry volunteer programs are not without controversy. Professional urban foresters may feel that volunteers aren't capable of highly skilled work, that they can't consistently count on volunteers, or that volunteers may challenge their professional authority and judgment.

Source: Westphal and Childs, 1994: 28.

Guest Statement

Public participation in environmental management

Ali Memon, New Zealand

Environmental equity and market liberalism are, in a sense, competing ethics regarding public participation in environmental management. We are currently witnessing a contest between these two ethics, with the future direction of environmental management depending on the outcome.

The justification for public participation in environmental management has been vigorously contested by Neo-Liberal critics whose approach to environmental management has gained political ascendancy in many parts of the world. This approach is based on assumptions about environmental management as objective and value free. The ultimate aim is to put a tight rein on public participation in resource allocation and environmental management decision making in favour of the market. Thus, wide ranging economic restructuring programmes have been implemented in many developed and developing countries with minimal public consultation and, in several instances, with major undesirable ecological and social impacts.

An environmental ethics perspective is necessary to reassert the case for partnerships and participatory approaches. The term *ethics* denotes values, principles or rules which either do or should underlie the moral behaviour of human beings, including their practices, relationships and responsibilities. A key question, in terms of the rationale for participation and who should participate, is the scope or extent of our moral responsibilities to others with whom we interact regarding the use, management or protection of the environment.

The extent of this moral responsibility can be conceptualized in terms of four dimensions:

(1) restricted to fellow human beings in our own *immediate* vicinity.
     (However, the extent of our moral relationships and responsibilities can be extended to include:)
(2) human beings in other communities, other regions or other parts of the globe,
(3) future generations, and
(4) non-human life, or the world of living and non-living nature as a whole rather than just to fellow human beings.

The first two dimensions relate progressively to wider interpretations of *intragenerational* environmental equity, defined here as the fair distribution of environmental well being in terms of environmental quality and risk among those living now. Environmental quality comprises both "good" and "bad" elements, distributed across communities, nations and the globe. The "good" elements include access to resources to satisfy basic human needs (e.g. to food and shelter in developing countries; access by indigenous minorities in Western societies to natural resources such as fisheries) as well as access to environmental attributes such as amenities, cleanliness, etc. in more affluent parts of the world. Many environmental conflicts in large cities relate to the location of NIMBYs ("not in my backyard") because local communities feel threatened by the perceived undesirable impacts on their neighbourhoods.

The third dimension relates to *inter-generational* equity, or the fair distribution of environments among successive generations of humans. Typically, many of us tend to think of inter-generational obligations in restrictive kinship terms (e.g. passing on the family inheritance). But concerns such as population increase, declining stocks of renewable and non-renewable resources, and implications of climate change have recently led us to question our moral obligations to future generations in broader community, national and international or global contexts.

The fourth dimension relates to equity in relationship between humans and the rest of the natural world (*ecological environmental equity*). Conventionally, our relationship with nature tends to be primarily instrumental. Economists, for example, conceptualize this relationship in terms of the source and sink functions of the environment, or its value to humankind. In contrast, the emphasis in ecological equity is on the meaning of environment in a deeper sense, the sense of our moral relationship with the non-human world. It is about acknowledging and respecting the integrity of the natural environment, apart from the values that humans may put on it.

I believe environmental equity is a central aspect of well being for individuals and communities. The attributes of environmental equity defined here (intergenerational, intra-generational, ecological) underpin the principle of sustainable development. The goal of sustainable development should be pursued through an open, consultative approach to policy formulation and implementation at all levels of government.

Ali Memon was appointed to a personal chair in environmental management and planning in the Division of Environmental Management and Design at Lincoln University, New Zealand, in 1999. Prior to that, he was an Associate Professor of Geography and Director of the Postgraduate Regional and Resource Planning Programme at the University of Otago. His primary research interest is environmental management from a social science perspective, with the objective of understanding societal responses to environmental problems and adequacy of institutional arrangements for formulating, implementing and evaluating plans and policies. Structured within the broader historical context of the experience of capitalist societies and based on fieldwork in different parts of the world, his work has been published widely.

improve service. Ali Memon elaborates on some of these ideas in his guest statement.

8.2.2 Kinds of partners and partnerships

Partnerships are applicable to many management functions. They can be used regarding policy development; data collection; research; analysis and planning; programme development, design and delivery, evaluation; monitoring; enforcement; administration; and fund raising. Depending upon the situation,

Box 8.4 Effective participation

Often, the effectiveness of a public participation exercise is judged on the basis of how many people show up at a public meeting. However, more than attendance is involved in an effective public participation process. Trust, communication, opportunity and flexibility are the crucial elements that ultimately determine the effectiveness of a public participation program . . .

Source: Law and Hartig, 1993: 32.

partnerships can be developed with client groups, volunteer associations, community groups, non-governmental organizations, educational institutions, business or industry, aboriginal people and other levels of government.

Partnerships can be of many different kinds. They can range from the personal or informal through to voluntary or legally binding arrangements. They may be short term and project specific, or long term and broad in scope. They may involve sharing of work or financial costs, or the sharing only of information.

8.2.3 Key elements for successful participation and partnerships

Many elements for successful participation and partnerships are the same as for effective conflict resolution (Box 8.4; Chapter 11). However, some key elements are:

- *Compatibility* between participants. Such compatibility often is based on *respect* and *trust*, even when legitimately different expectations or needs exist. With respect and trust, differences can often be overcome, and indeed can be used to help each participant to broaden his or her outlook.
- *Benefits* to all partners. If there are no real benefits to all the participants, and if they are not perceived to be shared fairly, then a sustained partnership will be difficult to achieve.
- *Equitable representation and power* for participants need to be agreed upon and established. Even though some partners may have fewer resources or capacity than others, means must be found to ensure that all partners are involved.
- *Communication* mechanisms. There is a need both to facilitate communication internally between the partners, and with groups external to the partnership.
- *Adaptability*, especially given the uncertainty and changing circumstances that often are encountered in resource and environmental issues. A willingness to be flexible and to learn from experience, as outlined in Chapter 6, usually is a strong advantage.
- *Integrity, patience and perseverance* by partners. Obstacles often will be encountered, frustration will occur, progress will be slow or slowed down, and signs of progress may not appear for some time. These elements, combined with trust and respect, allow partners to get through the difficult times which inevitably occur.

186

Box 8.5 Basis for effective partnerships

. . . partnering must be based on an understanding that the missions, legislative mandates, and administrative policies among partners may be very different. It requires that differences in view be identified and accepted, and that commonalities in interest be sought as the building blocks for consensus. The goal should be to ensure that there are no real losers, that all receive some spoils in pursuing a common target. Partners must recognize that trade-offs must be made to improve the collective whole. A necessary condition for establishing mutual trust is that partnering arrangements be open, frank and honest. Unless that condition is met, there will be little incentive for meaningful cooperation.

Source: Viessman, 1993: 14.

The above elements are not essential for successful partnerships, but the more that are present the greater is the likelihood that a partnership will endure and be effective (Box 8.5).

8.2.4 Degree of involvement through partnerships

The degree or amount of public involvement which is desirable and feasible must be determined. Arnstein (1969) observed that a participatory approach can represent a redistribution of power from managers to the public. On that basis, she argued that degrees of involvement could be identified, ranging from non-participation, to tokenism, to actual sharing of power (Table 8.1).

Table 8.1 Arnstein's eight rungs on the ladder of citizen participation

Rungs on the ladder of citizen participation	Nature of involvement	Degree of power sharing
1. Manipulation	Rubberstamp committees	
2. Therapy	Power holders educate or cure citizens	Non-participation
3. Informing	Citizens' rights and options are identified	
4. Consultation	Citizens are heard but not necessarily heeded	Degrees of tokenism
5. Placation	Advice is received from citizens but not acted upon	
6. Partnership	Trade-offs are negotiated	
7. Delegated power	Citizens are given management power for selected or all parts of programmes	Degrees of citizen power
8. Citizen control		

Source: Arnstein, 1969.

Traditional managers are often hesitant to go beyond the categories of non-participation or tokenism, in the belief that the general public is usually ignorant or apathetic, that the time required is disproportionate to the benefits, that the managers have a responsibility to exert professional judgement, and that public agencies have legally based obligations which cannot be transferred to another party. In contrast, citizens increasingly are expecting what they consider to be "meaningful" participation, which in their view usually means sharing some of the power. The sharing or re-allocating of power raises the issue of *accountability*, in the sense of to whom a group given power can be held accountable regarding decisions taken.

Various degrees of participation are illustrated by the four types of *strategic alliances* (Ontario Ministry of Natural Resources, 1995) identified in Table 8.2. *Contributory partnerships* involve an arrangement in which a public or private organization has agreed to provide sponsorship or support, normally through actual funding, for some activities in which it will have little or no direct operational participation. While the financial contribution is often essential for the success of the activity, this type of arrangement is a weak type of partnership since not all partners are actively involved in decision making.

Table 8.2 Strategic alliances identified by the Ontario Ministry of Natural Resources (1995)

Type of strategic alliance	Purpose	Extent of power sharing
(1) Contributory	*Support sharing*: to leverage new resources or funds for programme/service delivery	Government retains control, but contributors may propose or agree to the objectives of the strategic alliance
(2) Operational	*Working sharing*: to permit participants to share resources and work, and exchange information for programme/service delivery	Government retains control. Participants can influence decision making through their practical involvement
(3) Consultative	*Advisory*: to obtain relevant input for developing policies and strategies, and for programme/service design, delivery, evaluation and adjustment	Government retains control, ownership and risk, but is open to input from clients and stakeholders: the latter may also play a role in legitimizing government decisions
(4) Collaborative	*Decision making*: to encourage joint decision taking with regard to policy development, strategic planning, and programme/service design, delivery evaluation and adjustment	Power, ownership and risk are shared

Operational partnerships have partners sharing work rather than decision-making power. The emphasis here is upon reaching agreement on mutually desirable or compatible goals, and then working jointly to achieve them. Collaboration may be very high, in that the partners share non-financial resources to a considerable extent. Power is retained primarily or exclusively by the partner which provides the financial resources, and this is usually the public sector partner.

Consultative partnerships are those in which the resource management agency actively seeks advice from individuals, groups and other organizations outside government. The mechanism is usually a committee or council which is primarily designed to provide advice to the public agency about a specified policy field or issue. Control is clearly retained by the public agency, which has the discretion to decide the extent to which it will respond to the advice received. However, the partners can exert significant influence on decisions, because the public agency recognizes the political costs of ignoring advice that it has actively sought. Out-of-pocket expenses or daily payments are often made to members of the advisory group, based on an agreement reached at the beginning of the process.

Real decision-making power is shared in *collaborative partnerships*. The intent is to achieve mutually compatible objectives, and the resources to be shared may involve information, labour or money. This is the only one of the four partnerships in which each partner explicitly gives up some autonomy. More specifically, in this arrangement, a public agency turns over some power to groups or organizations outside of the government. Normally, such re-allocation does not include any responsibilities for which the public agency is legally accountable. In the best form of collaborative partnership, decisions are reached through consensus. Such consensus-building is usually most effective when the issue or problem is one that no partner can resolve unilaterally. Financially, there may be a mutual sharing, and indeed a two-way flow, of expenses and revenues.

The main implication of this discussion is that there is not one best "model" for partnerships. Many choices exist. The kind of partnership and the nature of participation have to be determined by the various people or groups involved.

8.2.5 Stakeholders

In designing partnerships, an issue can arise as to who are genuine *stakeholders*. In contrast to the view expressed in Box 8.6, a stakeholder is generally considered to be a person or group directly affected by or with an interest in a decision, or with legal responsibility and authority relative to a decision.

Regarding people or groups who might be affected by a decision, a distinction should be made between the *active* and the *inactive* publics. The *active* public involves those people who are organized into interest groups, such as Friends of the Earth, Sierra Club, Pollution Probe and Greenpeace. The largest of these groups are well organized and articulate, and often have financial resources and full-time staff to monitor activities, conduct research and make submissions to government. In contrast, the *inactive* public, or the silent majority, are those people who do not usually become actively involved in social or

Box 8.6 Who is a stakeholder?

... the context in which it is currently being used is inconsistent with the definition by which it found its way into the language in the first place. In fact it is being used in exactly the opposite sense to that for which it was intended.

... What is a stakeholder? ... Any reputable dictionary defines a stakeholder as just what is implied, someone who holds the stakes during the course of a wager; the person selected by the opponents in a wager, deemed by both of the betting parties, to possess the required honesty and impartiality to be trusted to hold the money or valuables being wagered until the uncertainty is resolved. Therefore, quite contrary to the intended current application, a stakeholder is someone who has no stake in the particular issue in question.

Since a stakeholder is, in fact, a disinterested third party, as opposed to someone who has a vested interest, continued use of the term in current context shows a flagrant disdain for the integrity of the language. Most definitely, it should not be used in professional presentations, written or oral.

Source: Wiens, 1995: 3, 7.

environmental issues, being more focused on coping with issues at work and at home. The reality is that many of the organized groups which form the core of the active public make it their business to become involved in environmental and resource issues, whether or not they are invited to become members in a partnership. Their voices normally are heard, and the public managers do not have to make special efforts to hear from them since they view part of their function as commenting upon and participating in planning.

The challenge for managers is to determine if active public interest groups reflect a reasonable cross section of the stakeholders to be affected by decisions. There often has been concern that the members of the active public do not in fact fairly represent all of the stakeholders. As a result, resource and environmental managers often have made substantive effort to interact with members of the inactive public, even if that may be viewed by some as falling into Arnstein's categories of non-participation or tokenism, shown in Table 8.1. However, in fairness to environmental and resource agencies, it should be said that many people do not want to become actively involved. Their lives are full and complicated enough by day-to-day matters, and they are often content to rely on professionals to do what they were hired to do – plan and manage.

8.2.6 Timing for public input

Partnerships may be established at varying times during analysis and planning. Smith (1982: 561–3) suggested that planning occurs at three levels: *normative*, or determining what ought to be done; *strategic*, determining what can be done; and, *operational*, determining what will be done. He concluded that many public participation programmes are used in the *operational* stage. However, Smith and others have argued that partnerships need to be established earlier in the planning process, so that the public can become involved at the *normative* and *strategic* stages. Otherwise, the public may conclude that their participation is little more than cosmetic, or tokenism in Arnstein's language, because many of the key decisions are taken before the *operational* phase is reached.

To illustrate, for energy planning a number of issues must be addressed. At an early phase, consideration should be given to what is an appropriate mix of strategies involving various sources of energy (conventional thermal, nuclear, hydroelectric, solar, wind) and changing patterns of energy use, through conservation and other initiatives to reduce demand. Having decided upon the supply sources, decisions then must be made regarding where the new sources will be located. Often it is at this stage that the public is invited to become involved to help identify acceptable sites. However, the public may wish to re-open questions related to whether or not there is a need for new sources of energy supply, if actions were taken to reduce demand. If people arrive at a public hearing or meeting with different expectations as to whether the issues being discussed are *normative* (mix of supply and demand management strategies) or *operational* (sites for new power sources), there can be a high level of frustration created.

Critics of partnerships or public participation may charge that people came wanting to discuss issues that were already resolved, and therefore conclude that a participatory approach generates excessive costs and delays. In contrast, advocates of a participatory approach may conclude that too frequently the participation is superficial or tokenism, with the most important decisions already having been made before the public was invited to become involved. The implication is that it is important to recognize the different stages or phases of planning, and to ensure that partners or public participation exercises are designed so that those involved understand the stages and agree regarding the purpose of the partnership exercise.

8.2.7 Components of partnership programmes

As noted earlier, partners can be drawn into a management process in many ways. However partnerships are organized, normally five key functions should be included.

First, information must be provided to the public. The most important aspect for this function is that the resource and environmental management agency must make an honest effort to determine both the needs and wants for information by stakeholders. Providing information becomes a credible partnership function when an agency makes a conscious effort to determine not just the information which is convenient to supply, but also what information the stakeholders want; establishes a systematic process to provide the information; and presents the information in a timely manner. Thus, attention is required regarding not just the *content* of information but also the *form* in which it is provided.

Second, following the provision of information (information out phase), opportunity must be provided for the partners or the general public to provide their perspectives, whether related to the nature of the problem, the range of possible solutions, or their role in implementation and monitoring of results. This often is referred to as the "information in" or "gathering information" function. This is an important partnership function, since it signals that the public agencies do not have all the information or understanding, and are explicitly seeking input from others. For this function, it is important that the lead agency has a clear purpose regarding what information it seeks and how it will use the information. Furthermore, to achieve ongoing stakeholder participation, provision should be made to communicate with the stakeholders about the agency's interpretation of the information collected.

Third, since a number of iterations are usually required, such as to define the problem, to develop alternative solutions and to design an implementation strategy, provision should be made for continuous exchange or interaction among the representatives of the resource and environmental management agencies and the other partners. This function is often referred to as one of "promoting dialogue". The main objective is to be able to manage systematically and effectively the process of stakeholder dialogue, but not necessarily to obtain a consensus among the stakeholders. At this stage, an agency need not commit itself in advance to any particular response to the results of the consultation process. This function is well suited to a situation in which an agency is committed to a general direction or action, but has not finalized the preferred direction or action. Not only should attention be given to creating dialogue between the agency and stakeholders, but also to facilitate dialogue among stakeholders themselves. And, as with the second function, the consultation should be designed to produce a specific result or end point.

The fourth function involves facilitating consensus, and the fifth deals with negotiating consensus. These will be considered in detail in Chapter 11. These are longer-term functions, and success in regarding each of them usually should not be expected in the early stages of a public participation initiative.

8.2.8 Mechanisms for participation

Having agreed upon the functions to be included in the partnership programme, it then is important to determine the mix of mechanisms to be used. As Table 8.3 indicates, many mechanisms exist. The challenge is to custom design from the alternatives to meet conditions and needs of a particular situation. *Lobbying* often is not recognized as a form of participation, but it definitely is one method used by interest groups to represent their views to decision makers. Advisory bodies may take many forms, but usually involve a group established to investigate a problem. Mediation and negotiation are methods which have emerged to identify different interests and to find mutually satisfactory solutions. They will be considered in more detail in Chapter 11.

8.2.9 Balancing fairness and efficiency

As noted in Chapter 4, sustainability emphasizes the concepts of equity and empowerment (Box 8.7). Creation of partnerships is usually justified on the

Table 8.3 Public participation mechanisms

	Representativeness	Information in	Information out	Continuous exchange	Ability to make decisions
Public meetings	Poor	Poor	Good	Poor	Poor–fair
Task force	Poor	Good	Good	Good	Fair–good
Advisory groups	Poor–good	Poor–good	Poor–good	Good	Fair
Social surveys	Good	Poor	Fair	Poor	Poor
Individual/ group submissions	Poor	Good	Poor	Poor	Poor
Litigation	Poor–fair	Good	Good	Poor	Good
Arbitration	Poor–fair	Good	Good	Poor	Good
Environmental mediation	Poor–fair	Good	Good	Fair	Good
Lobbying	Poor–fair	Good	Fair	Good	Fair

Source: Mitchell, 1989: 119.

Box 8.7 Achieving empowerment

The essence of critical EA [environmental assessment] education is education and learning that facilitates public involvement in resource management and, thereby, empowers local communities to take greater control of resource use decisions that directly affect them.

Source: Diduck, 1999: 87.

---

basis that they provide for a more open and transparent management process, and therefore for greater equity. Furthermore, by being involved in defining the problem and identifying solutions, the partners are more likely to accept or "buy in" to proposed recommendations. There is no doubt that, in the short term, a participatory approach often extends the time required for analysis and planning. If adequate time is to be allowed for "information out" and "information in" over a number of iterations, the process will be longer than if technical resource and environmental managers worked on their own. However, it is commonly accepted that in the long run a participatory approach is also often efficient in that it results in less challenging of findings and solutions toward the end of the planning process. In that regard, time "lost" in the early part of the analysis and planning is usually recaptured by the time of the implementation phase. Furthermore, if the investment of time in a participatory approach leads to less argument and opposition to recommendations, then it often seems to be the case during the overall life of an issue or problem that a participatory approach is both more efficient and equitable compared to an approach that does not incorporate participation.

8.2.10 Monitoring effectiveness of partnerships

If experience from partnerships and public participation is to be helpful in improving future initiatives, it is important that we build in the capacity to monitor effectiveness. In that manner, lessons from past and ongoing experience can become part of the *social learning* approach advocated in adaptive environmental management, as discussed in Chapter 6.

Smith (1983) has suggested that monitoring and evaluation could usefully focus upon three aspects: *context*, *process* and *outcome*. Today, many people who advocate a results-based management approach would split his third category into three separate categories: *outputs*, *outcomes* and *impacts*. The *context* category reminds us that any partnership or public participation exercise occurs with reference to previous events and decisions; historical relationships between partners; changing interests; objectives and expectations; and shifting ideological, economic and political circumstances. If the effectiveness of a partnership initiative is to be determined, there must be awareness of such contextual aspects.

The *process* associated with a partnership arrangement is often crucial for its success, and indeed it is this aspect which has attracted the most attention in evaluation. As noted in previous sections, there are many choices in designing a process, and it is important to be aware of what the range of choice is, before judging the adequacy of the choices actually taken. Emphasis here usually concentrates upon the goals and objectives for participation, the number and type of stakeholders involved, and the methods used.

The final aspects are the coupled elements of *outputs*, *outcomes* and *impacts*. Outputs are normally measurable aspects, such as number of options considered, number of interests accommodated in the selected solution, and satisfaction of participants. Outcomes relate to the significance of the outputs, in the short and medium term. Here, interest is not just upon the number of interests accommodated in the solution, but also upon the capacity for them to be implemented, and their ability to meet the ongoing needs and expectations of the participants. *Impacts* relate to results or changes in the longer term. The issue of monitoring will be revisited in Chapter 13 when a specific experience of monitoring public participation will be reviewed.

8.2.11 Overview

Many aspects require attention in designing a partnership arrangement or a public participation programme. As illustrated in the above sections, consideration should be give to the following: (1) rationale, (2) kinds of partners and partnerships, (3) elements for success, (4) degrees of involvement, (5) types of stakeholders, (6) timing, (7) programme components, (8) mechanisms, (9) balancing fairness and efficiency, and (10) monitoring and evaluation. With these in mind, we turn now to examples of partnerships and participation which illustrate some of these aspects.

8.3 Evolving patterns of participation and partnerships

Experience from Montana, USA, illustrates the manner in which public participation programmes can evolve as a result of resource and environmental managers adjusting on the basis of earlier experiences. This example is based on a report by McMullin and Nielsen (1991).

8.3.1 Example of the Missouri River, Montana

South of Great Falls, the Missouri River is a high-quality and heavily fished trout stream. The Montana Department of Fish, Wildlife and Parks (MDFWP) had been monitoring the brown trout population in the stream and became concerned about the low numbers, which they believed were due to heavy fishing pressure. The MDFWP proposed that the daily limit of five brown trout of any size be dropped to one brown trout longer than 56 cm, along with a continuing limit of up to five of the more abundant rainbow trout. The proposal also stated that fishing gear would be restricted to flies and lures. Baited hooks would be disallowed, based on evidence that the mortality of trout released after being caught with a baited hook was much higher than when released from flies and lures. Other than offering the opportunity to keep a trophy-sized fish (over 56 cm), brown trout fishing would become a catch-and-release activity.

Several public meetings were used to obtain public input during development of the modified catch regulations. Those attending had the opportunity to provide comments, and to answer a questionnaire. The conclusion of the managers from this input was that the majority of respondents supported the proposal. However, the fishers were polarized. Fly fishers were strongly in agreement. Bait fishers were opposed, believing that the brown trout could be protected without a restriction on bait.

During the formal open meeting at which the Montana Fish and Game Commission was expected to adopt the proposal, emotions were high. Bait anglers protested strongly and bitterly, and argued effectively that not all options to protect the brown trout population had been examined systematically. Fly and lure fishers spoke equally fervently in support of the proposal. The Commission members became uneasy about the sharp polarization, and about a proposal which was a "take-it-or-leave-it" kind. They indicated that they wanted to have more than one option to consider, and rejected the proposed change of catch regulations. The outcome was dissatisfaction by all participants. The MDFWP believed that a special interest group had derailed its proposal to protect the fishery. Fly fishers were unhappy that the Commission had apparently chosen to ignore the majority of views, which favoured the proposal, from the public. Bait anglers felt that the MDFWP was favouring the fly fishers. And, Commission members concluded that the MDFWP had not adequately addressed the polarized positions prior to the Commission's public hearing. However, the one thing that everyone agreed was the image and credibility of the MDFWP, and the perceived value of "public participation", had been damaged. The next example provides a contrast, and indicates how the MDFWP adjusted its approach to public participation based on the experience with the proposed changes to catch regulations on the Missouri River.

8.3.2 Example of the Bighorn River, Montana

The Bighorn River is close to Billings, the largest city in Montana, and also is within a one-day drive of Denver. Construction of a dam in 1967 changed a warm and silty river into a cool and clear-water trout stream. The Bighorn River soon gained a reputation for having a world-class trout fishery, and angling pressure grew. Furthermore, the river fishery was closed in the mid 1970s, as a result of a legal challenge from the Crow Tribe, native people who claimed that only tribe members could fish. In 1981, the US Supreme Court ruled that the state of Montana had authority for management, and the river was re-opened for recreation fishing. Sport fishing then grew quickly, and an associated guide and outfitter industry developed.

Concern peaked in 1986, when numbers angling that summer doubled from those in the previous year. Fishers complained that the fish population was being damaged, and that crowding by fishers was ruining the experience for many. Guides and outfitters were particularly outspoken, and called for the MDFWP to introduce rigorous restrictions. However, studies of the Bighorn River by the MDFWP indicated that angler pressure was much less a stress on the fish than were habitat problems. The managers believed that water releases from the dam were inadequate for the needs of the large brown trout so prized by fishers, that quality of the water from the dam was detrimental to the fish, and that the food supply for the brown trout was inadequate. Thus, in their view, the priority need was to improve the habitat, not change fishing regulations. Despite these findings from the MDFWP, the fishers focused on the merits and problems of regulations, and became polarized in the same way as had occurred at the Missouri River.

The MDFWP realized that it needed to modify its approach to participation and partnerships. Not only public participation was required, but also capacity for *conflict resolution*. Indeed, public participation and involvement of partners was deemed essential for resolution of the conflict, a general issue to be considered further in Chapter 11. Based on the experience with the Missouri River, the MDFWP developed a five-stage management process:

(1) *Involving concerned citizens to establish management goals for the fishery.* Public meetings were held both in Billings, from where many of the fishers came, and in Fort Smith, a small community along the Bighorn River in which many of the guides and outfitters were based. Because of the controversy, the local media had provided detailed coverage of the issue and the upcoming meetings, which led to a large number of people attending who represented diverse interests. Attendance rates were well above what normally would be expected for such events. After the MDFWP outlined its understanding of the fishery and the habitat, a *facilitator* chaired the

discussions and kept the focus on management goals. As a result, the meetings concentrated on *ends* (management goals) rather than on *means* (specific regulations). The meetings at Billings and Fort Smith generated the same conclusions. The management goal should be to manage the river for a high-quality, wild trout fishing experience with emphasis upon catching large, trophy-sized trout.

(2) *Preparing and distributing a draft management plan.* The draft plan accomplished several purposes. First, it provided a means to organize the concerns which had emerged during the goal-setting meetings. In addition to the concerns, the draft plan included measurable objectives which would allow achievement of the agreed upon goal. Second, the draft plan presented data in lay terms, which helped to inform people about the status of the fish populations, and the variables affecting them. The draft plan was distributed widely, and included people who had contacted the MDFWP with concerns about the fisheries, as well as guides and outfitters, sports club officers and elected officials. Speaking engagements were arranged by the MDFWP at a number of fishing clubs and civic clubs over a three-month period.

(3) *Obtaining feedback to the draft plan through a brief, self-addressed survey.* People were given the opportunity to express agreement or disagreement regarding the draft plan through a survey form. The draft plan had accomplished its information and education functions, as 93% of respondents indicated that it had improved their understanding about both the river and the fishery. Agreement regarding the recommended objectives and strategies varied from 69 to 83%. For MDFPW, a striking result from the survey was that 50% of respondents indicated that they had changed their views about how the fishery should be managed after having read the plan. A number of people stated that they initially had believed that regulations were the best approach, but after reading the plan agreed that priority should be given to habitat issues.

(4) *Submitting the final plan to the Montana Fish and Wildlife Commission.* Following modifications to the plan based on the survey responses, the MDFWP made a formal presentation of the plan to the Commission, in which the implications for Commission action were highlighted. The Commission accepted the plan.

(5) *Following up with the public.* The MDFWP committed itself to report to the public when major milestones of the plan were met, or when conditions changed enough that modifications or departures were required to the plan. The MDFWP staff prepared news releases which were used by the local media, and staff members accepted speaking invitations both to give updates and to respond to questions.

Lessons

McMullin and Nielsen (1991) indicated that the revised management process did not eliminate all conflict and controversy. However, it did allow the MDFWP managers to address and resolve most of the difficulties during the 90-day period designated for review of the draft proposal. This more systematic and reflective approach to receiving input regarding a plan was viewed as much more useful by both the MDFWP and the Commission, compared with use of one pressure-packed Commission meeting in which short notice and limited options were involved. The effectiveness of the revised process was striking, particularly since the decision for the Bighorn River followed immediately after the disapproval by the Commission of the Missouri River proposal. A state-wide conservation group did present an option for a restrictive regulation for the brown trout fishery on the Bighorn River, but that was not accepted because that option (and others) had already been systematically assessed during the planning process. The overall implications of this experience are presented in Box 8.8.

Box 8.8 Lessons from the Montana experience

Commonly cited problems of the public involvement process are: (a) a "representative" public is difficult to reach, (b) citizen participation may promote conflict rather than resolve it, (c) participatory democracy in resource allocation diminishes the role and stature of the professional manager, and (d) the public is often not sufficiently informed to make good resource allocation decisions.

The key to solving the first two problems is having reasonable goals for citizen participation programs and employing adequate techniques to achieve them. . . . Managers who seek consensus on controversial issues will invariably be disappointed because the public has real, deeply held differences in desires and outlooks. Instead of consensus, the goal should be informed consent.

The third problem is clearly an illusion. Montana managers found their professional credibility was enhanced rather than diminished by emphasizing their role in formulating management options and explaining the implications of each. When ecological conditions dictate that only one course of action is reasonable, resource managers should assert their authority. However, if multiple strategies will achieve an objective, resource clients should be equal partners in the decision making process. Moreover, many resource allocation decisions are based on incomplete ecological information or include significant, value-based elements. As stewards of publicly owned resources, resource managers have no more right to make these value-based decisions than any other member of the public.

In regard to the fourth problem, the Bighorn experience demonstrated that users can make informed decisions if data are presented to them in understandable form. . . . good communication between resource managers and resource users helps the public make an informed choice that benefits the resource even when it causes them economic hardship.

Resource managers must continue to evolve to be creators of management options, disseminators of information and facilitators in a publicly-oriented decision making process.

Source: McMullin and Nielsen, 1991: 557–8.

8.4 Stakeholders and the private sector

The experience in Montana reflects many of the issues encountered in a participatory approach designed by a public resource management agency. However, involvement of partners and stakeholders also can be initiated by the private sector. For example, Eckel, Fisher and Russell (1992) argued that involvement of stakeholders was critically important when designing a system of environmental performance measurement for a firm. They argued that a system for environmental performance measurement (SEPM) was desirable due to the following realities in the business world:

- as well as having economic impacts, business activity also has environmental and social impacts;
- businesses are being held liable for environmental costs, as reflected by the increasing number of regulations, incentives and penalties; and
- incorporating environmental management into a firm is "good business" as it can result in direct cost reduction or indirect increases in goodwill.

The products of a good SEPM should include disclosure of (1) environmental obligations and contingencies; (2) environmental risks inherent in the operations of an organization, an item increasingly being expected by stakeholders; (3) financial risk to the organization; as well as (4) separate disclosure of expenditures for environmental management, an item increasingly being asked for by regulators, investors and analysts.

One of the key tasks in preparing a corporate policy and objectives for environmental management is to identify environmental issues pertinent to the firm. Two methods are available to identify such issues: initial environmental audits and consulting with stakeholders. Environmental audits will be discussed in Chapter 13 when considering monitoring and evaluation. Here, attention focuses on consultation with stakeholders. The gains from stakeholder consultation are identified in Box 8.9.

Eckel, Fisher and Russell (1992: 19) emphasized that "consulting with stakeholders should not be viewed as an act of altruism or public relations". In their view, stakeholder consultation should be considered as an integral input into development of corporate policy, as it helps to develop direction and focus for a firm's short- and long-term environmental objectives and responsiveness.

Box 8.9 Consulting with stakeholders

Consultation will clarify stakeholders' expectations about corporate environmental performance, information requirements, and preferred environmental solutions. It will help identify risk reduction projects that boost the firm's reputation with stakeholders. Stakeholders include: creditors, shareholders, regulators, employees, customers, suppliers, communities in which the company operates, and local, provincial and federal governments.

Source: Eckel, Fisher and Russell, 1992: 19.

Stakeholders can help a firm to determine whether it will simply seek to "stay out of trouble" regarding environmental matters, or if it will strive to become a leader in developing "cutting-edge" practices and procedures regarding environmental monitoring and reporting standards. In that context, they argue that stakeholder consultation provides many benefits. Discussions with customers may identify unhappiness with a firm's environmental record that, without concerted attention, could result in decreasing demand for the firm's products or services. Discussions with government officers may assist the firm to become aware of impending regulatory changes. Perhaps more importantly, it may allow the firm to become a partner with the government in helping to design the nature and timing of such changes.

Thus, consulting with stakeholders can be as useful for the private sector as for the government sector. With increasing privatization and commercialization of resource and environmental management functions, more and more resource and environmental managers will be employed in the private sector. It will be important for them to remember that developing partnerships, and consulting with stakeholders, is as important in the private as in the public sector.

8.5 Stakeholder consultation in developing countries

Development of partnerships and involvement of stakeholders are increasingly occurring in the Third World. For example, in northwestern Botswana, Van der Sluis (1994) explained that a land-use zoning plan developed for Planning Zone 6, or the Ngamiland Western Communal Remote Zone, used a participatory approach (Figure 8.1). Zone 6, a fragile area in the Kalahari Sandveld of Botswana, is primarily a hunter-gatherer region and a cattle post. Many conflicts have arizen among users. Crops and veld products have been damaged or destroyed by livestock. Ploughing is occasionally done in grazing areas, which reduces the fodder for livestock. Wild dogs and lions from the nearby Kaudom Game Reserve in Namibia kill livestock, and a growing tourism industry poses a threat to the integrity of national monuments.

Against the above background, a land-use plan was created to try and minimize the conflicts among different uses and users (Box 8.10). Van der Sluis (1994: 8) indicated that "to the extent possible, the residents of the area were involved in the planning process". Meetings were held in all the villages in the area to be covered by the plan. This created some major logistical challenges, as the population in the area is highly dispersed, with some of the remote cattle posts being more than a two-hour drive from a village. Since few of the local people had any means of transportation other than walking, they were collected and driven to village meeting places. Land-use conflicts were identified and discussed, and alternative land-use zoning arrangements were debated. Such meetings were not without controversy, as the BaMbukushu and Bushmen argued for their land to be zoned for grazing, ploughing, gathering of veld foods and wildlife use. In contrast, the BaHerero had quite different interests because of their highly mobile pastoralist land-use activities. Suggestions were

Figure 8.1   Ngamiland Western Communal Remote Zone, Botswana (Van der Sluis, 1994: 8)

Box 8.10 Consultation in Ngamiland West

Consultations with the people were considered crucial in the planning process, since experiences in other parts of the district showed that limited consultations can lead to serious delays in acceptance of the plans, both at the local and regional levels. Also, a zoning plan developed in cooperation with the people stands a better chance to be implemented and adhered to.

Source: Van der Sluis, 1994: 9.

---

sought as how best to reconcile land degradation problems, and conflicting needs and activities. Records were kept for all of the meetings.

The outcome was a land-use zoning system based on the suggestions of the land users, as well as from evaluation of both the capacity of the land and present land use (Box 8.11). The land evaluation indicated that the present land use was "more or less optimal" for the area. All land use was marginal, with little gain to be achieved by modifying land-use patterns other than the potential for improved dryland farming in valleys. The consultations revealed that land-use conflicts could partially be avoided through zoning, but it would be almost impossible to remove cattle posts from areas used by hunter-gatherers and pastoralists. Van der Sluis (1994: 9) remarked that "the people's opinion was decisive in the planning: where no interest existed in zoning of land, the

---

Box 8.11 Outcome of the Ngamiland West land-use plan

Although the preparations and consultations took up to three years, the adoption and acceptance of the plan by the authorities was quick. The results have thus been quite satisfactory, especially when compared to the experience with some other land use zoning plans, which had yet to be accepted by the authorities after more than two years.

The socio-economic research has shown the importance of combining different activities in Zone 6, or the "multi-stranded" economy. From the land evaluation, it becomes clear why such large areas are "empty", without physical developments, cattle posts or settlements: the land users do not want to risk investing large sums in drilling boreholes, since the chances of finding water are very low. However, they do recognise the existing land degradation problems and welcome the assistance of the . . . zoning . . . to halt their deterioration.

Source: Van der Sluis, 1994: 10.

[Planning Team] zoned all land for 'mixed land use' ". The Planning Team "did not attempt to change present land use or existing land rights, which would have been very difficult. . . . Instead, it prepared guidelines to assist the Land Board in future land allocations".

8.6 Implications

Participation and partnerships. These concepts are increasingly being incorporated into resource and environmental management initiatives. The rationale for using them is not only to be "politically correct". There is growing and persuasive evidence that public participation often leads to more effective resource and environmental management in the medium and long term, even if some extra time is required in the short term.

Many kinds of partnerships and participation methods are possible. As a result, the resource and environmental manager has the opportunity to craft an approach that draws upon a mix of methods that best meets the conditions and needs in a given situation. Increasingly, managers are striving to involve partners and the public in *normative* and *strategic* stages of planning, rather than confining involvement only to the *operational* phase. Expectations by the public are growing for greater power sharing, which in some instances will cause a dilemma for resource and environmental agencies legally responsible for aspects of management. Nevertheless, there is a trend toward delegation of more power to local groups and people, as will be shown in Chapter 9 when discussing the concept of *co-management*.

The examples in this chapter, ranging from public sector initiatives in the USA and Botswana, and private sector strategies in North America, indicate that managers in many countries and cultures can anticipate encountering the same generic issues regarding partnerships and participation. As a result, resource and environmental managers need to develop a carefully thought out rationale for their use of partnerships and public participation, and also need to be aware of the range of methods and techniques that can be applied. It appears that in the future we can anticipate more pressure and demands for increased involvement of the public in resource and environmental management.

References and further reading

Arnstein S 1969 A ladder of citizen participation. *Journal of the American Institute of Planners* 35: 216–24

Bakker K and D Hemson 2000 Privatising water: BoTT and hydropolitics in the new South Africa. *South African Geographical Journal* 82: 3–12

Bartlett A G, M C Nurse, R B Chhetri and S Kharel 1993 Towards effective community forestry through forest user groups. *Journal of World Forest Resource Management* 7: 49–69

Batterbury S 1998 Local environmental management, land degradation and the *"gestion des terriors"* approach in West Africa: policies and pitfalls. *Journal of International Development* 10: 871–98

Bennett J 1998 Development alternatives: NGO–government partnership in Pakistan. *Development* 41(3): 54–7

Bhatt C P 1990 The Chipko Andolan: forest conservation based on people's power. *Environment and Urbanization* 2: 7–18

Boiko P E, R L Morrill, J Flynn, E Faustman, G van Belle and G S Omenn 1996 Who holds the stakes? A case study of stakeholder identification at two nuclear weapons productions sites. *Risk Analysis* 16: 237–49

Bramwell B and A Sharma 1999 Collaboration in local tourism policy making. *Annals of Tourism Research* 26: 392–415

British Columbia Commission on Resources and Environment 1995 *Public Participation*. Volume 3, Victoria, Commission on Resources and Environment

Brown J M and E A Campbell 1991 Risk communication: some underlying principles. *International Journal of Environmental Studies, A* 37: 271–83

Buchy M and S Hoverman 2000 Understanding public participation in forest planning: a review. *Forest Policy and Economics* 1: 15–25

Bullard J E 2000 Sustaining technologies? Agenda 21 and UK local authorities' use of the World Wide Web. *Local Environment* 5: 329–41

Burroughs R 1999 When stakeholders choose: process, knowledge and motivation in water quality decisions. *Society and Natural Resources* 12: 797–809

Cardskadde H and D J Lober 1998 Environmental stakeholder management as business strategy: the case of the corporate wildlife habitat enhancement programme. *Journal of Environmental Management* 52: 183–202

Cecil B P 1998 The psyche of alliance management: policy implications for strategic alliance formation in the Great Lakes region. *Great Lakes Geographer* 5: 77–90

Chatterjee N 1995 Social forestry in environmentally degraded regions of India: case-study of the Mayurakshi Basin. *Environmental Conservation* 22: 20–30

Chess C 2000 Evaluating environmental public participation: methodological questions. *Journal of Environmental Planning and Management* 43: 769–84

Choguill M B G 1996 A ladder of community participation for undeveloped countries. *Habitat International* 20: 431–44

Cleaver F 1999 Paradoxes of participation: questioning participatory approaches to development. *Journal of International Development* 11: 597–612

Connor D M 1988 *Constructive Citizen Participation: A Resource Book*. Revised. Victoria, BC, Development Press

Curtis A 1998 Agency–community partnership in Landcare: lessons for state-sponsored citizen resource management. *Environmental Management* 22: 563–74

Curtis A and M van Nouhuys 1999 Landcare participation in Australia: the volunteer perspective. *Sustainable Development* 7: 98–111

Curtis A, A Britton and J Sebels 1999 Landcare networks in Australia: state-sponsored participation through local organizations. *Journal of Environmental Planning and Management* 42: 5–21

Curtis A, M van Nouhuys, W Robinson and J Mackay 2000 Exploring Landcare effectiveness using organisational theory. *Australian Geographer* 31: 349–66

Dale A P 1992 Aboriginal Councils and natural resource use planning: participation by bargaining and negotiation. *Australian Geographical Studies* 30: 9–26

Daniels S E and G B Walker 1997 *Rethinking Public Participation in Natural Resource Management: Concepts from Pluralism and Five Emerging Approaches*. On line at http://www.fao.org/WAICENT/faoinfo/forestry/FREE/PLURAL/1/DANWAL.htm

Datta S K and K J Virgo 1998 Towards sustainable watershed development through people's participation: lessons from the Lesser Himalaya of Uttar Pradesh, India. *Mountain Research and Development* 18: 213–33

Davies A R 1999 Where do we go from here? Environmental focus groups and planning policy formation. *Local Environment* 4: 295–316

Diduck A 1999 Critical education in resource and environmental management: learning and empowerment for a sustainable future. *Journal of Environmental Management* 57: 85–97

Donald B J 1997 Fostering volunteerism in an environment stewardship group: a report on the Task Force to Bring Back the Don, Toronto, Canada. *Journal of Environmental Planning and Management* 40: 483–505

Eckel L, K Fisher and G Russell 1992 Environmental performance measurement. *CMA Magazine* March: 16–23

Edwards-Jones E S 1997 The River Valleys project: a participatory approach to integrated catchment planning and management in Scotland. *Journal of Environmental Planning and Management* 40: 125–41

Ewing S 1999 Landcare and community-led watershed management in Victoria, Australia. *Journal of the American Water Resources Association* 35: 663–73

Flader S L 1998 Citizenry and the State in the shaping of environmental policy. *Environmental History* 3: 8–24

Gbadegesin A and O Ayileka 2000 Avoiding the mistakes of the past: towards a community oriented management strategy for the proposed national park in Abjua-Nigeria. *Land Use Policy* 17: 89–100

Goodwin P 1998 "Hired hands" or "local voice": understanding and experience of local participation in conservation. *Transactions of the Institute of British Geographers* 23: 481–99

Griffin C B 1999 Watershed Councils: an emerging form of public participation in natural resource management. *Journal of the American Water Resources Association* 35: 505–18

Grigg N S 1999 Integrated water resources management: who should lead, who should pay? *Journal of the American Water Resources Association* 35: 527–34

204

Grimble R and M-K Chan 1995 Stakeholder analysis for natural resource management in developing countries. Some practical guidelines for making management more participatory and effective. *Natural Resources Forum* 19: 113–24

Grimble R and K Wellard 1997 Stakeholder methodologies in natural resources management: a review of principles, contexts, experiences and opportunities. *Agricultural Systems* 55: 173–93

Hampton G 1999 Environmental equity and public participation. *Policy Sciences* 32: 163–74

Hanna K S 1999 Integrated resource management in the Fraser River estuary: stakeholders' perceptions of the state of the river and program influence. *Journal of Soil and Water Conservation* 54: 490–8

Hartig H N and N Law 1994 Institutional framework to direct development and implementation of Great Lakes Remedial Action Plan. *Environmental Management* 18: 855–64

Hartup B K 1994 Community conservation in Belize: demography, resource use, and attitudes of participating landowners. *Biological Conservation* 69: 235–41

Haughton G 2000 Information and participation within environmental management. *Environment and Urbanization* 11: 51–62

Hill K A 1991 Zimbabwe's wildlife conservation regime: rural farmers and the state. *Human Ecology* 19: 19–34

Hobson J 2000 Sustainable sanitation: experiences in Pune with a municipal–NGO–community partnership. *Environment and Urbanization* 12: 53–62

Hordijk M 1999 A dream of green and water: community based formulation of Local Agenda 21 in peri-urban Lima. *Environment and Urbanization* 11: 11–29

Hughey K F D, R Cullen and G N Kerr 2000 Stakeholder groups in fisheries management. *Marine Policy* 24: 119–27

Innes J E and D E Boohr 1999a Consensus building as role playing and bricolage: towards a theory of collaborative planning. *Journal of the American Planning Association* 65: 9–26

Innes J E and D E Boohr 1999b Consensus building and complex adaptive systems: a framework for evaluating collaborative planning. *Journal of the American Planning Association* 65: 412–23

Jakes P and J Harms 1995 *Report on the Socioeconomic Roundtable convened by the Chequamegon and Nicolet National Forests*. St Paul, Minnesota, US Department of Agriculture, Forest Service, North Central Forest Experiment Station, General Technical Report NC-177

Johnson H and G Wilson 2000 Institutional sustainability: "community" and waste management in Zimbabwe. *Futures* 32: 301–16

Johnson T R 1999 Community-based forest management in the Philippines. *Journal of Forestry* 97(11): 26–30

Jorbay S A 2000 Local Agenda 21 in practice – a Swedish example. *Sustainable Development* 8: 201–14

Kakonge J O 1996 Problems with public participation in EIA process: examples from Sub-Saharan Africa. *Impact Assessment* 14: 309–20

Kattenborn B P, H Riese and M Hundeide 1999 National park planning and local participation: some reflections from a mountain region in southern Norway. *Mountain Research and Development* 19: 51–61

King S, M Conley, B Latimer and D Ferrari 1989 *Co-design: A Process of Design Participation*. New York, Van Nostrand Reinhold

Landre B K and B A Knuth 1993 Success of citizen advisory committees in consensus-based water resources planning in the Great Lakes basin. *Society and Natural Resources* 6: 229–57

Larritt C 1995 Taking part in Mutawintjii: aboriginal involvement in Mootwingee National Park. *Australian Geographical Studies* 33: 242–56

Law N and J H Hartig 1993 Public participation in Great Lakes Remedial Action Plans. *Plan Canada* March: 31–5

Leiss W 1993 *Guide to Consultation Processes*. Institute for Risk Research Paper No. 31, Waterloo, Ontario, University of Waterloo

Lerner S (ed.) 1993 *Environmental Stewardship: Studies in Active Earthkeeping*. Department of Geography Publication Series No. 39, Waterloo, Ontario, University of Waterloo

Lobster D J 1992 Using forest guards to protect a biological reserve in Costa Rica. *Journal of Environmental Planning and Management* 35: 17–41

Loker C A, J Shanahan and D J Decker 1999 The mass media and shareholders' beliefs about suburban wildlife. *Human Dimensions of Wildlife* 4: 7–26

Makombe K (ed.) 1994 *Sharing the Land: Wildlife, People and Development in Africa*. IUCN/ROSA Environmental Issues Series No. 1, Harare, Zimbabwe, International Union for the Conservation of Nature and Natural Resources (IUCN) Regional Office for Southern Africa, and Washington, DC, IUCN Sustainable Use of Wildlife Programme

Margerum R D 1999 Getting past yes: from capital creation to action. *Journal of the American Planning Association* 65: 181–92

Martin P and H Ritchie 1999 Logics of participation: rural environmental governance under neo-liberalism in Australia. *Environmental Politics* 8: 117–35

McClaran M P and D A King 1999 Procedural fairness, personal benefits, agency expertise, and planning participants' support for the Forest Service. *Natural Resources Journal* 39: 443–58

McClaran M P and D A King 2000 Procedural fairness, personal benefits, agency expertise, and planning participants' support for the Forest Service. *Natural Resources Journal* 39: 443–58

McDaniels T L, R S Gregory and D Fields 1999 Democratizing risk management: successful public involvement in local water management decisions. *Risk Analysis* 19: 497–510

McMullin S L and L A Nielsen 1991 Resolution of natural resource allocation conflicts through effective public involvement. *Policy Studies Journal* 19: 553–9

Mercer D, M Keen and J Woodfall 1994 Defining the environmental problem: local conservation strategies in metropolitan Victoria. *Australian Geographical Studies* 32: 41–57

Michaels S, R J Mason and W D Sidecki 1999 Motivations for ecostewardship partnerships: examples from the Adirondack Park. *Land Use Policy* 16: 1–9

Miller A 1993 The role of citizen scientist in nature resource decision-making: lessons from the spruce budworm problem in Canada. *The Environmentalist* 13: 47–59

Mitchell B 1989 *Geography and Resource Analysis*, Second edition. Harlow, Longman Scientific and Technical

Mullen M W and B E Allison 1999 Stakeholder involvement and social capital: keys to watershed management success in Alabama. *Journal of the American Water Resources Association* 35: 655–62

Mwangi S W 2000 Partnerships in urban environmental management: an approach to solving environmental problems in Nakuru, Kenya. *Environment and Urbanization* 12: 77–92

Olsen B 1999 The Forest Service, water yield and community stability: defining the contours of an agency commitment to include Land Grant communities in the timber management process. *Natural Resources Journal* 39: 819–43

Ontario Ministry of Natural Resources (1995) *Memorandum: MNR Guide to Resource Management Partnerships – Administrative Considerations*. Toronto, Ontario Ministry of Natural Resources, 25 July

O'Riordan T and R Ward 1997 Building trust in shorelines management: creating participatory consultation in shoreline management plans. *Land Use Policy* 14: 257–76

Palerm J R 1999 Public participation in environmental impact assessment in Spain: three case studies evaluating national, Catalan and Balearic legislation. *Impact Assessment and Project Appraisal* 17: 259–71

Palerm J R 2000 An empirical–theoretical analysis framework for public participation in environmental impact assessment. *Journal of Environmental Planning and Management* 43: 581–600

Perera L A S R and A T M N Amin 1996 Accommodating the informal sector: a strategy for urban environmental management. *Journal of Environmental Management* 46: 3–15

Petts J 1995 Waste management strategy development: a case study of community involvement and consensus-building in Hampshire. *Journal of Environmental Planning and Management* 38: 519–36

Pinkerton E (ed.) 1989 *Co-operative Management of Local Fisheries: New Directions in Improved Management and Community Development*. Vancouver, University of British Columbia Press

Plummer R and C Stacey 2000 A multiple case study of community-based water management initiatives in New Brunswick. *Canadian Water Resources Journal* 25: 293–307

Porto M, R La Laina Porto and L G T Azevedo 1999 A participatory approach to watershed management: the Brazilian system. *Journal of the American Water Resources Association* 35: 675–83

Reed M G 1994 Locally responsive environmental planning in the Canadian hinterland: a case study in northern Ontario. *Environmental Impact Assessment Review* 14: 245–69

Reed M G 1995 Cooperative management of environmental resources: a case study from northern Ontario, Canada. *Economic Geography* 71: 132–49

Richards E M 1993 Lessons from participatory natural forest management in Latin America: case studies from Honduras, Mexico and Peru. *Journal of World Forest Resource Management* 7: 49–69

Roberts I 2000 Leicester environmental city: learning how to make Local Agenda 21 partnerships and participation deliver. *Environment and Urbanization* 12: 9–26

Robertson M, P Nichols, P Horwitz, K Bradby and D MacKintosh 2000 Environmental narratives and the need for multiple perspectives to restore degraded landscapes in Australia. *Ecosystem Health* 16: 119–33

Rocha E M 1997 A ladder of empowerment. *Journal of Planning Education and Research* 17: 31–44

Rowe G and L J Fewer 2000 Public participation methods: a framework for evaluation. *Science, Technology and Human Values* 25: 3–29

Rudel T K 2000 Organizing for sustainable development: conservation organizations and the struggle to protect tropical rainforests in Esmeraldas, Ecuador. *Ambio* 29: 78–82

Rydin Y and M Pennington 2000 Public participation and local environmental planning: the collective action problem and the potential of social capital. *Local Environment* 5: 171–89

Selin S and D Chavez 1995 Developing a collaborative model for environmental planning and management. *Environmental Management* 19: 189–95

Sexton K, A A Marcus, K W Easter and T D Burkhardt 1999 Introduction: integrating government, business, and community perspectives. In K Sexton, A A Marcus, K W Easter and T B Burkhardt (eds) *Better Environmental Decisions: Strategies for Governments, Businesses and Communities.* Washington, DC, Island Press, pp. 1–14

Short C and M Winter 1999 The problem of common land: towards stakeholder governance. *Journal of Environmental Planning and Management* 42: 613–30

Simonsen W and M D Robbins 2000 *Citizen Participation in Resource Allocation.* Boulder, Colorado, Westview Press

Smith L G 1982 Mechanisms for public participation at a normative planning level in Canada. *Canadian Public Policy* 8: 561–72

Smith L G 1983 The evaluation of public participation in water resources management: a Canadian perspective. In J W Frazier, B J Epstein, M Bardecki and H Jacobs (eds) *Papers and Proceedings of Applied Geography Conferences.* Vol. 6, Toronto, Ryerson Polytechnical Institute, Department of Geography, pp. 235–44

Smith P D, M H McDonough and M T Mang 1999 Ecosystem management and public participation: lessons from the field. *Journal of Forestry* 97(10): 32–8

Spyke N P 1999 Public participation in environmental decisionmaking at the new millennium: structuring new spheres of public influence. *Boston College Environmental Affairs Law Review* 26: 263–313

Stedman R C and D J Decker 1996 Illuminating an overlooked hunting stakeholder group: non-hunters and their interest in hunting. *Human Dimensions of Wildlife* 1: 29–41

Stein T V, D H Anderson and T Kelly 1999 Using stakeholders' values to apply ecosystem management in an upper Midwest landscape. *Environmental Management* 24: 399–413

Tesh S N 1999 Citizen experts in environmental risk. *Policy Sciences* 32: 39–58

Thapa G B 1998 Issues in the conservation and management of forests in Laos: the case of Sangthong District. *Singapore Journal of Tropical Geography* 19: 71–91

Thomas-Slayter B 1992 Implementing effective local management of natural resources: new roles for NGOs in Africa. *Human Organization* 51: 136–43

Thompson J 1994 Kenya's catchment approach – lessons for the SADC region. *Splash* 10(1): 15–16

Timoth D J 1999 Participatory planning: a view of tourism in Indonesia. *Annals of Tourism Research* 26: 371–91

Tuler S and T Webler 1999 Voices from the forest: what participants expect of a public participation process. *Society and Natural Resources* 12L: 437–53

Ungate C 1996 Tennessee Valley Authority's Clean Water Initiative: building partnerships for watershed improvement. *Journal of Environmental Planning and Management* 39: 113–22

Van der Sluis T 1994 Community-based land use planning – Ngamiland West. *Splash* 10(1): 8–10

Vari A and J Caddy 1999 *Public Participation in Environmental Decisions: Recent Developments in Hungary.* Budapest, Akademai Kiado

Viessman W 1993, The water management challenge. *Water Resources Update* 90 (Winter): 13–15

Vizayakumar K and P K J Mohapatra 1991 Coal mining impacts and their stakeholders: a SIAM approach. *International Journal of Environmental Studies, A* 37: 297–303

Wacker C, A Viaro and M Wolf 1999 Partnerships for urban environmental management: roles of urban authorities, researchers and civil society. *Environment and Urbanization* 11: 113–25

Walesh S G 1999 DAD is out, POP is in. *Journal of the American Water Resources Association* 35: 535–44

Warner M 1997 "Consensus" participation: an example for protected areas planning. *Public Administration and Development* 17: 413–32

Wernstedt K 2000 Terra firma or terra incognita? Western land use, hazardous waste, and the devolution of US federal environmental programs. *Natural Resources Journal* 40: 157–83

Westphal L and G Childs 1994 Overcoming obstacles: creating volunteer partnerships. *Journal of Forestry* 92(10): 28–32

Wiens L H 1995 Stakeholders misrepresented. *Water News*, Canadian Water Resources Association, 14, June: 3, 7

Wild A and R Marshall 1999 Participatory practice in the context of Local Agenda 21: a case study evaluation of experience in three English local authorities. *Sustainable Development* 7: 151–62

Williams B L, S Brown, M Greenberg and M A Kahn 1999 Risk perception in context: the Savannah River site stakeholder study. *Risk Analysis* 19: 1019–35

Wittayapak C and P Dearden 1999 Decision-making arrangements in community-based watershed management in Northern Thailand. *Society and Natural Resources* 12: 673–91

Wood G 1999 Contesting water in Bangladesh: knowledge, rights and governance. *Journal of International Development* 11: 731–54

World Commission on Environment and Development 1987 *Our Common Future*. New York and Oxford, Oxford University Press

Chapter 9

Local Knowledge Systems

9.1 Introduction

In Chapter 8, attention focused upon the concepts of *partnerships* and *stakeholders*. The main implication of that chapter was the advantages from a participatory approach in resource and environmental management. Such an approach recognizes that professionally trained experts can usually learn and benefit from the experiential knowledge of people who live and work in an area. Such knowledge has been called *traditional*, *indigenous* or *local*, to differentiate it from knowledge based upon science or formal study. In this chapter, some of the characteristics of what will be called *local knowledge systems* are reviewed. Then, consideration is given to the method of *participatory local appraisal*, a method increasingly being used to analyse local understanding. Section 9.2 focuses on *co-management*, an approach that explicitly seeks to incorporate local and scientific understanding, and that often results in a re-allocation of authority for management. Section 9.3 provides some examples of integration of local knowledge into resource and environmental management situations.

Box 9.1 Scientific and local knowledge

Modern scientific knowledge, with its accompanying world view of humans as being apart from and above the natural world has been extraordinarily successful in furthering human understanding and manipulation of simpler systems. However, neither this world view nor scientific knowledge have been particularly successful when confronted with complex ecological systems. These complex systems vary greatly on spatial and temporal scales rendering the generalizations that positivistic science has come up with of little value in furnishing practical prescriptions for sustainable resource use. Science-based societies have tended to overuse and simplify such complex ecological systems, resulting in a whole series of problems of resource exhaustion and environmental degradation.

It is in this context that the knowledge of indigenous societies accumulated over historical time, is of significance. The view of humans as a part of the natural world and a belief system stressing respect for the rest of the natural world is of value for evolving sustainable relations with the natural-resource base.

Source: Gadgil, Berkes and Folke, 1993: 151.

9.2 Local knowledge systems

The concept of *local knowledge systems* has its roots in the idea of indigenous or traditional knowledge and management systems. Indigenous, or native or tribal peoples, today are found on every continent and in many countries. Definitions of indigenous peoples vary. Nevertheless, common elements usually include: (1) descendants of original inhabitants of an area which has been occupied by more powerful outsiders; (2) distinctly different language, culture or religion compared with the dominant group; (3) often associated with some type of subsistence economy; (4) frequently descendants of hunter-gatherers, fishers, nomadic or seasonal herders, shifting farmers or cultivators; (5) social relations which emphasize kinship, group decision making by consensus, and collective sharing and management of natural resources (Durning, 1992: 8). Durning has suggested that if spoken language is used as a measure, the people of the world belong to 6,000 cultures, 4,000–5,000 of these being indigenous.

Because of their close ties to the environment and resources, indigenous people developed, by trial and error, understanding of the ecosystem in which they lived (Box 9.2, Figure 9.1). Such people did not always live in harmony with their environment and resources, and did and could cause degradation. At the same time, since their survival depended on maintaining the integrity of the ecosystem from which they derived their food and shelter, any major mistakes were usually not repeated. Their accumulated understanding of their environment was often transmitted in oral rather than in written form, and often could not be explained in scientific terms.

In many instances, their practices mimicked the patterns and behaviours of natural systems. For example, the practice of mixed cropping as an element of shifting cultivation replicated much of the complexity and diversity of sub-tropical or tropical vegetation systems. Different food such as maize, plantain, taro and groundnuts are often grown side-by-side on the same plots. To the Western-trained scientist, such an approach appears primitive and inefficient. However, the different rates of development of the crops ensure that the soil is kept under permanent cover. This reduces exposure to the sunlight and heating

---

Box 9.2 Emergence of indigenous knowledge and management

It thus appears plausible that over the course of human history, there have been human groups whose interests were strongly linked to the prudent use of their resource base, and that such groups did indeed evolve appropriate conservation practices. These practices were apparently based on some simple rules of thumb that tended to ensure the long term sustainability of the resource base. These rules were necessarily approximate. They would have been arrived at through a process of trial and error, with the continued acceptance of practices which appear to keep the resource base secure, coupled with the rejection of those practices which appear to destroy the resource base.

Source: Gadgil and Berkes, 1991: 136.

Figure 9.1 Through centuries of experimentation, Balinese farmers have developed sophisticated irrigation systems to support growing of sawah (rice). Western civil engineers, surveying the terraced rice fields with laser equipment, have been unable to identify significant changes to improve the effectiveness of how the water is collected, stored or moved (Bruce Mitchell)

of the surface. Continuous cover also protects against soil erosion, especially during the wet season when rainfall can be intense. The various root systems result in effective use of the soil volume. The mix of crops also minimizes the vulnerability of the plot to infestation by weeds or pests, as it is unlikely that any such infestations will harm the entire range of crops.

Western, science-based resource management has provided many useful concepts and methods for resource management and use. In many cases, "productivity" has been multiplied significantly, and higher population densities of

---

Box 9.3 Limitations of science for resource management

To many, it is a paradox that with all its power, modern science seems unable to halt and reverse the depletion of resources and the degradation of the environment. Part of the reason for this paradox may be that scientific resource management, and Western reductionist science in general, developed in the service of the utilitarian, exploitive, "dominion over nature" world view of colonialists and developers. It is best geared to the efficient utilization of resources as if they were boundless. . . .

Thus, modern resource management science is well suited, by design, for conventional exploitive development, but not for sustainable use. The task then is to re-think and re-construct a new resource management science that is better adapted to serve the needs of ecological sustainability and the people who use these resources.

Source: Gadgil and Berkes, 1991: 138.

---

people can be supported. However, as indicated in Box 9.3, the scientific approach has not always been able to avoid or restore degradation, or to sustain productivity.

The growing recognition that indigenous people who live in an area have understanding and insights about resources, environment and ecosystems as a result of observation over various seasons and many years has been extended to recognize that any people, indigenous or otherwise, living in an area may be aware about aspects that a scientist could miss (Figure 9.2). Such awareness has led to the acceptance of the participatory approach outlined in Chapter 8, and in growing interest to combine *local knowledge systems* with science-based knowledge. In the balance of this chapter, attention is given to ways of understanding such local knowledge systems, how they may be used in co-management arrangements, and to some experiences related to local knowledge.

9.3 Participatory local appraisal

Participatory local appraisal is used here to describe a method for studying local or indigenous knowledge systems. It has emerged from what has become known as *rapid rural appraisal* and *participatory rural appraisal*. Each is described below.

9.3.1 Rapid rural appraisal

Rapid rural appraisal (RRA) has been defined as "a systematic, but semi-structured, activity carried out in the field by a multidisciplinary team and designed to acquire quickly new information on, and new hypotheses about, rural life" (Conway and McCracken, 1990: 223). Conway and McCracken (1990) conclude that two central characteristics of RRA are:

- **Pursuit of *optimal ignorance*.** When collecting information about rural systems in a limited time frame, costs should be minimized. Thus, any

Figure 9.2 Experiential knowledge, based on years of interaction with local environments such as by this fadama farmer in Sokoto state in northwestern Nigeria, can produce insights which are not obtained through scientific investigation (Bruce Mitchell)

approach should facilitate rapid collection of information by focusing on selected key variables and appreciating what is not worth knowing. A corollary is *"appropriate imprecision"*, or not measuring what need not be measured or more accurately than is appropriate. Appropriate imprecision follows the belief that it is better to be approximately right than precisely wrong.

Guest Statement

Traditional ecological and technical knowledge for sustainable livelihoods and development

T. Vasantha Kumaran, India

Traditional ecological and technical knowledge is rooted in the past and intricately connected to the culture and values of a community of people in the present. As culture is dynamic, the traditional ecological and technical knowledge evolves over time and encompasses the new and the innovative. It is for this reason that it is important to distinguish between ancient and modern traditional knowledge, which so blend as to make a unified whole and genuine, practical knowledge. The main issue is simply whether or not the traditional knowledge is still relevant or useful in the current local or regional context.

Working with village communities in the rain shadow of the Western Ghats, in southern India, whose lives are intertwined with the monsoon winds and sands piled by them over an area of 120 km^2, I am convinced that the people of the villages: (1) given space and time and position of equality, would be able to posit and think of their life situation collectively in a manner more practical and competent than any of us could; and (2) do not need to plan their growth on a model designed by outsiders. In fact, urban India needs to learn a lot from villages as much or more than it needs to give, because the village abounds in traditional knowledge, both ecological and technical. There are techniques, such as *theppam*, devised by the people for irrigation water management, which are traditional and still find uses. A key element of the community action plan, in the making through participatory approaches, is that the traditional knowledge of the local people becomes a necessary and valuable input into the process. In particular, traditional strategies of natural resource use and conservation are critical components in developing a viable plan of community action. This is possible because for the people every cause is both personal and collective. Both men and women are knowledgeable. There is wisdom amid them. There is common sense. There is politeness. There is concern for others. Traditional knowledge and expertise promote the inherent dignity and agency of the rural people.

There is also an extraordinary strength in working class men and women. They are not afraid of acting openly on atrocities perpetrated and on corruption in high places within the community. There are women in each of the villages who are committed to issues and action, despite the added challenges they face in trying to take action. Many women are fiercely independent, endowed with the gift of logic and a love for mainstream participation. The villagers are fearless, especially women, and clear in their position, even though they are not formally educated. Social experience provides a valuable education. There is a strong moral fibre in all of them. And they know what they are doing. Collective thinking is natural to people. There is a good exchange of information among village members. Although faceless, they have a vision. Although helpless, for reasons of caste, class and gender, they have the strength to carry out action. Exploratory thinking is often shared in small public fora, in meetings at the social spaces of the villages. Hence,

215

most of the answers to questions of day-to-day life and living, even the questions of development, lie with the people themselves.

Given these characteristics, sometimes natural and other times paradoxical, the people do possess knowledge of practical significance such as biological and technical insights, systems of resource management (cropping pattern, prevention of soil erosion, *theppam*, prevention of soil erosion and improving soil quality), conservation (community forests known as *kuduvals*), development planning (improving livelihoods and initiating social welfare activities), and environmental assessment (reclamation, effectiveness of certain species of trees, state of sand dunes, change over from traditional agriculture to cash crop farming practices). All they miss, however, is collaborative problem solving through shared decision making. That is the reason why a community action plan is a vital instrument, designed by the people but through catalytic work of academics, NGOs, government and institutions. With the Community Action Plan in the making, I am sure we could promote environmentally sustainable resource reclamation (land and water), and address desertification and collaborative problem solving.

T. Vasantha Kumaran is a Professor in the Department of Geography at the University of Madras, Chennai, India. He obtained his MA (1973) and PhD (1980) from the University of Mysore. He has been involved in many research projects, most significantly in diffusion of innovations (population planning), tribal ecology (agricultural development and traditional ecological knowledge), contemporary health care (location–allocation problems and provider–user spatial behaviour), medical geography (hepatitis, malaria and filariasis), and community economic development (rural communities). His current research examines the options for restoration of the environment in five degraded /desertified villages in South India in a rain shadow region through creation of a Community Action Plan, with village resource management plans as constituents. The project assesses options for sustainable livelihoods through traditional ecological and technical knowledge. Gender and development is the main focus in all of this research.

- **Use of *triangulation*.** Triangulation emphasizes a diversity of sources and means for gathering data, and analytical methods. Accuracy and completeness will be greater if each aspect of a problem is investigated using various means. Understanding is sought by using a diversity of information, rather than through statistical replication. There can be tension between completing the data collection rapidly, and wanting to use a wide variety of sources and methods, but at a minimum it is important not to over-rely on a single source or method.

Given the above two characteristics, Conway and McCracken concluded that effective RRAs have five notable features, shown in Box 9.4. It should be clear from these features that RRA is intended to facilitate collection of data as quickly as possible, but to do so by interacting informally with local people in their own environments. The emphasis is qualitative rather than quantitative.

Box 9.4 Five features of RRAs

Iterative. The goals and process of the study are not rigidly set at the outset; they can be and often are modified as the investigators learn what is and is not relevant.

Innovative. No standardized methodology exists. Techniques are custom chosen for specific situations, and depend significantly upon available knowledge and skills.

Interactive. Normally, RRAs are conducted by a team. The team members are selected to encourage a diversity of interdisciplinary insights, and to foster synergy.

Informal. In contrast to the formality of many "scientific" methods, emphasis is upon informal interviews and discussions, and partly structured interviews.

In the field. Learning occurs in the field, often by helping or working with local people. Learning occurs "as you go". Field work often is followed by a workshop to share and discuss findings. Reports are written quickly after the field work.

Source: After Conway and McCracken, 1990: 224.

The five key features shown in Box 9.4 emerged in response to what have been viewed as three aspects of much development work in Third World countries (Chambers, 1994a: 956–7). These include:

(1) *Anti-poverty bias.* Much research in developing countries is conducted by urban-based professionals. Often, the work of such people has numerous biases: (a) *spatial*, with visits usually focusing on areas near cities or other easily accessible areas, to the neglect of peripheral areas; (b) *temporal*, with visits normally in the dry and cool seasons, whereas most problems are exacerbated in the wet and hot seasons; (c) *people*, with attention frequently on officials or elites rather than the poor, and on men rather than women; (d) *project*, with emphasis on officially supported activities, rather than on local, informal initiatives; and (e) *diplomatic*, with outside experts avoiding issues or questions considered to be offensive or sensitive to officials of the host country.

(2) *Over-reliance on questionnaire surveys* (Box 9.5). Too often questionnaire surveys have been designed by people with inadequate understanding of local cultures and languages. As a result, ideas are raised that are difficult or impossible to understand by local people, and the meaning of answers often becomes distorted or changed during translation.

Box 9.5 Problems with questionnaire surveys

Again and again, over many years and in many places, the experience had been that large-scale surveys with long questionnaires tended to be drawn-out, tedious, a headache to administer, a nightmare to process and write up, inaccurate and unreliable in data obtained, leading to reports, if any, which were long, late, boring, misleading, difficult to use, and anyway ignored.

Source: Chambers, 1994a: 956.

---

(3) *Emergence of better alternatives*. Unlike the other two aspects, this one was positive. It reflected a growing awareness about more cost-effective methods. A core idea was recognition by professionals that local people often were very knowledgeable about matters which affected their lives, including the behaviour and patterns of local ecosystems. This led to interest in what was termed *indigenous technical knowledge* or *indigenous ecological knowledge*, and acceptance that it had much practical value for problem solving. As Matowanyika (1991: 89) explained, such indigenous knowledge systems were based on the following characteristics related to resource use: (a) primarily rural; (b) relying on primary production based on a local physical environment; (c) integration of economic, social and cultural values and institutions, with kinship being a central unifying characteristic and the household being the base for the division of labour; (d) distributive systems that encourage mutual support and reciprocity; (e) a wide variety of resource ownership systems, but always including major communal ownership; and (f) relying mainly on local knowledge, information and experience.

Agroecosystem analysis, discussed in Chapter 5, has been applied in several countries in Southeast Asia, and helped to create credibility for RRAs. By the mid 1980s, work at the University of Khon Kaen, Thailand, was viewed by many as the global leader in developing and applying RRAs, and in providing training for RRAs. In 1985 an international conference on RRA was hosted at Khon Kaen, and the proceedings from that conference became the standard reference (University of Khon Kaen, 1987). The outcome was different types of RRA, as shown in Box 9.6, each designed for different purposes but often able to be used in combination or sequence.

With RRA well established, people began to consider how it could be strengthened. RRA had been designed to facilitate the rapid collection of data about rural systems, or ecosystems. The intent also was to be sensitive to the

---

Box 9.6 Kinds of RRA

Exploratory. Obtain initial information about a problem or ecosystem. The goal is to identify preliminary key questions or hypotheses.

Topical. Investigate a particular topic, often emerging from one or more questions or hypotheses identified during the exploratory RRA. The product is a detailed proposition to be used as the basis for management decisions, or for further research.

Participatory. Include local resource users and officials in decisions about new initiatives based on the findings from exploratory or topical RRAs. The output becomes locally managed experiments or activities in which local people have central roles.

Monitoring. Track progress of the experiments, and the implementation of the activities. The output is a revised hypothesis, along with changes in the experiments or activities.

Source: After Conway and McCracken, 1990: 225.

---

local context, especially the culture, traditions and languages. However, RRA was viewed by some as too "extractive" in that the process remained one-sided, similar to the questionnaire survey it had been designed to replace. Considerable time of the respondents was still taken up by an outsider. Little or nothing was returned to the respondents, as the outsiders had controlled the information which had been assembled. Thus, in the mid 1980s, two new words began to be used: *participation* and *participatory*. This led to RRA being renamed as *participatory rural appraisal*.

9.3.2 Transition to participatory rural appraisal

During the Khon Kaen conference in 1985, different types of RRA were identified, one of which was labelled "participatory". As noted in Box 9.6, the orientation of participatory rural appraisal was to facilitate or stimulate community awareness and capability regarding a problem or issue. Particular attention was given to enabling local people to conduct their own analyses of problems, and to share their findings. The role of the outsider became one of catalyst, rather than one of expert. Stimulation of community awareness and capability was also intended to reduce the extractive nature of the RRA, and to help local people to empower themselves. In that regard, what became known as *participatory rural appraisal* (PRA) is consistent with some of the basic aspects of sustainable development and transactive planning (local empowerment, equity, social justice).

RRA and PRA have some common elements, but the main distinction has been that RRA was designed to allow outsiders to collect data quickly and efficiently about a problem or ecosystem using a mix of informal methods and a variety of sources. In contrast, PRA was focused more on enabling local people to undertake their own investigations, to develop solutions and to implement action. Tables 9.1 and 9.2 highlight the similarities and differences

Table 9.1 A comparison of RRA and PRA

	RRA	PRA
Period of major development	Late 1970s, 1980s	Late 1980s, 1990s
Major innovators based in	Universities	NGOs
Main users at first	Aid agencies, universities	NGOs, government field organizations
Key resource earlier undervalued	Local people's knowledge	Local people's analytical capabilities
Main innovations	Methods, team management	Behaviour, experiential training
Predominant mode	Elicitive, extractive	Facilitating, participatory
Ideal objectives	Learning by outsiders	Empowerment of local people
Longer-term outcomes	Plans, project publications	Sustainable local action and institutions

Source: After Chambers 1994a: 958.

Table 9.2   RRA and PRA as a continuum

Nature of process	RRA .... PRA
Mode	Extractive ..... Elicitive ...... Sharing .... Empowering
Oursiders' role	Investigator .... Facilitator
Information owned, analysed and used by	Outsiders .... Local people
Methods used	Mainly RRA plus sometimes PRA .... Mainly PRA plus sometimes RRA

Source: Chambers, 1994a: 959.

between these two approaches. It is suggested here that PRA should more properly be called *participatory local appraisal* (PLA), to indicate that it can be applied in either rural or urban settings, incorporates the participatory ideas discussed in Chapter 8, and focuses upon appraisal that leads to action. An important concern continues to be the relatively rapid and systematic collection of information related to key questions and variables, rather than a comprehensive analysis. In this manner, PLA is consistent with the ideas expressed in Chapter 5 regarding *integrated resource management*.

Chambers (1994b: 1254) highlighted common principles shared by RRA and PRA. These include:

(1)   *A reversal of learning*. Learning occurs from local people, directly, from on-site and face-to-face meetings.

(2)   *Learning rapidly and progressively*. Through deliberate exploration, flexibility in choice of methods, being opportunistic, improvising, iterating, cross checking and adapting.

(3)   *Offsetting biases*. Achieved by being relaxed and not rushing, listening rather than lecturing, probing rather than moving quickly from topic to topic, being humble rather than important, and searching out the poorest and most vulnerable rather than those of highest status.

(4)   *Optimizing trade-offs*. The costs of learning are related to the usefulness of information, and attention is given to trade-offs among quantity, relevance, accuracy and timeliness.

(5)   *Triangulation*. Cross checking and progressive learning through use of multiple sources and methods.

(6)   *Seeking diversity*. Rather than focusing on convergence and common patterns, also looking for and learning from exceptions, oddities, dissenters and outliers in any distribution. Another way of expressing this view is to focus upon variability rather than averages. Seeking diversity goes beyond triangulation, as it systematically is intended to search for and analyse contradictions, anomalies and differences.

While the above six principles are common to both RRA and PRA, Chambers (1994b: 1254–5) argued that another set of principles was specific to PRA, and these are:

(1) *They do it.* The intent is to facilitate investigation, analysis, presentation and overall learning by local people, in order that they generate and own the results, and learn from them. This principle often leads to a facilitator starting a process of participatory analysis, but then withdrawing and leaving the community to pursue and complete the work.

(2) *Self-critical analysis.* Facilitators continuously assess their own behaviour, acknowledging and accepting that they will make mistakes, and seek to learn from failures. This is often characterized as "falling forward".

(3) *Personal responsibility.* Practitioners take personal responsibility for decisions and actions, and do not look to the authority of manuals or rules for PRA. This view emphasizes the importance of using best judgement in the context of local conditions.

(4) *Sharing.* Information and ideas are shared openly among local people, among them and external facilitators, and among different PRA practitioners.

9.3.3 Methods for participatory local appraisal

Chambers (1994a; 1994b; 1994c) and Conway and McCracken (1990) have identified the range of methods that often are used in different combinations for PLA. Some of the more important are summarized below.

(1) Secondary sources

This source includes more than books and journal articles. It includes reports, maps, aerial photographs, satellite imagery, files, memoranda, annual reports, survey results, computerized data records, census information, project documents and newspaper stories. Scanning as many secondary sources as possible will help to avoid unnecessary duplication of work already completed, will help to identify important issues and potential data sources, and may help to identify key people to contact.

(2) Semi-structured interviews

These have been viewed as a central method for PLA. These can be completed with individuals or groups, and can cover resource users (farmers, hunters and gatherers), officials (government officers) or local elites (school teachers, village leaders). Interviews are always conducted in an informal manner, preferably in the usual surroundings of the informant. A written questionnaire is not used, but notes are taken of key ideas and information. The interview is usually based on some key questions, but is deliberately left open ended, and there is willingness, even anticipation, to explore unexpected topics or issues. Any semi-structured interview would not normally last more than one hour, but there might be more than one interview with a respondent over a period of time.

(3) Direct observation

Direct observation occurs while the outsider is conducting other work, and can involve the systematic observation of events, processes, relationships or patterns.

221

This type of data collection is similar to what social scientists refer to as *participant observation*. Direct observation is often used to verify the insights obtained from secondary sources and from semi-structured interviews. The danger, of course, is that there can be a *reactive effect*. That is, normal processes and relationships may not be seen, as the presence of an outside observer may change "normal" behaviour. The longer the analyst is able to observe people or an ecosystem, the greater are the chances that what is seen will be patterns not influenced by a reactive effect.

In PLA, useful methods can involve asking to be, and being, taught by local resource users, and helping them with their tasks. In that way, barriers between the outsider and local people will be gradually broken down. The outsider will also gain a new appreciation for skills which, on the basis of superficial examination, may appear to be rudimentary or simple but turn out to require considerable skill.

(4) Visual models

These are increasingly being used, particularly in cross cultural situations, or ones in which formal education of respondents is minimal. Visual models refer to the use of diagrams such as sketch maps (on paper or on the ground), transects, seasonal calendars, bar diagrams, time trends, Venn diagrams and decision trees. It is not just the researcher or outsider who uses these methods. Local people can complete transects, sketch maps, seasonal calendars and time trends. These are frequently used methods in PLA.

- *Participatory modelling.* Local people use the ground, floor or paper to construct social, demographic or natural resource maps, showing existing patterns of use, capacity for different uses, ownership, shared uses, etc.
- *Transect walks.* The outsider or local people walk through an area, systematically making observations about problems, solutions and opportunities. Transects are often done along a slope, or involve "combing" or "looping". While walking with others, it is possible to observe, ask questions and discuss. Figure 9.3 is an example of a diagram prepared on the basis of a transect walk.
- *Seasonal calendars.* Changes and/or activities by month, season or other meaningful time period can be recorded. These can include patterns of resource use, distribution of rain and prices for ecosystem products.
- *Time lines and trend/change analysis.* A chronology of events with approximate dates can be created. Attention can be directed to how things have changed; ecosystem histories can be created; changes in customs, practices and activities can be recorded; changes in resource-use patterns can be noted. These records can be established by groups, and often become a source of considerable interest, particularly when it almost inevitably emerges that people have different perceptions about the timing or significance of changes.
- *Institutional Venn diagrams.* Individuals and organizations, important to the community, can be identified, and their relationships can be depicted diagrammatically.

Figure 9.3 Agroecosystem Transect, Dusun Nating, South Sulawesi, Indonesia (Mackenzie, 1993)

Dusun: Nating Kabupaten: Enrekang Sub-watershed: Tabang AE Zone: Montane

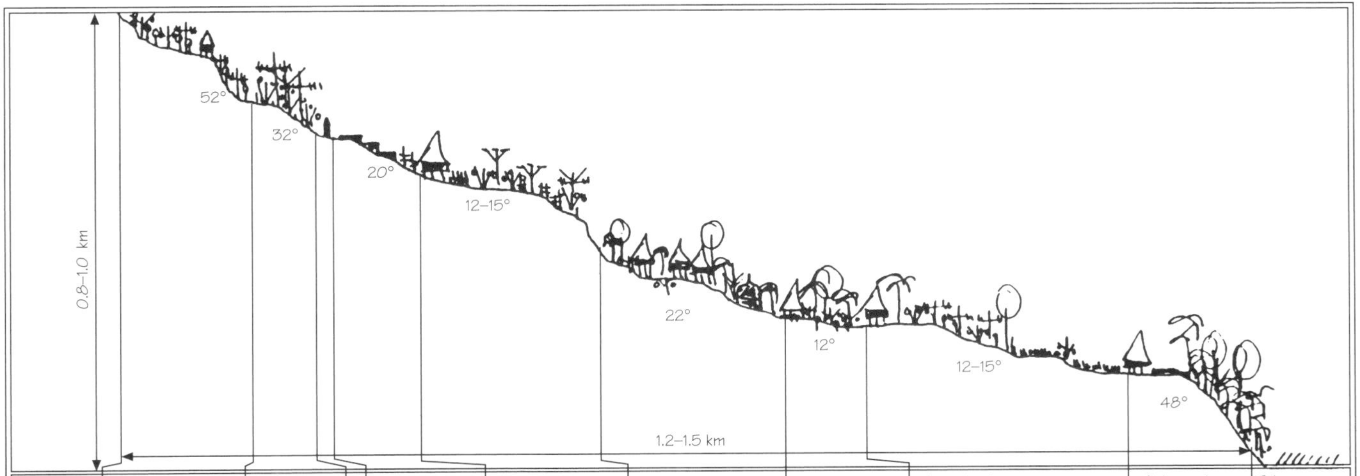

Land use	• farm land • fuelwood	• farm land	• fish ponds	• farm land • housing	• village proper • homegardens	• farm land • homegarden	• farm land • some housing	• fish ponds • homegarden
Crops and trees	• arabica coffee • maize • suren 1 yr.	• v. mature coffee w. lg. suren 30 yrs	• gold fish • cassava	• coffee/cassava/ kani-suren • sm. clove tree nursery	• jackfruit/banana/ coffee/cocoa/veg.	• sugar pala • jackfruit • coffee • banana	• coffee/maize/suren • banana/jackfruit	• gold fish • banana/ jackfruit
Forest	• young, simple secondary forest	• secondary forest • mature						• traditional protection forest
Livestock				• chickens	• chickens • goats	• chickens		
Soils	• "tanah putik" • violet podsolic	• violet podsolic	• violet w. black podsolic	• red w. violet podsolic	• red podsolic	• red podsolic	• violet podsolic • soil exhausted	• black podsolic
Problems	• rill erosion • slumps • alang alang (imaskata grass) • wild boars		• no dry season fish crop. • poor soil	• poor fencing • wild boars			• rill erosion • lg. clearings w. alang alang • wild boars	

0.8–1.0 km   1.2–1.5 km

Protection Forest Marker — est. 1992

52° 32° 20° 12–15° 22° 12–15° 12° 48°

Tabang River

(5) Workshops

Workshops may occur throughout the period of data collection, but most usually are used toward the end of the data collection period. The outsiders meet with local people with whom they have been working to examine the information collected, share analysis and interpretations, consider opportunities and possible actions, and search for preferred initiatives. The number of participants can be small or large; but if the latter, then usually it is helpful to combine full group sessions with other sessions involving smaller discussion groups. The workshops may be short (less than a half day) or may extend over several days.

Many of the methods outlined above will be familiar to the social scientist; some will not be familiar. As indicated earlier, there is no formula, menu or recipe to determine the right mix of methods. The choice will be governed by judgements about needs and circumstances related to the system being studied, and by the creativity of the analyst. In a later section in which experience with local knowledge systems is considered, there will be opportunity to consider how PLA methods could be used to understand what has been and is happening.

9.3.4 Contributions or "discoveries" from PLA

PRA or PLA has led to four "discoveries" (Chambers, 1994b). The first is that local people often have a greater ability to observe, estimate, quantify, rank, compare, diagram, map and model than outsiders generally thought was possible. Particularly revealing has been the remarkable capacity of local people to map and model.

A second discovery is the significance of the behaviour of outsiders, and of establishing rapport from the outset. What has been referred to as achieving "relaxed rapport" and trust between outside investigators or facilitators and local people is now recognized as essential preconditions for effective PLA. Again and again, empirical findings have shown that if initial behaviour and attitudes of the outsiders are relaxed and appropriate, the methods of PLA create and reinforce further rapport.

The power and popularity of participatory diagramming and visual sharing is the third major discovery. In traditional questionnaire surveys, information is collected and appropriated by the outsider, and the meanings are transformed into concepts familiar to the outsider. With visual sharing of a diagram, map or model, everyone participating can see what has been developed, can comment on, and offer suggestions to modify it. Triangulation occurs as a result of local people cross checking and correcting one another. The learning is progressive, and the information is ultimately owned by the participants.

The fourth discovery is value of sequencing when applying participatory methods. While some participatory methods (such as participatory observation) have been well known for decades, and others are newer, the striking insight has been the power achieved through using methods in different combinations and sequences. For example, in participatory mapping, local people often draw a set of maps, with each additional one becoming more detailed and useful by including new and complementary information. Or, a map prepared during discussions can then be used to plan a transect walk during which the local people who prepared the map act as guides for the outsiders. The transect walks in turn can lead to identification of issues, which then can be used for a ranking or scoring exercise.

9.3.5 Cautions about PLA

While PLA has offered new insights for resource and environmental management, it is important to be aware of limitations. Key ones have been identified by Chambers (1994c) and Mohan (1999), and are considered here.

First, there is a danger of PLA becoming "instantly fashionable", and therefore being used uncritically. In some instances, traditional survey approaches have been modified ever so slightly, and then have been referred to as being based on a participatory appraisal approach. If PLA is misused or abused, then it will be discredited. PLA can provide new and useful insights, but by itself is not a panacea for all the difficulties encountered when seeking information from local people.

A second danger is using PLA approaches, but in a rushed manner. The word "rapid" was included in RRA approaches to highlight an option relative to the time-consuming field work usually relied upon by anthropologists or to counter the use of large-scale questionnaire surveys. However, at some point, preoccupation with the idea of "rapid" can become liability, and result in the loss of the critical elements of "relaxed rapport" and trust. However, Mohan's (1999: 45) views about what constitutes "relaxed rapport" should be given attention. He noted that outsiders often make inappropriate cultural assumptions about what constitutes being relaxed and informal. Furthermore, he has argued that sometimes outsiders treat a situation as informal when local people perceive it to be formal.

A third potential problem is formalizing of PLA concepts and methods into guidelines and manuals, with the result that practitioners over-rely on such protocols, and lose sight of the need to use judgement and to custom design with regard to local conditions (Box 9.7). There cannot be a recipe or cook book regarding the best mix of tools or techniques for PLA. Indeed, Chambers (1994c: 1441) has argued that a key lesson is that practitioners learn best in the field, through experience, learning from doing, and taking responsibility for what they do and the mistakes that they make.

A fourth danger relates to Mohan's (1999: 46) view that PLA techniques are usually "culturally loaded". Although they have been designed to connect with authentic "life worlds" of local people, they usually do this by using concepts

---

Box 9.7 Learning from doing

It has been not books of instructions, but personal commitment, critical awareness, and informed improvisation, which have best assured quality and creativity.

Source: Chambers, 1994c: 1441.

---

and categories created in Western minds. Thus, while clipboards and survey questionnaires have been set aside, the underlying interests, concepts and analytical categories are usually mainly Western and also primarily academic.

A fifth and related difficulty is the general view that local communities are harmonious, leading to many PLA methods promoting the belief of a consensual view within them (Mohan, 1999: 46). Local communities often are more heterogeneous than homogeneous, and differences exist between the well- and less-well off, men and women, and older and younger people. The romanticized view that consensus exists and can be found, and that individuals and groups live in harmony with each other and their environment, may predispose outsiders to focus on certain aspects of evidence, and to ignore or not "to see" other aspects. To counter this problem, investigators are encouraged to look for differences and variability rather than averages and commonality.

A sixth concern is based on what Chambers (1994c) has called "routinization and ruts". Repetition of the use of PLA can lead practitioners into regular habits and over-reliance on selected methods. Even for specific techniques such as participatory mapping or modelling, many alternative ways exist to apply them. The reliance on unvarying standard practices can lead to limitation on obtaining data and to confining insights that may be gained.

Lack of sensitivity regarding who is being empowered at the local level is a seventh danger. Whether empowerment is equitable depends very much on who is being empowered. A naïve view that participation is good, regardless of who participates or gains, can contribute to entrenchment of the power of local elites (usually older, better off, higher status males), leaving the poor and vulnerable no better off, and maybe worse off. Thus, a challenge is to apply PLA so that the weaker or more vulnerable members of groups are identified, and empowerment and equity are served.

An eighth danger is to become over committed to PLA, and not be willing to innovate further. Mohan (1999: 44) has observed that participatory rural appraisal methods can take a community only so far in understanding problems and opportunities, and argued that other approaches are needed. He suggested that *Participatory Action Research* (PAR) or *Participatory Learning and Action* (another acronym using PLA) offer an additional benefit by consciously using generation of knowledge to lead to action about and transformation of conditions in local areas.

The above eight limitations or problems with PLA are reminders that PLA should not be viewed as the ideal way to tap into local knowledge systems related to resource and environmental management. PLA does offer useful ways to gain insights and understanding, and creates capacity for action to change conditions, but it also has limitations which need to be recognized.

9.4 Co-management

Co-management arrangements reflect one means of achieving the partnership approach reviewed in Chapter 8. However, it is included in this chapter

Box 9.8 Co-management defined

Co-management arrangements in general involve genuine power sharing between community-based managers and government agencies, so that each can check the potential excesses of the other.

Source: Pinkerton, 1993: 37.

There is no widely accepted definition of co-management. The term broadly refers to various levels of integration of local- and state-level management systems. Co-management [is defined here] to mean the sharing of power and responsibility between the government and local resource users. A more precise definition is probably inappropriate because there is a continuum of co-management arrangements from those that merely involve, for example, some local participation in government research being carried out, to those in which the local community holds all the management power and responsibility.

Source: Berkes, George and Preston, 1991: 12.

because co-management has often been explicitly designed to recognize and incorporate local knowledge systems from resource users.

Co-management is an approach which incorporates local- and state-level management systems (Box 9.8). The state-level responsibilities are conducted by a government agency with a legal mandate. The management style is often characterized as centralized and hierarchial, with a headquarters determining overall policy, and regional offices implementing the policy. Its approach is normally based on "science" or "scientific data", and on calculations or estimates based on such knowledge or data. Enforcement is based on the authority provided through legislation and regulations. Local-level management systems, in contrast, are based on experiential knowledge, cultural traditions, customary practices and self-regulation. The approach is highly decentralized. Decisions are based on development of consensus, and enforcement occurs through social sanctions.

Given the striking differences between the two approaches, there often is lack of understanding and respect by followers of one system for the other. In addition, problems can occur because they often represent fundamentally different ideas about the concept of *rights* to resource access and use. State-level management systems are normally based on the idea that there is ownership of the resource, or that rights to access or use of the resource are allocated by the government which owns the resource on behalf of the society. For local-level systems, the concept of ownership is often a foreign one. No-one "owns" the resource, but people have the right of access to, or use of, the resource, as determined by customary practices and arrangements. The challenge, therefore, is to determine how to incorporate each system's strengths, many of which are complementary.

As noted above, one key issue is to determine the extent to which control of, or power over, the resources or environment will be shared. Berkes, George and

Preston (1991: 12–13) suggested that a parallel can be made between co-management arrangements and Arnstein's "ladder of citizen participation", discussed in Chapter 8. Their view is that different types of co-management can be represented as rungs on a ladder, each reflecting different degrees of power sharing between citizens and the government. To illustrate, for situations in which resources can be managed locally, such as harvesting of beaver, nearly all the management power can be delegated to the local people. With full community control, given credibility by the state government, local people can make decisions about allocations and patterns of access and use. On the other hand, situations exist in which resources cannot readily be managed locally, such as for migratory species of birds or mammals. In these situations, local people can participate as partners, and can share their local knowledge about the behaviour of the species in their region. However, the state government also has an important role to play since the species move across a number of regions each of which contains local users.

Co-management agreements most frequently have been developed with reference to renewable resources, especially fish, birds and forests. Box 9.9 outlines the rationale for, and benefits from, a co-management agreement between a government and fishers. It usually is agreed that co-management agreements contribute to encouraging community-based development, decentralizing

---

Box 9.9 Co-management for fisheries

Co-management agreements between government and fishing interests have arisen out of crises caused by rumoured or real stock depletion or from political pressure resulting from claims that the government's ability to manage is insufficient to handle specific problems. Co-management agreements are a creative way to break the impasse in government/fishermen conflicts over the most effective solutions to such crises. Typically in such cases, fishermen demand a real voice in decision-making because they have lost faith in government's ability to solve management problems: they point to government's lack of adequate data and to its role in making the problems worse. Government officials, who may equally distrust fishermen, whom they see as unrelenting predators who will eliminate the fish unless more strictly regulated, become willing to surrender some power in exchange for fishermen's co-operation and assistance in management. A balance of power is struck so that fishermen do not believe they have been simply co-opted into government's convenience. For its part, government can act as a check to any local violations which do not conserve fish stocks or fairly share the benefits of fish production.

Co-management regimes work by altering the relationships among the actors in the fishery – primarily between fishermen and government, but also those among individual fishermen and among fishermen's groups. Basically, by instituting shared decision-making among these actors, co-management systems set up a game in which the pay-offs are greater for co-operation than for opposition and/or competition, a game in which the actors can learn to optimize their mutual good and plan co-operatively with long-term time horizons.

Source: Pinkerton, 1989: 4–5.

regulatory power, and reducing conflict through consensus building and participatory principles.

9.4.1 Aspects which contribute to effective co-management

Pinkerton (1989) and Ostrom (1999) have each reviewed co-management arrangements, and developed a set of "preconditions" which support successful agreements for co-management related to fisheries. The following points, based on their work, have been modified so as to apply to most resources.

(1) *Most favourable preconditions.* (a) The presence of a real or imagined crisis in the depletion or degradation of the resource, such as the crash of a fish species. (b) The resource has not deteriorated so much that people believe it would be useless to organize themselves for a cooperative effort, or the resource is not so under-used that little advantage would occur from organizing and using it. (c) The availability of resource units is relatively predictable. (d) Reliable and valid information regarding the general condition of the resource is available at reasonable cost. (e) Willingness of local resource users to contribute financially or in kind, or to obtain other sources of support, for the rehabilitation of the resource or to contribute to management tasks. (f) Opportunity for negotiation and/or introduction of a co-management experiment focused on a specific management task or function, which subsequently could be expanded to other tasks.

(2) *Most favourable mechanisms and conditions.* (a) Agreements are reached which are formalized, legal and multiple-year. (b) A mechanism exists to return back into local communities some of the wealth produced by co-management. (c) A mechanism exists which both conserves and enhances the resource and the integrity of the local cultural system. (d) External support (such as from universities, NGOs) can be recruited, and external forums for discussion (technical advisory committees) which include more than the immediately affected stakeholders can be created.

(3) *Best spatial scale.* (a) A relatively small area, meaning an area such as a watershed in which benefits can be readily identified and appreciated by participants. The small area also allows users to have an accurate knowledge about the resource, external boundaries and internal microenvironments. (b) The number of participants and local communities is relatively small so that effective communication can occur. (c) The government bureaucracy is relatively small, and its mandate can be related specifically to the local or regional scale.

(4) *Groups most predisposed toward co-management.* (a) The existing group(s) already has a cohesive social system, often based on kinship, ethnicity or similar method of resource use. (b) Users have a common or shared perception of the resources and how their actions (individually and collectively) affect each other and the resource itself. (c) Users trust each other to keep promises and relate to one another with reciprocity. (d) Users are able to determine arrangements for access and use, without the likelihood that

229

Box 9.10 Importance of people

. . . the motivations and attitudes of key individuals can make or break co-management, no matter how much legal backing or supportive arrangements an agreement has.

Source: Pinkerton, 1989: 29.

---

external authorities will challenge or over-ride their rules. (e) The group is dependent on the resource for a major portion of its livelihood. (f) Users with higher economic and political assets are similarly affected by the present pattern of resource access and use. (g) The group(s) can effectively define its boundaries, leading to clear identification of membership in the group and ease of allocating access to the resource and applying sanctions.

(5)   *The human factor.* While preconditions, conditions, mechanisms and spatial scale are all important, successful co-management agreements ultimately depend upon the relationships among people (Box 9.10). Co-management is therefore most likely to be successful when a dedicated person or core group creates ongoing and consistent pressure to move forward the provisions in the agreement. Furthermore, co-management is more likely to be successful if new and sustainable relationships are created among the people, and this occurs when: (a) cooperation is fostered among individuals and groups in the local community of resource users; (b) commitment is generated among local people to share both the benefits and the costs of their efforts to enhance and conserve the resource; (c) allocation decisions generate an approach to conflict resolution which motivates people to negotiate shared and equitable access to the resource; (d) a negotiating relationship is established which includes people who are directly and indirectly affected by the resource allocation decisions (e.g. regarding a fishery, other water users are included); (e) a willingness exists or is created to share data about the resource; and (f) trust and respect grow between resource users and government officials, leading government to turn over more power to local people.

In an ideal form, co-management arrangements reflect many aspects of sustainable development. However, as with participatory local appraisal approaches discussed earlier, it is important not to overly idealize the co-management approach. As Reed (1995) has observed, based on her study of a co-management initiative in northern Ontario, Canada, the preconditions and conditions are not always met. In the Ontario example, which involved a local community in partnership with a provincial government agency, she found that some local people coalesced to skew participation toward those who shared their economic values (Box 9.11). Key participants maintained close control of the participation process within the local community, and they also became a conduit to promote particular values and views to the provincial government. In Reed's view, the outcome was a set of proposals for modest change which lacked an

---

Box 9.11 Distinguishing between co-management in name and in practice

. . . the procedures adopted by the committee were not consistent with the premises of co-management. The Township Council, with members of the Business Association, acted to avoid local division by influencing initial membership of the committee toward local business interests. Once the committee was operating, it exerted close control over information provided to the public and restricted opportunities for the broader community to become involved in its deliberations. Committee members maintained close control over aspects such as access to information, rules for participation, mechanisms used, timing of participation, and feedback to the public. Thus, the co-management committee . . . simply replaced the paternalistic approach previously associated with the MNR [Ministry of Natural Resources] with paternalism initiated by the local committee.

Source: Reed, 1995: 145.

---

overall vision, or a commitment for implementation. Unequal access to power, both within the local community, and between the local community and the provincial government, resulted in the co-management initiative being one more in name than in deed. Reed's analysis is a helpful caution not to become uncritically enthusiastic about co-management agreements.

9.5 Local knowledge for resource and environmental management

9.5.1 Cod fisheries in Atlantic Canada

In July 1992, the Government of Canada announced a fishing moratorium for northern cod in the waters off Atlantic Canada because of depleted stocks. It was anticipated that up to 20,000 fishers and fish process workers would lose their jobs, many of whom lived in small coastal communities with few alternative employment opportunities. In subsequent decisions up to December 1993, other fisheries were closed, with the outcome being a total loss of jobs for 40,000–50,000 people. Many reasons were offered for the depletion of the fisheries, including domestic and foreign fleet overfishing, changing environmental conditions and predation by seals. Others suggested that fishery managers needed to share some of the blame, as they apparently had badly miscalculated the extent of the fish stocks.

How could the fishery scientists and managers have been so wrong? Many reasons can be offered (Box 9.12). First, their data about the fishery were poor, due to the challenges of sampling a migratory resource whose habitat extends over a vast area. Second, the scientists' models contained many assumptions, based on theory of fish behaviour. The scientists used their models, and data taken from research vessels, but rarely had experience fishing or talking to

---

Box 9.12 Scientists ignore early warnings from fishers

Newfoundland's inshore fishery continued to fail. To the inshore fishermen, the reason was clear: fish were scarce, and offshore draggers were catching them before they came closer to shore.

Maybe not, said the scientists. Maybe the fish were staying offshore. Maybe fewer crews were actually fishing. Maybe it was the water temperature, or food cycles, or . . .

No said the fishermen. There are not enough fish. And they were right.

Source: Cameron, 1990: 31–3.

---

fishers who were seeing patterns that did not always show up in their data. It is the latter point that is of interest here.

What insights did the local fishers have, and, if they had been heeded, how might that have helped the scientists? Could the local knowledge of fishers have provided an early warning which the "scientific" data and computer models did not capture? One expert who was appointed to conduct an inquiry into the status of the fishery expressed doubt about the adequacy of the scientific understanding of the fishery, as highlighted by his comments in Box 9.13. Given all of the problems in studying fish stocks and behaviour, it is difficult to understand why a potentially valuable source of information would be over-looked or discounted, if it could be accessed at reasonable cost.

Neis (1992) examined the ecological knowledge of local fishermen in New-foundland. She provided some insights regarding their understanding, and why scientists have been slow or reluctant to consider their knowledge. Based on a study of fishers in one area (Petty Harbour), she found that they were among the first groups to argue that the scientific assessments regarding the health of the northern cod stocks were incorrect. Their questioning of the assessments was based on their experience of catching smaller fish, requiring longer fishing days, and needing a longer fishing season to maintain catches of previous years. They subsequently took a leading role in creating the Newfoundland Inshore Fisheries Association, a group which consistently questioned the accuracy and validity of the scientific assessments of the cod stocks. At the same time, however, Neis (1992: 156) noted that "the scope and nature of the ecological

---

Box 9.13 Ignorance about the fishery

"The state of our ignorance is appalling. We know almost nothing of value with respect to the behaviour of fish. We don't even know if there's one northern cod stock, or many, or how they might be distinguished. We don't know anything about migration patterns or their causes, of feeding habits, or relationships in the food chain. I could go on listing what we don't know."

Source: Dr Leslie Harris, as quoted in Cameron, 1990: 35.

Box 9.14 A scientist's view about local knowledge

... to separate out testable elements of this [scientific] view of the fishery system from the fishermen's point of view is really difficult. I imagine there is probably integration of all kinds of variables going on simultaneously in any particular fisheries situation on any given day, and also over the years, as people modify the traditional lore. You can't really do a controlled experiment under these situations to say, "we falsified the null hypothesis so now we can move on to the next step in the method." That reductionist approach would seem to me to be different from what you would consider to be traditional lore that integrates a lot of different observations and people's intuitions and gut feelings and is kind of tested but you don't know what kind of testing it's undergone from generation to generation. Have the conditions remained constant over time, or have they been changing? If they have, then how do you know what you are seeing is really the result of the causal mechanism that is attributed to it? So it's basically at odds with scientific method because traditional knowledge has so much more information in it that is unspoken or already subsumed and the scientific method says reduce it and test it at each point and control for all of the other co-current variables. It is hard to integrate those two views of the system.

Source: As quoted in Neis, 1992: 166.

---

knowledge of Newfoundland inshore fishers has never been the direct focus of either scientific or social science research".

Franklin (1990: 40) has observed that "it should be the experience that leads to a modification of knowledge, rather than abstract knowledge forcing people to perceive their experience as being unreal or wrong". That is, if the facts do not match the theory, analysts have two choices. One is to assume that the theory or model is correct and the data are flawed, and therefore to renew efforts to obtain better data. During the 1980s, this was the approach of many fishery scientists regarding their computerized models of fish behaviour in the northwest Atlantic waters. The other choice is to conclude that the model is flawed, and return to the drawing board to determine what changes should be made to the model or theory. The local fishers were arguing that the model was flawed, not their experiential evidence. However, as the comments in Box 9.14 from one scientist reveal, there often are different values and mindsets held by scientists and local resource users, resulting in hesitation by scientists to place much weight on "evidence" provided by local people. This situation is changing, as local knowledge systems are increasingly being accepted. However, the experience with the Atlantic fisheries of Canada confirms the observation of Reed that we must not take a romanticized view of local knowledge and co-management agreements, and assume that there will always be trust, respect and cooperation from the outset.

Thus, a major challenge for local indigenous knowledge can be scepticism from scientists because such knowledge is "only" anecdotal. Such vernacular knowledge is often not given much weight or respect from scientists. The irony

---

Box 9.15 Questionable data

. . . to say we are going to do our survey in the first three weeks of October every year and that creates a constant for us is not correct because it's not constant. It's constant in terms of time, our calendar, but that's meaningless to fish who don't use our calendar. They use another calendar entirely, which is based on temperature, food availability, and salinity and a number of other environmental circumstances.

Source: Dr Leslie Harris, quoted in Neis, 1992: 169.

---

is that "scientific" data often can be viewed as equally anecdotal. Scientific information can be badly flawed, and often isolated from a broader context which is critical if it is to be interpreted sensibly. For example, the expert who conducted the investigation about the decline of the northern cod stocks indicated little confidence in sampling methods to assess the stock, as shown by his comments in Box 9.15. As Neis concluded, only after many crises related to fishery stocks have scientists began to recognize the neglect in their computer models of ecological relationships, such as those among species, between stocks and oceanographic conditions, and between catch levels and human and marine ecology. In her view, such relationships are the ones that provide the framework for traditional ecological knowledge.

The experience with the Atlantic Canada fisheries is not unique. Other examples demonstrate that often governments have developed policies or practices, which have turned out to be very much less than perfect, as a partial result of either ignoring or misunderstanding local knowledge or management systems. Thus, Dove (1986) convincingly showed that the Indonesian government pursued poorly conceived agricultural policies in South Kalimantan due to misunderstanding of swidden agricultural practices. Even when confronted with evidence showing that their policies and practices were based on misunderstanding, the government was unwilling to make any changes. Thus, it may still be some time before local knowledge systems and participatory approaches are widely accepted and endorsed. Nevertheless, it appears that both of them should become mainstream elements in resource and environmental management, and in some places this is already occurring, as shown in the guest statement by Robson Mutandi.

9.6 Implications

The existence of local knowledge systems reinforces the importance of incorporating partnerships and participatory approaches more systematically in resource and environmental management. As the example of the northern cod fishery in the northwest Atlantic Ocean emphasized, scientists and their models can often be incorrect. In many situations, local knowledge systems can provide

Guest Statement

Forging effective partnerships for sustainable community-based natural resources and enviromental management in Zimbabwe

Robson Mutandi, Zimbabwe

Emerging rural development theory and practice indicate that problems faced by developing countries' rural poor are far too big for any single player or stakeholder. Functional and effective partnerships at various levels of development intervention need to be created where they do not exist and re-enforced where they exist in one form or another. Strategic partnerships, alliances and collaboration among key stakeholders have been forged at the global, national, regional and sub regional (grassroots) level to effect positive changes in the livelihoods of the rural poor in developing countries. At the global level, leading agencies for development assistance and governments are joining forces though "co-financing" arrangements to address the needs of the rural poor. At the national level, these agencies are joining forces with government agencies, industry, academic institutions and NGOs to ensure a more effective delivery system for the rural population. At the local level, frameworks are being created to ensure that target groups are included in decision-making and implementation processes that affect their lives. At a grassroots level, functional and effective partnerships are particularly critical for community-based natural resources management, especially in African dryland ecosystems. This involves, among other interventions, processes that empower rural communities to seek and implement strategies that improve their livelihoods. The sustainability of these empowerment processes hinges on two key elements: (1) participation of the community, and (2) incorporation of local or indigenous knowledge systems into the development process.

The South Eastern Dry Areas Project (SEDAP) in Zimbabwe is one attempt to forge partnerships at all three levels. At the global level, the project is jointly funded by the International Fund for Agricultural Development (IFAD) through a loan to the Government of Zimbabwe, by a grant from the Netherlands Government, and through the Public Sector Investment Program (PSIP) of the Government of Zimbabwe. At the national level, collaboration has been forged among key stakeholders, including the government, NGOs, universities and the private sector, the last through rural credit facilities with commercial banks and agro-industrial firms. Through its *rural institutional strengthening* component, the project is able to address the question of effective and functional partnerships through a community empowerment process that ensures full participation by the target groups and inclusion of key elements of their indigenous knowledge systems into the management of natural resources in their areas. This process not only ensures that communities are fully incorporated into the project, but that they take leading roles in decision making, project implementation and management processes at the local level. This is one way that the project ensures sustainable management and utilization of natural resources in these arid regions.

The project covers eleven districts in the southeastern border areas of Zimbabwe. Low and erratic rainfall, associated with high levels of unpredictability and variability, characterize these areas. Thus, drought coping strategies and risk aversion are central to the livelihood systems in the project area. In this regard, local knowledge systems play a critical role in sustaining the livelihood system. SEDAP recognizes that many environmental problems are localized and specific, and thus require local and ecologically particular responses. In this respect, the project also recognizes that indigenous knowledge systems are more likely to provide the missing link in formal natural resources and environmental management strategies and processes among rural communities in developing countries. Hence the need to forge partnerships at grassroots levels with communities taking the leading role and outsiders playing the role of facilitator.

Robson Mutandi is an environmental management and rural development practitioner with over 17 years' experience working with rural communities, international and local development agencies and policy-making institutions in southern Africa. His work and publications focus mainly on indigenous knowledge systems, resource management in the commons, gender roles and relations in resource management, the application of participatory methodologies in rural development, and dryland ecosystems management among rural communities in Southern Africa. He is the National Project Facilitator for SEDAP.

In January 2001, he began working for the United Nations Development Program as Chief Technical Advisor to the Ministry of Environment, Gender and Youth Affairs in Lesotho regarding sustainable livelihood systems and biodiversity conservation in communal areas.

the "ground truthing" to help verify basic assumptions or building blocks in analysts' and managers' models. The track record of scientific understanding and predictions regarding resource and environmental matters is not so good that scientists can dismiss local knowledge as "only" vernacular or anecdotal.

If local knowledge is to be understood by those with scientific backgrounds, many of the traditional social science methods of collecting information, especially the use of questionnaire surveys, need careful review and assessment. Concern about the problems with questionnaires and social surveys has led to the emergence of what has here been called *participatory local appraisal*. The intent is to obtain data that will be useful to planners and managers, but to do so in a less extractive or exploitive way. Indeed, participatory local appraisal has a central goal to help empower local people, and to assist them to help define, understand and resolve their own problems. The techniques associated with participatory local appraisal should not be considered to be applicable only in developing countries. Their techniques offer considerable potential for work in developed countries as well.

Perhaps one of the most impressive examples of incorporating local knowledge systems into resource and environmental management has occurred through *co-management* initiatives. Furthermore, Mitlin and Thompson (1994: 11) have suggested that for two reasons co-management is an excellent way to link local level development planning with higher level planning structures through applying participatory approaches: (1) cooperative arrangements give local groups space to manoeuvre, regularize land rights, fund local initiatives and employ participatory approaches to understand local needs and priorities better; and (2) community development groups can use participatory approaches as a means to create local awareness and to mobilize local resources for community action. When co-management exercises have worked well, both sides have explicitly accepted the value of local knowledge and have incorporated it into management strategies. Furthermore, such initiatives also have been most successful when some genuine power or authority has been allocated to the local users or managers. Co-management offers one of the exciting opportunities for creating effective partnerships for resource and environmental management in the future.

References and further reading

Agrawal A 1995 Dismantling the divide between indigenous and scientific knowledge. *Development and Change* 26: 413–39

Agrawal A and C C Gibson 1999 Enchantment and disenchantment: the role of community in resource conservation. *World Development* 27: 629–49

Akis S, N Peristianis and J Warner 1996 Residents' attitudes to tourism development: the case of Cyprus. *Tourism Management* 17: 481–94

Amalric F 1999 Natural resources, governance and social justice. *Development* 42(2): 5–12

Appiah-Opoku S 1999 Indigenous economic institutions and ecological knowledge: a Ghanaian case study. *The Environmentalist* 19: 217–27

Bebbington A 1999 Capitals and capabilities: frameworks for analyzing peasant viability, rural livelihoods and poverty. *World Development* 27: 2021–44

Berkes F (ed.) 1989 *Common Property Resources: Ecology and Community-based Sustainable Development*. London, Belhaven

Berkes F 1999 *Sacred Ecology: Traditional Knowledge and Resource Management*. Philadelphia, Taylor and Francis

Berkes F, P George and R J Preston 1991 Co-management: the evolution in theory and practice of the joint administration of living resources. *Alternatives* 18: 12–18

Bignal E M and D I McCracken 2000 The nature conservation value of European traditional farming systems. *Environmental Reviews* 8: 149–71

Blackburn J with J Holland (eds) 1998 *Who Changes? Institutionalizing Participation in Development*. London, Intermediate Technology Publications

Bocco G and V M Toledo 1997 Integrating peasant knowledge and geographic information systems: a spatial approach to sustainable agriculture. *Indigenous Knowledge and Development Monitor* 5(2) [online]. Available on the Internet: http://www.nufficcs.nl.circan/ikdm/

Brokensha D W, D M Warren and O Werner 1980 *Indigenous Knowledge Systems and Development*. Lanham, Maryland, University Press of America

Brosius J P, A L Tsing and C Zerner 1998 Representing communities: histories and politics of community-based natural resource management. *Society and Natural Resources* 11: 157–68

Brouwe R 1995 *Baldios* and common property resource management in Portugal. *Unasylva* 46: 37–43

Brown D N and R S Pomeroy 1999 Co-management of Caribbean Community (CARICOM) fisheries. *Marine Policy* 23: 549–70

Burgess J, J Clark and C Harrison 2000 Culture, communication, and the information problem in contingent valuation surveys: a case study of a Wildlife Enhancement Scheme. *Environment and Planning C* 18: 505–24

Caldwell W, S Bowers, H Greening, L Norman and O Williams 1999 A community development approach to environmental management. *Environments* 27: 63–78

Cameron S D 1990 Net losses: the sorry state of our Atlantic fishery. *Canadian Geographic* 110(2): 28–37

Chambers R 1994a The origins and practice of participatory rural appraisal. *World Development* 22: 953–69

Chambers R 1994b Participatory rural appraisal (PRA): analysis of experience. *World Development* 22: 1253–68

Chambers R 1994c Participatory rural appraisal (PRA): challenges, potentials and paradism. *World Development* 22: 1437–54

Cheptstow-Lusty A and M Winfield 2000 Inca agroforestry: lessons from the past. *Ambio* 29: 322–8

Cizek P 1993 Guardians of Manomin: aboriginal self-management of wild rice harvesting. *Alternatives* 19: 29–32

Clark A L and J C Clark 1999 The new reality of mineral development: social and cultural issues in Asia and Pacific nations. *Resources Policy* 25: 189–96

Cleaver F 1999 Paradoxes of participation: questioning participatory approaches to development. *Journal of International Development* 11: 597–612

Cleveland D A, D Soleri and S E Smith 1994 Do folk crop varieties have a role in sustainable agriculture? *BioScience* 44: 740–51

Colchester M 1994 Sustaining the forests: the community-based approach in South and South-East Asia. *Development and Change* 25: 69–100

Collier G A, D C Mountjoy and R B Nigh 1994 Peasant agriculture and global change. *BioScience* 44: 398–407

Conservation of Arctic Flora and Fauna 1997 *Recommendations on the Integration of Two Ways of Knowing: Traditional Indigenous Knowledge and Scientific Knowledge* [online]. Available on the Internet: http://www.grida.no/caff/inuvTEK.htm

Conway D, K Bhattari and N R Shrestha 2000 Population–environment relations at the forested frontier of Nepal: Tharu and Pahari survival strategies in Bardiya. *Applied Geography* 20: 221–42

Conway G R and J A McCracken 1990 Rapid rural appraisal and agroecosystem analysis. In M A Altieri and S B Hect (eds) *Agroecology and Small Farm Development*. Boca Raton, CRC Press, pp. 221–35

Costa-Neto E M 2000 Sustainable development and traditional knowledge: a case study in a Brazilian artisanal fishermen's community. *Sustainable Development* 8: 89–95

Cream K 1999 Centralised and community-based fisheries management strategies: case studies from two fisheries dependent archipelagos. *Marine Policy* 23: 243–57

Critchley W R S, C Reij and T J Willcocks 1994 Indigenous soil and water conservation: a review of the state of knowledge and prospects for building on traditions. *Land Degradation and Rehabilitation* 5: 293–314

Dove M R 1986 Peasant versus government perception and use of the environment: a case-study of Banjarese ecology and river basin development in South Kalimantan. *Journal of Southeast Asian Studies* 17: 113–36

Dudley E 1993 *The Critical Villager: Beyond Community Participation*. London, Routledge

Duffield C, J S Gardner, F Berkes and R B Singh 1998 Local knowledge in the assessment of resource sustainability: case studies in Himachal Pradesh, India, and British Columbia, Canada. *Mountain Research and Development* 18: 35–49

Durning A T 1992 *Guardian of the Land: Indigenous Peoples and the Health of the Earth*. Worldwatch Paper 112, Washington, DC, Worldwatch Institute

Dwivedi R 1999 Displacement, risks and resistance: local perceptions and actions in the Sardar Sarovar. *Development and Change* 30: 43–78

Franklin U 1990 *The Real World of Technology*. Toronto, CBC Enterprises

Fraser I and T Chisholm 2000 Conservation or cultural heritage? Cattle grazing in the Victoria Alpine National Park. *Ecological Economics* 33: 63–75

Gadgil M and F Berkes 1991 Traditional resource management systems. *Resource Management and Optimization* 8: 127–41

Gadgil M, F Berkes and C Folke 1993 Indigenous knowledge for biodiversity conservation. *Ambio* 22: 151–6

Gibson C C and T Koontz 1998 When "community" is not enough: institutions and values in community-based forest management in Southern Indiana. *Human Ecology* 26: 621–47

Gill N 1994 The cultural politics of resource management: the case of bushfires in a conservation reserve. *Australian Geographical Studies* 32: 224–40

Gillingham M E 1999 Gaining access to water: formal and working rules of indigenous irrigation management on Mount Kilimanjaro, Tanzania. *Natural Resources Journal* 39: 419–41

Grenier L 1998 *Working with Indigenous Knowledge: A Guide for Researchers*. Ottawa, International Development Research Centre

Grundy I, J Turpie, P Jagger, E Witkowski, I Guambe, D Semwayo and A Solomon 2000 Implications of co-management for benefits from natural resources for rural households in north-western Zimbabwe. *Ecological Economics* 33: 369–81

Guevara J R 1996 Learning through participatory Action research for community ecotourism planning. *Convergence* 29: 24–39

Hannah L, B Rakotosamimanana, J Ganzhorn, R A Mittermeier, S Olivieri, L Iyer, S Rajaobelina, J Hough, F Andreamialisoa, I Bowles and G Tilkin 1998 Participatory planning, scientific priorities and landscape conservation in Madagascar. *Environmental Conservation* 25: 30–6

Hartup B K 1994 Community conservation in Belize: demography, resource use and attitudes of participating landowners. *Biological Conservation* 69: 235–41

Holland J with J Blackburn (eds) 1998 *Whose Voice? Participatory Research and Policy Change*. London, Intermediate Technology Publications

Horowitz L S 1998 Integrating indigenous resource management with wildlife conservation: a case study of Batang Ai National Park, Sarawak, Malaysia. *Human Ecology* 26: 371–403

Hviding E and G B K Baines 1994 Community-based fisheries management, tradition and the challenges of development in Marovo, Solomon Islands. *Development and Change* 25: 13–39

Ite U and W Adams 2000 Expectations, impacts and attitudes: conservation and development in Cross River National Park, Nigeria. *Journal of International Development* 12: 325–42

Kaplan I M 1999 Suspicion, growth and co-management in the commercial fishing industry: the financial settlers of New Bedford. *Marine Policy* 23: 227–41

Khan N A 1998 Interviews with the Sahibs: bureaucratic constraints on community forestry programmes in Bangladesh. *Journal of World Forest Resource Management* 9: 73–93

Khan N A and S K Khisa 2000 Sustainable land management with rubber-based agroforestry: a Bangladeshi example of uplands community development. *Sustainable Development* 8: 1–10

Kimmerer R W 2000 Native knowledge for native ecosystems. *Journal of Forestry* 98(8): 4–9

Kleemeier E 2000 The impact of participation on sustainability: an analysis of the Malawi rural piped scheme program. *World Development* 28: 929–44

Klooster D 2000 Institutional choice, community and struggle: a case study of forest co-management in Mexico. *World Development* 28: 1–20

Klooster D and O Masera 2000 Community forest management in Mexico: carbon mitigation and biodiversity conservation through rural development. *Global Environmental Change* 10: 259–72

Lane M B 1999 Regional Forest Agreements: resolving resource conflicts or managing resource politics? *Australian Geographical Studies* 37: 142–53

Leach M, R Mearns and I Scoones 1999 Environmental entitlements: dynamics and institutions in community-based natural resource management. *World Development* 27: 225–47

Li T M 2000 Articulating indigenous identity in Indonesia: resource politics and the tribal slot. *Comparative Studies in Society and History* 42: 149–79

Litke S and J C Day 1998 Building local capacity for stewardship and sustainability: the role of community-based watershed assessment in Chilliwack, British Columbia. *Environments* 25: 91–109

Loomis T M 2000 Indigenous populations and sustainable development: building on indigenous approaches to holistic, self-determined development. *World Development* 28: 893–910

Mackenzie A K 1993 *Exploring Rapid Rural Appraisal for Community-Based Watershed Planning: The Bila River Watershed, South Sulawesi, Indonesia*, unpublished Master's thesis, Guelph, University of Guelph, Faculty of Graduate Studies

Martin P and S Lockie 1993 Environmental information for total catchment management: incorporating local knowledge. *Australian Geography* 24: 75–85

Matowanyika J Z Z 1991 In pursuit of proper contexts for sustainability in rural Africa. *The Environmentalist* 11: 85–94

McKean M A 1992 Success on the commons: a comparative examination of institutions for common property resource management. *Journal of Theoretical Politics* 4: 247–81

Mehta J N and S R Kellert 1998 Local attitudes toward community-based conservation policy and programmes in Nepal: a case study in the Makalu–Barun Conservation Area. *Environmental Conservation* 25: 320–33

Mitlin D and J Thompson (eds) 1994 *Special Issue on Participatory Tools and Methods in Urban Areas*. RRA Notes Number 21, London, International Institute for Environment and Development, Sustainable Agriculture Programme and Human Settlements Programme

Mohan G 1999 Not so distant, not so strange: the personal and the political in participatory research. *Ethics, Place and Environment* 2: 40–54

Moote M A, M P McClaren and D K Chickering 1997 Theory in practice: applying participatory democracy to public land planning. *Environmental Management* 21: 877–89

Morin-Labatut G and S Akhatoar 1992 Traditional environmental knowledge: a resource to manage and share. *Development* 4: 24–30

Mosse D 1994 Authority, gender and knowledge: theoretical reflections on the practice of participatory rural appraisal. *Development and Change* 25: 497–526

Motteuz M, E Nel, K Rowntree and T Binns 1999 Exploring community environmental knowledge through particpatory methods in the Kat River valley, South Africa. *Community Development Journal* 34: 227–31

Neis B 1992 Fishers' ecological knowledge and stock assessment in Newfoundland. *Newfoundland Studies* 8: 155–78

Nickerson-Tietze D J 2000 Community-based management for sustainable fisheries resources in Phang-nga Bay, Thailand. *Coastal Management* 28: 65–74

Noble B F 2000 Institutional criteria for co-management. *Marine Policy* 24: 69–77

Notzke C 1995 A new perspective in aboriginal natural resource management: co-management. *Geoforum* 26: 187–209

O'Connell-Rodwell C E, T Rodwell, M Rice and L A Hart 2000 Living with the modern conservation paradigm: can agricultural communities co-exist with elephants? A five-year case study in East Caprivi, Namibia. *Biological Conservation* 93: 381–91

Ohmagari K and F Berkes 1997 Transmission of indigenous knowledge and bush skills among the Western James Bay Cree women of Subarctic Canada. *Human Ecology* 25: 197–222

Ostrom E 1987 Institutional arrangements for resolving the commons dilemma. In B J McKay and J M Acheson (eds) *The Question of the Commons: The Culture and Ecology of Communal Resources*. Tucson, University of Arizona Press, pp. 250–65

Ostrom E 1990 *Governing the Commons: The Evolution of Institutions for Collective Action*. Cambridge, Cambridge University Press

Ostrom E 1992 The rudiments of a theory of the origins, survival and performance of common-property institutions. In D W Bromley (ed.) *Making the Commons Work: Theory, Practice and Policy*. San Francisco, Institute for Contemporary Studies, pp. 293–312

Ostrom E 1999 *Self-governance and Forest Resources*. Center for International Forestry Research, Occasional Paper No. 20, Bogor, Indonesia, CIFOR

Ostrom E, R Gardner and J Walker 1994 *Rules, Games, and Common-pool Resources*. Ann Arbor, University of Michigan Press

Ostrom E, J Walker and R Gardner 1993 Covenants with and without a sword: self-governance is possible. In T L Anderson and R T Simmons (eds) *The Political Economy of Customs and Culture: Informal Solutions to the Commons Problem*. Lanham, Maryland, Rowma and Littlefield, pp. 127–56

Osunade M A A 1994a Community environmental knowledge and land resource surveys in Swaziland. *Singapore Journal of Tropical Geography* 15: 157–70

Osunade M A A 1994b Indigenous climate knowledge and agricultural practice in southwestern Nigeria. *Malaysian Journal of Tropical Geography* 25: 21–8

Pinkerton E 1989 Attaining better fisheries management through co-management prospects, problems and propositions. In E Pinkerton (ed.) *Co-operative Management of Local Fisheries: New Direction in Improved Management and Community Development*. Vancouver, University of British Columbia Press, pp. 3–33

Pinkerton E 1993 Co-management efforts as social movements: the Tin Wis Coalition and the drive for forest practices legislation in British Columbia. *Alternatives* 19: 33–8

Pinkerton E 1996 The contribution of watershed-based multi-party co-management agreements to dispute resolution: the Skeena Watershed Committee. *Environments* 23: 51–68

Pomeroy R S and F Berkes 1997 Two to tango: the role of government in fisheries co-management. *Marine Policy* 21: 465–80

Power J, J McKenna, M J MacLeod, A J G Cooper and G Connie 2000 Developing integrated participatory management strategies for Atlantic dune systems in County Donegal, northwest Ireland. *Ambio* 29: 143–9

Quiggin J 1993 Common property, equality and development. *World Development* 21: 1123–38

Rai S C, E Sharma and R C Sundriyal 1994 Conservation in the Sikkim Himalaya: traditional knowledge and land-use of the Mamlay watershed. *Environmental Conservation* 21: 30–34, 56

Reed M G 1995 Cooperative management of environmental resources: a case study from Northern Ontario, Canada. *Economic Geography* 71: 132–49

Rhoades R 2000 Integrating local voices and visions into the global mountain agenda. *Mountain Research and Development* 20: 4–9

Rodon T 1998 Co-management and self-determination in Nunavut. *Polar Geography* 22: 199–235

Rodríguez-Navarro G E 2000 Indigenous knowledge as an innovative contribution to the sustainable development of the Sierra Nevada of Santa Marta, Colombia. The elder brothers, guardians of the "Heart of the World". *Ambio* 29: 455–8

Saleh M A E 2000 Environmental planning and management for the Assarawat highland region of south-western Saudi Arabia: the traditional versus the professional approach. *The Environmentalist* 20: 123–39

Sekhar N U 2000 Decentralized natural resource management: from state to co-management in India. *Journal of Environmental Planning and Management* 43: 123–38

Singleton S 2000 Co-operation or capture? The paradox of co-management and community participation in natural resource management and environmental policy-making. *Environmental Politics* 9: 1–21

Singleton S and M Taylor 1992 Common property, collective action and community. *Journal of Theoretical Politics* 4: 309–24

Songorva A N 1999 Community-based wildlife management (CWM) in Tanzania: are the communities interested? *World Development* 27: 2061–79

Songorva A N, T Buhrs and K F D Hughey 2000 Community-based wildlife management in Africa: a critical assessment of the literature. *Natural Resources Journal* 40: 603–43

Stevens T H, D Dennis, D Kittredge and M Rickenbach 1999 Attitudes and preferences toward co-operative agreements for management of private forestlands in the North-eastern United States. *Journal of Environmental Management* 55: 81–90

Taconi L 1997 Property rights and participatory biodiversity conservation: lessons from Malikula Island, Vanuatu. *Land Use Policy* 14: 151–61

Taiepa T, P Lyver, P Horsley, J Davis, M Bragg and H Moller 1997 Co-management of New Zealand's conservation estate by Maori an Pek eha: a review. *Environmental Conservation* 24: 236–50

Takasaki Y, B L Barham and O T Coomes 2000 Rapid rural appraisal in humid tropical forests: an asset position-based approach and validation method for wealth assessment among forest peasant households. *World Development* 28: 1961–77

Tapela B N and P H Omara-Ojungu 1999 Towards bridging the gap between wildlife conservation and rural development in post-apartheid South Africa: the case of the Makuleke community and the Krueger National Park. *South African Geographical Journal* 81: 148–55

Thomson J T and C Coulibaly 1995 Common property forest management systems in Mali: resistance and vitality under pressure. *Unasylva* 46: 16–22

Thorborn C C 2000 Changing customary marine resource management practice and institutions: the case of Sasi Lola in the Kei Islands, Indonesia. *World Development* 28: 1461–79

Tosun C 2000 Limits to community participation in the tourism development process in developing countries. *Tourism Management* 21: 613–33

Twyman C 1998 Rethinking community resource management: managing resources or managing people in western Botswana? *Third World Quarterly* 19: 745–70

Twyman C 2000 Participatory conservation? Community-based natural resource management in Botswana. *Geographical Journal* 166: 323–35

Ulluwishewa R 1995 Traditional practices of inland fishery resources management in the dry zone of Sri Lanka: implications for sustainability. *Environmental Conservation* 22: 127–32

United Nations, Food and Agriculture Organization 1989 *Community Forestry: Participatory Assessment, Monitoring and Evaluation.* Rome, Food and Agriculture Organization

University of Khon Kaen 1987 *Proceedings of the 1985 International Conference on Rapid Rural Appraisal.* Rural Systems Research and Farming Systems Research Projects, Khon Kaen, Thailand, University of Khon Kaen

Unruh J D 1995 The relationship between indigenous pastoralist resource tenure and state tenure in Somalia. *GeoJournal* 36: 19–26

Usher P J 1987 Indigenous management systems and the conservation of wildlife in the Canadian north. *Alternatives* 14: 3–9

Vadya A P 1983 Progressive contextualization: methods for research in human ecology. *Human Ecology* 11: 265–81

Warner G 1997 Participatory management, popular knowledge, and community empowerment: the case of sea urchin harvesting in the Vieux-Fort area of St Lucia. *Human Ecology* 25: 29–46

Wiles A, J McEwen and M H Sadar 1999 Use of traditional ecological knowledge in environmental assessment of uranium mining in Saskatchewan. *Impact Assessment and Project Appraisal* 17: 107–14

Wilson P I 2000 Wolves, politics, and the Nez Perce: wolf recovery in Central Idaho and the role of native tribes. *Natural Resources Journal* 39: 543–64

Wollenberg E 1998 A conceptual framework and typology for explaining the outcomes of local forest management. *Journal of World Forest Resource Management* 9: 1–35

Yakuba J M 1994 Integration of indigenous thought and practice with science and technology: a case study of Ghana. *International Journal of Scientific Education* 16: 343–60

Zimmerer K S 1994 Local soil knowledge: answering basic questions in highland Bolivia. *Journal of Soil and Water Conservation* 49: 29–34

Chapter 10

Gender and Development

10.1 Introduction

The important role of women in economic and social development in both their communities and countries, relative to men, is increasingly being recognized. Numerous formal "events" have helped to enhance recognition of this role. For example, the United Nations Decade for Women from 1975 to 1985 drew global attention to women, especially the trying conditions under which many of the poorest lived. The United Nations Conference on Environment and Development (UNCED) in Rio de Janeiro during 1992, which was discussed in Chapter 4, presented a global action plan to institutionalize the role of women in environment and development. Some of the key objectives and actions related to women and environment in *Agenda 21*, one of the main documents to emerge from UNCED, are shown in Boxes 10.2 and 10.3.

The United Nations Fourth World Conference on Women, held in Beijing, China, during September 1995, further highlighted important issues if the gap between men and women is to be narrowed. This conference followed earlier major meetings in Mexico City (1975), Copenhagen (1980) and Nairobi (1985). The Beijing Conference was attended by representatives from 189 countries, and involved an official conference held in Beijing as well as a parallel conference for non-governmental organizations in nearby Huariou. Together the two conferences attracted some 50,000 people.

Box 10.1

Studies show that because of their responsibilities for securing food, fuel, and water – and the labor burdens imposed on them when the resources needed to produce these goods become scarce – women tend to have a greater interest in preserving and conserving croplands, forests and other natural resources for perpetual use, whereas men are more often concerned with converting these resources into cash. Development programs that vest control over natural resources solely within the hands of men, or profit-making enterprises in general, are in effect explicitly supporting short term consumption at the expense of long term sustainability.

Source: Jacobson, 1992: 13.

Box 10.2 Selected objectives proposed for national governments in Chapter 24 (Global Action for Women Towards Sustainable and Equitable Development) in Agenda 21

(1) to implement the Nairobi Forward-looking Strategies for the Advancement of Women [1985], particularly with regard to women's participation in national ecosystem management and control of environmental degradation;

(2) to increase the proportion of women decision makers, planners, technical advisers, managers and extension workers in environment and development fields;

(3) to consider developing and issuing by the year 2000 a strategy of changes necessary to eliminate constitutional, legal, administrative, cultural, behavioural, social and economic obstacles to women's full participation in sustainable development and in public life;

(4) to establish by the year 1995 mechanisms at the national, regional and international levels to assess the implementation and impact of development and environment policies and programmes on women and to ensure their contributions and benefits.

Source: United Nations Conference on Environment and Development 1992 Agenda 21. New York, United Nations.

The spate of conferences, formal declarations and intentions over the past 25 years suggest that the role of women is changing, and its importance is receiving growing recognition. Yet, deeds do not always match words, nor does action necessarily follow from good intentions. Furthermore, many people are arguing that the issue is not one that can be simplified into a question of women versus men. Instead, they argue that the more fundamental issue is that of *gender*. In this chapter, attention will be given to the differences between women's and gender issues, their implications for resource and environmental management, and some experiences in which attempts have been made to institute gender-focused change.

10.2 Multiple roles for women

For gender planning, one of the greatest challenges to be overcome is the frequent assumption about gender-based divisions of work within and outside the household. Especially in Third World countries, women's work usually involves three components. *Reproductive* work relates to child bearing and raising, as well as nurturing all family members to ensure their health and well being. Such reproductive work extends beyond biological reproduction to include those domestic tasks necessary to maintain and reproduce the labour force for a society. While child bearing is a biological function unique to women, there is no particular reason or logic as to why child rearing, and nurturing and caring for the family, should be women's work.

Box 10.3 Aspects requiring urgent action, as stipulated in Chapter 24 of Agenda 21

Countries should take urgent measures to avert the ongoing rapid environmental and economic degradation in developing countries that generally affects the lives of women and children in rural areas suffering drought, desertification and defor-estation, armed hostilities, natural disasters, toxic waste and the aftermath of the use of unsuitable agro-chemical products.

In order to reach these goals, women should be fully involved in decision-making and in the implementation of sustainable development activities.

Research, data collection and dissemination as stipulated in Chapter 24 of Agenda 21

Countries should develop gender-sensitive databases, information systems and participatory action-oriented research and policy analyses with the collaboration of academic institutions and local women researchers on the following:

(1) knowledge and experience on the part of women of the management and conservation of natural resources for incorporation in the databases and information systems for sustainable development;

(2) the impact on women of environmental degradation, particularly drought, desertification, toxic chemicals and armed hostilities; and

(3) the integration of the value of unpaid work, including work currently design-ated "domestic", in resource accounting mechanisms in order to better rep-resent the true value of the contribution of women to the economy, . . . ;

(4) measures to develop and include environmental, social and gender impact analyses as an essential step in the development and monitoring of pro-grammes and policies.

Source: United Nations Conference on Environment and Development 1992, Agenda 21, New York, United Nations.

Productive work involves activity by both women and men, for payment in cash or in kind. Such work can be "market-based production" which results in earning of money. Or, it can be subsistence or home production which gener-ates an in-kind rather than a monetary value (Figure 10.1). Much of the product-ive work of women, especially in rural areas, generates in kind returns which are essential for the well being of their families. However, because there is no exchange value, such activity normally is invisible in the regional or national economy, and therefore does not get the same recognition as market-based activity. In many societies, men dominate market-based economic activity, whereas women feature more frequently in non-market-based activity. As with most aspects of *reproductive* work, there is no reason why there should be such a gender division of labour. However, those involved in market-based activity tend to have more power because of their income-earning capacity.

Community managing involves time allocated to participating in activities within the local community to help further the welfare of its members. In

Figure 10.1 Women often have "productive" work which does not generate income, such as this woman drawing water for household use, in Jin Ma, Chengdu, China (Bruce Mitchell)

many ways, this kind of activity is an extension of the *reproductive* role. The focus here normally is to ensure both provision and maintenance of facilities for collective needs, such as water, health care and education. The *community managing* work usually is voluntary, and is done in "free time" after reproductive and productive tasks have been attended to. In contrast, men tend to allocate their time to *community politics*, which involves participation at the formal political level. Their community work often is paid, either directly or indirectly, by earning wages or by achieving enhanced status and power. There is no good reason why women should concentrate upon *community managing*, and men upon *community politics*, but traditional gender divisions of labour often result in this split.

The implications of the multiple roles of women are significant regarding issues of empowerment, social justice and equity. First, the triple role for women in many Third World countries means that they are the first to begin working during the day, and often the last to finish at the end of the day. As a result, little time is available for self-improvement, or to pursue interests of their own. Second, much of their *productive* and *community managing* activities are invisible in any economic accounting, so their contribution to the household, community, region and country is often undervalued by family members and political leaders. Third, the reduced opportunity to earn monetary income contributes to reduced overall status and power in the household and community. Fourth, the lack of involvement in *community politics* reinforces a gender bias in many decisions, and helps contribute to maintenance of a status quo in which the role of and opportunities for women are significantly less than for male counterparts. All of these implications have fundamental ramifications for resource and environmental management, as will be illustrated later in Sections 10.4, 10.5, and 10.6, which consider initiatives about forestry, water and agriculture, respectively.

10.3 From "women in development" to "gender and development"

10.3.1 Women in development

Moser (1993) has explained that the concept of *women in development* appeared during the early 1970s in the USA through a Women's Committee of the Washington DC chapter of the Society for International Development. The term was then adopted by the US Agency for International Development (USAID) in its development activities. *Women in Development*, or WID, was based on the belief that women were an untapped and under-used resource who could and should contribute directly to economic development.

Box 10.4 Difference between biological sex and gender

Gender differs from biological sex in important ways. Our biological sex is given; we are born either male or female. But the way in which we become masculine or feminine is a combination of these basic biological building blocks and the interpretation of our biology by our culture. Every society has different "scripts" for its members to follow as they learn to act out their feminine or masculine role, much as every society has its own language. . . .

Gender is a set of roles which, like costumes or masks in the theatre, communicate to other people that we are feminine or masculine. This set of particular behaviours – which embrace our appearance, dress, attitudes, personalities, work both within and outside the household, sexuality, family, commitments and so on – together make up our "gender roles".

Source: Mosse, 1993: 2.

As Mosse (1993: 158) noted, however, WID focused upon initiatives such as development or transfer of better (and hopefully locally appropriate) technologies to reduce workloads for women. In that manner, WID emphasized the *productive* role of women in the economy, particularly their capacity for generating income, but neglected their *reproductive* and *community managing* roles. Nevertheless, WID was a first step to begin sensitizing people about a need to alter their thinking about the role of women.

10.3.2 Gender and development

The shift in attention from women to gender occurred as a result of unhappiness that the difficulties of women were being interpreted with reference to their *sex* or biological differences from men, rather than with regard to their *gender*, of their social relationships with men. It was the gender relationship through which women systematically had been subordinated. A gender and development (GAD) perspective emphasizes that while sex is biologically determined, gender is socially or culturally determined. As a result, while a person's sex is difficult to change, gender roles can be altered if societal values can be modified. Thus, gender roles need not be static. They can vary among cultures at any given time, and can vary with a culture over time. The different emphases between WID and GAD are outlined in Box 10.5.

GAD is also preferred over WID because it does not treat women as a homogenous group. The notion that women have a "position" in a society implies some universal slot for women. However, it is not sensible to consider women as one group with common values and interests. Rich and poor women may have less in common than poor women and poor men. In white-dominated societies, a black woman may believe she has more in common with black men than with white women. In addition to gender, therefore, the status of and opportunities for women will also be influenced by their financial, ethnic, class and other characteristics.

Box 10.5 Differences between WID and GAD

The WID approach, despite its change in focus from one of equity to one of efficiency, is based on the underlying rationale that development processes would proceed much better if women were fully incorporated into them (instead of being left to use their time "unproductively"). It focuses mainly on women in isolation, promoting measures such as access to credit and employment as the means by which women can be better integrated into the development process. In contrast, the GAD approach maintains that to focus on women in isolation is to ignore the real problem, which remains their subordinate status to men. In insisting that women cannot be viewed in isolation, it emphasizes a focus on gender relations, when designing measures to "help" women in the development process.

Source: Moser, 1993: 3.

Gender is, therefore, very much focused on empowering women in their relationships with men. It emphasizes a bottom–up rather than a top–down approach to management. It seeks to facilitate women becoming more self-reliant, through changing and transforming practices and structures – such as labour codes, civil codes, religious and cultural customs, and property rights – that have been disadvantageous to them. Not surprisingly, some government, religious and cultural leaders have been unsettled by GAD since it often challenges basic values and traditional customs of a society. For example, at the UN Fourth World Conference on Women in Beijing (1995), representatives from the Vatican and from some conservative Islamic nations opposed recommendations regarding various sexual rights and freedoms for women.

In the following sections, some specific experiences related to resource and environmental management are examined. The examples are taken from forestry and water.

10.4 Gender and forestry

10.4.1 Chipko movement, India

"Perhaps the most famous example [of a women's grassroots movement] of this resistance is the Chipko – or tree-hugging – movement of the Indian Himalayas" (Jacobson, 1992: 14). During the 1970s, the Chipko movement emerged when local women demonstrated to protect stands of forests from commercial harvesting which had been endorsed by the government of India. Forest ecosystems play many roles. They serve to stabilize soil conditions by retarding runoff, and thereby reducing soil erosion. Wetlands also act as sinks to collect runoff after snowmelt or rainfall, and then release the water more slowly than would occur if they or other vegetation were not there to slow down the runoff. Forests also provide fuel wood, and fodder for animals, as well as traditional medicines. On the other hand, forests also offer commercial products such as timber and resin.

The deforestation of the Himalayas, primarily driven by the interests of governments and commercial forestry companies interested in the marketable products from forest ecosystems, had been a major contributor to landslides and flooding, and associated erosion of soil. Growing recognition of this problem led to a protest by women who were concerned about the over-emphasis on the commercial uses of forests, increasing environmental degradation and the loss of non-commercial functions (source of domestic fuel wood, fodder, traditional herbal medicines, berries and other food).

The literal interpretation of the word "Chipko" is "to embrace", and this was the tactic used by the women, first throughout villages in the Garhwal Himalaya in the north during the 1970s, and later in states such as Karnataka in the extreme south of India. The women successfully stopped the cutting down of trees in the forests by using their own bodies to block the loggers (Box 10.6). Once the cutting was stopped, the women organized themselves to protect the

---

Box 10.6 Gender differences illustrated by the Chipko movement

In some of the conflicts . . . , the different gender issues of women and men were painfully highlighted, when women embraced the trees that were to be cut down by their own husbands, employed by the forest contractors. The demands of the women, for a supply of fuel, fodder and water, conflicted absolutely with the demands of the men for a cash income.

Source: Mosse, 1993: 147.

---

forests on an ongoing basis. This usually involved a group of women taking turns to watch a forest during the day, and even at night, to stop goats or cattle from grazing, and people from cutting firewood or collecting fodder. Once the forests began to recover, traditional uses were re-started. As Jacobson (1992: 14) remarked, Chipko evolved into a full-fledged ecological movement and became an outstanding example of female empowerment. The Chipko movement often has been used to point out what women in the Third World can accomplish when they organize and challenge traditional gender-defined patterns of resource use. However, such initiatives are not always so successful, as shown next.

10.4.2 Community forestry management, India

Sarin (1995) has suggested that decentralized and "participatory" forest management has long been considered an appropriate long-term solution to deforestation and environmental degradation in India. Decentralization and participation can be achieved in numerous ways, but since the late 1970s emphasis has been on two approaches at the local level. Each has had implications for the role of women relative to men.

- **Joint forest management (JFM).** Several state forestry departments have introduced what are called joint forest management initiatives as ways to involve local people, especially women and tribals who live in villages dependent on forests. JFM arrangements involve a negotiated partnership between local institutions (LIs) of forest users, and the state departments responsible for forests. The basic principle is sharing of the duties for protecting the forests, and of the income from forest products. The largest and most well-established JFM programme is in the state of West Bengal, with more than 2,500 forest protection committees (FPCs) involved in regenerating 300,000 hectares of forest land.
- **Village initiatives.** In areas in which the state forestry department has not taken the initiative, thousands of villages have organized themselves to regenerate adjacent forests in the states of Orissa, Bihar, Madhya Pradesh, Andhra Pradesh, Rajasthan, Gujarat and Himachal Pradesh. The villagers have developed rules and ways to enforce them, have allocated responsibilities among their groups, and have stopped outsiders from using forests.

Box 10.7 Role of women in community-based forestry

Women remain as marginalised at the community level as in national policies because of their disadvantaged position defined by patriarchal gender relations, which traditionally exclude women from the arena of political participation, even at the community level.

Source: Sarin, 1995: 27.

The last task often has required men to go into the forest during the night to confront outside cutters. Such confrontations sometimes have resulted in physical violence.

Implications for women

Women are often more dependent on the forests for subsistence needs (fuel, fodder, food), and hence are often affected more than men when degradation occurs. However, Sarin (1995) has suggested that frequently the women's needs have received no more consideration under community-oriented initiatives than under the unilateral programmes of forestry departments (Box 10.7).

Much of the marginalization of women can be related to traditional gender roles related to forest responsibilities and uses. Men are normally responsible for cutting wood for timber to be used in house construction and for agricultural implements. Such needs occur relatively infrequently for a household. In contrast, women normally have responsibility for collecting daily fuel wood, fodder and water (Figure 10.2). While these roles are not absolutely fixed, and certainly vary depending upon caste, tribe and village, role differentiation is fairly common, and is especially so in the villages most dependent on forests. The joint programmes and locally initiated programmes have been contolled by men who tend to be dominant in local politics, and hence the programmes usually emphasize extraction of timber for house and implement building. As Sarin (1995: 28) noted, "women continue to lack a voice in community decision-making fora because of the unquestioned assumption that their household men will automatically take care of their interests". Unfortunately, this assumption is not often reflected in what happens, with negative consequences for the women.

To illustrate, in a number of tribal villages in the state of Gujarat, Sarin (1995) explained that the male leaders of numerous tribal villages organized the men to protect some seriously degraded forests. Rigorous rules were established, which were more restrictive than a state forestry department would have considered. They were also enforced. Entry into designated forests with any tools was completely banned, and people were appointed as watchmen to ensure compliance. Each village household contributed a portion of its grain harvest to pay the watchmen for their time. In the early years, even grazing was disallowed. No women were present when the objectives and rules were established.

Figure 10.2 Women carrying forest fodder back to their homes, in the Kullu Valley, Himachal Pradesh, India (John Sinclair)

The purpose of these restrictions was to allow the regeneration of a teak forest from which timber could be cut for house construction and agricultural implements. No consideration was given to the implications for the women's tasks of collecting firewood or green fodder. Women from higher income families were able to purchase alternatives for firewood and fodder, usually by buying agricultural residuals from landless or marginal households. Poorer households experienced humiliation when their female members were stopped from collecting wood and fodder in the restricted forests by the village watchmen. Arguments often ensued, and the confrontations between the men and women became so unpleasant that some men began to refuse to serve as watchmen. The resolution involved the women making the greatest concessions, by agreeing to obtain substitutes, either by using poorer quality cooking fuels (such as dung droppings or weeds), or by going to more distant but still open forests to collect wood (Box 10.8).

This experience illustrates how preoccupation with men's *productive* roles dominated decisions, and generated more effort for women in their *reproductive* and *community managing* roles. The implications are several, and include the necessity to recognize: (1) the different needs of women and men when developing resource-use policies and practices, (2) the desirability of including both men and women in decision making about resource allocation, and (3) the heterogeneity of women in a community. As Sarin (1995: 29) remarked, "the 'community' is not a homogenous, faceless entity but encompasses a diversity of needs and interests with gender differences being a major variable".

Box 10.8 Women's workload increases

For the women this meant an increase in the labour and time required for fetching cooking fuel as well as a worsening of their quality of life due to switching to poorer quality, smokier fuels. In terms of the sustainability of the forest protection efforts, the new arrangement meant shifting the pressure of unsustainable extraction of firewood to other, more distant, areas so that those closer by could be regenerated as timber reserves.

Source: Sarin 1995: 28.

10.5 Gender and water

Lynch (1991: 37) has observed that gender roles in Peru have changed significantly since the early 1980s. At a national level, women's issues have captured attention, and women's groups have often been in the vanguard of grassroots movements for decentralization and deconcentration of power. In resource and environmental management, however, Lynch concluded that women's role in water management had received relatively little attention either from women's groups or from government agencies. Given the importance of water in the highlands of Peru, and the growing role of women in irrigation, this lack of attention initially appears puzzling. However, a closer examination of gender roles indicates that this lack of attention is not surprising. Some significant societal changes will have to occur before women become more prominent in water management.

Experience with two government-supported irrigation systems, San Marcos and Santa Rita, in the department of Cajamarca, illustrates the issues. Cajamarca, in northern Peru, is populated by Spanish-speaking and *mesitzo* people. The two project areas are adjacent to district and departmental capitals, and have reasonable road connections to the coast. As with many areas in the Sierra Andes, these areas had been experiencing a transition to an open, cash economy, replacement of indigenous irrigation systems by ones introduced and managed to a considerable extent by government irrigation agencies, and an increase in women's participation in irrigation-based tasks because of out-migration and/or different work patterns by men. With regard to the last point, both rural men and women work outside the agricultural sector, but women are more likely to remain in agriculture and the men more likely to migrate for seasonal work. As a result, women are left to be both heads of households, and to perform the irrigation tasks normally handled by the household head.

According to Lynch, other research in the Central Sierra had found that irrigation tasks normally were allocated to men, as part of a patriarchal society in which a gender division of labour gave men control over critical resources (land, water, transportation, cash), which bestowed relatively more power to them. In Lynch's (1991: 39) words, "women are prevented from performing key tasks so that they depend on men for resource allocation. Thus, to the extent that women are structurally excluded from key institutions, their power is limited".

What has been the practice regarding irrigation agriculture in the Peruvian Andes? Lynch suggested that delivery of water from the farm gate to the crop, or the on-farm management, was traditionally men's work. Women usually only became involved in irrigation if they were members of small households (fewer than five people), or if the men left the farm to obtain more remunerative work. This pattern became more pronounced in poorer families which usually had smaller land holdings and therefore the man was pushed to work full time for wages away from the farm. With the exception of small land holdings in which women basically took on what was traditionally men's work, women normally would not be hired to perform on-farm irrigation work for others as labourers or share croppers.

Women normally did not participate in the construction and maintenance of irrigation systems. Indigenous irrigation systems are based on unpaid community labour for construction and maintenance. Women's roles normally involved preparing food for male workers, unless the irrigation system was very small and then women might contribute labour. If a full day's work was required, women would send a *peon* or pay some money as their contribution. Similar arrangements prevailed for maintenance. However, over time and as more men migrated to seek work in the market economy, women had become more involved in work crews for both construction and maintenance. Thus, with regard to construction and maintenance work, gender divisions of labour had become less distinct.

However, when water management is interpreted to involve the organization of labour, cash and materials to build, repair and maintain irrigation infrastructure, and to make decisions about the acquisition, allocation and distribution of water, women have been mainly absent. Several reasons account for this. First, at the community or local level, a group of irrigators sharing a common system elect a board of directors to be responsible for management decisions. The directors almost inevitably are men of high status, which includes wealthier landowners who often hold other community positions, and people with experience outside the local community. Women rarely have either of these characteristics. When Lynch did her work, the San Marcos project had no women members, and Santa Rita had one. Second, the state agency responsible for irrigation is dominated by men. According to Lynch (1991: 47), women employed in the state irrigation bureaucracy most often are extension specialists or sociologists who work primarily with women's groups and on women's issues. "Rarely, if ever, do they play central roles in system design, construction, or water allocation."

A number of characteristics of societal and government traditions thus impede a fuller role for women, as noted in Box 10.9. These patterns are reinforced by the fact that men are more likely than women to have experience in the market or cash economy, in interacting with government officials, and outside their communities. The experience of the Andean peasant women is usually restricted to the household and the community, emphasizing their *reproductive* and *community*

Box 10.9 Barriers to greater women's involvement in irrigation

First, societal norms in the highlands that define water management as men's work, as well as sexist stereotypes present in the larger society, make women seem invisible to government officials. Because irrigation institutions at all levels are male-dominated, irrigation is not a sector where the state has encouraged movement of women into a political sphere beyond the community. Nor has the state tried to use irrigation projects to mobilize peasant women.

Source: Lynch, 1991: 47.

Box 10.10 Opportunity for women must be expanded

Despite women's critical role in agriculture, their access to education and their representation in research, extension, and other support services is woefully inadequate. Women should be given the same educational opportunities as men. There should be more female extension workers, and women should participate in field visits. Women should be given more power to take decisions regarding agricultural and forestry programmes.

. . . In many countries women do not have direct land rights; titles go to men only. In the interests of food security, land reforms should recognize women's role in growing food. Women, especially those heading households, should be given direct land rights.

Source: World Commission on Environment and Development, 1987: 140 and 141.

---

managing roles, while the men's roles emphasize *productive* work. The outcome is that while the government bureaucracy does provide some role for women in irrigation and water management, the irrigation agencies mostly seek to co-opt women rather than to encourage them to have a central role in management. Where there has been a significant change in their roles within the local community, that often has been as a result of men leaving the local area to seek wage employment elsewhere, leaving the women to take on traditional men's work by default, as well as continuing their traditional women's work (Box 10.10).

10.6 Gender and agriculture

Jarosz' (1991) examination of the role of women as rice share croppers in Madagascar provides further insights into the *reproductive*, *productive* and *community managing* roles that women have in natural resource use. Sharecropping is a long-standing tradition within agriculture in the developing world. It can involve sharing of a crop, or of agricultural tasks. In many societies, such as Madagascar, sharecropping is a well-established way of organizing rural labour and agricultural production. Sharecropping provides young people with access to land and capital, and it also gives elderly peasant farmers access to labour. As a result, a fairly negotiated sharecropping agreement can be advantageous for both younger and older people, and can contribute to inter-generational equity. The disadvantage can be unfair agreements which increase the productivity and benefits for some, and lead to dispossession or economic stagnation for others. Jarosz was interested in the roles of class and gender in sharecropping arrangements in Madagascar as farmers used this system to gain access to resources.

Madagascar is an island country of 586,560 km^2, has a population of about 12 million people with about 75% living in rural areas, and is one of the 12 poorest nations in the world. Agricultural exports, especially coffee, vanilla and cloves, are the main source of foreign exchange. Sharply fluctuating prices for coffee, and increasing use of synthetic vanilla, have had major negative impacts on the national economy. The basis of the Malagasy diet is rice, which is grown as both a subsistence and cash crop.

Analysis was completed in the Alaotra region, a fertile basin at an elevation of about 800 m in the north central part of the island. Some 90,000 hectares of irrigated rice were under cultivation at the time of Jarosz' study. One-third of that area was devoted to an intensive irrigation project started in 1960, and was seeded with the high-yield rice varieties introduced during the Green Revolution. Dryland farming in the area also produces rice, along with maize, manioc, peanuts and vegetables. Sharecropping has been used in the Alaotra since the nineteenth century. When Madagascar achieved independence from France in 1960, about one-half of the people in that region were landless share croppers involved in many types of agreements with landholders. The normal arrangement is for the rice crop to be divided in half between the landowners and share croppers.

Regarding gender roles in sharecropping, Jarosz discovered that marital status and class were key variables influencing the kind of sharecropping in which women participated. Female heads of households almost invariably participate in sharecropping if they have one or more hectares of irrigated rice. In addition, women who inherited land near their ancestral homes, but who subsequently moved away to live closer to their husbands' lands, also sharecrop, normally in agreement with family members who live in the ancestral village. If a married couple divorce, the husband is allocated two-thirds, and the wife one-third, of all the property acquired during their marriage. Middle-class women normally sharecrop their land to obtain the needed male labour and, if necessary, also the capital needed for production. These differences highlight that women are not a homogeneous group, but that there are differences depending upon their marital and class situation.

Field work tasks, including clearing, ploughing and harrowing, are considered to be men's work. Ploughing requires considerable physical strength and team-work. One person guides the plough, and the other controls a team of either two or four zebu cattle. The latter task can be dangerous, because the zebu cattle are temperamental and unpredictable. Men have been badly gored by the zebu while pulling ploughs, and women are rarely seen ploughing. Other field work can be done by men or women, although transplanting and weeding are

---

Box 10.11 Vulnerability of poor rural women

One young women (sic) and her two tiny children live in a neighbor's tool shed. She has no direct access to a cooking fire, and she and her children are ill and malnourished. Single, landless men can earn $66 per year as permanent farm workers who can room and board on the farm of their employers. Landless women without children can earn $40 per year as domestic workers who room and board with their employers. Landless men can sharecrop dryland or marshland, options that are unavailable to landless female-headed households.

Source: Jarosz, 1991: 58.

---

normally viewed to be women's jobs. This characterization of tasks as customarily men's or women's work illustrates that a partial gendered division of labour exists.

Not surprisingly, Jarosz found that landless female-headed households were the most marginal and vulnerable in the Alaotra agricultural community. Such women found work on an irregular basis as seasonal labourers or laundresses. Incomes usually were inadequate to meet the basic needs of their families, and such families often lived from day to day. The poorest of these poor people usually are homeless (Box 10.11).

In Madagascar, Jarosz' findings indicated that the workloads for women had increased across all classes over the previous 20 to 30 years. In order to meet their *reproductive* responsibilities related to feeding their families, during the preharvest season (normally a relatively idle period for men), women would sell horticultural products, chicken, geese, fish, mats and baskets. The income from such subsidiary activities was not as great as from irrigated rice cultivation, but was still needed to provide for the needs of families. As farm people, the women also have been increasingly marginalized within the irrigated rice production system. Part of the cause for that has been the Green Revolution style of agriculture, with emphasis on high-yielding rice varieties, more technology-based irrigation and growing reliance on agrochemcials. All of these aspects of agricultural production are more men's than women's work. A particularly challenging time for the poorest female-headed households is the November to March period. Except for several weeks of transplanting work in the last two months of the year, very few income-earning tasks are available until April. As a result, the women and their children do not have enough to eat during this time. Furthermore, this period overlaps with the malaria season, a time when there is little money that can be used for anti-malarial medicine.

The experience in Madagascar led Jarosz (1991: 61–2) to conclude that not everyone gained from sharecropping. Her findings indicated that wealthy and middle class people did well both socially and economically as a result of share-cropping arrangements. However, "the young and landless, poor, female-headed households, and descendants of former slaves are at the greatest disadvantage and subject to the most severe exploitation". As a result, while gender by itself did not define the most vulnerable segment of the population, it along with class and kinship variables were central in influencing "who gains and who loses in cropping rice on shares in Madagascar". The poorest people, both men and women, were the most vulnerable and marginalized. However, the poor women were more vulnerable than the poor men. Their situation will not be improved until there is a shift in societal values regarding *reproductive*, *productive* and *community managing* needs, tasks and roles.

10.7 Integrating gender and environment

In some of the previous sections, attention focused on initiatives at the local or village level to increase the role of women in resource and environmental management, and thereby to soften or remove gender-based divisions. A common thread in the various experiences (forestry, water, agriculture) is the difficulty in getting women's issues recognized by the formal state agencies responsible for resource and environmental management. Changes in attitudes and values of senior managers will have to occur if women's issues are to be addressed systematically and on a sustained basis.

However, a structural or organizational issue must also be addressed. Levy (1992) has argued that both "environment" and "gender" have encountered the same dilemma regarding the best way to incorporate them into government bureaucracies. At least two choices exist. On one hand, custom-built women's or environmental agencies have been created. Certainly for the environment, following the Stockholm Conference in 1972 (see Chapter 4), there was a proliferation of new environmental ministries, departments and agencies in many countries. This approach made the "environment" more visible within government, but it also allowed traditional line or sectoral agencies to ignore or neglect environmental matters since they could argue that such a concern was the responsibility of the environmental ministry or department. On the other hand, women's or environmental considerations can be integrated into existing organizations and agencies. This option does not make gender or environmental issues as visible within a governmental system, but it does require each agency to consider the implications of its objectives and activities for them.

The choice between custom-built agencies, or integration within existing agencies, is a generic issue for managers, as indicated in the comments in Box 10.12 from the Brundtland Commission about alternative ways to incorporate environmental issues into management.

Levy (1992) indicated that, regarding women's issues, the choice usually has been to create special women's bureaus, ministries or departments. This particularly has been the situation in Asia, Africa and Latin America. Levy expressed concern that the focus of such special-purpose women's organizations was primarily at the project level, resulting in a very narrow and limited role. The outcome has been that the women's "sector" has often been a very weak one (Box 10.13). In Levy's (1992: 136) words, the women's sector is usually "characterized by a lack of any real political influence, and is therefore underfunded and under-staffed, both in numbers and qualifications. A key factor underlying these characteristics is the conceptualization of both the problems

---

Box 10.12 Incorporating environmental issues into management

The mandates of central economic and sectoral ministries are also often too narrow, too concerned with quantities of production or growth. The mandates of ministries of industry include production targets, while the accompanying pollution is left to ministries of environment. Electricity boards produce power, while the acid pollution they also produce is left to other bodies to clean up. The present challenge is to give the central economic and sectoral ministries the responsibility for the quality of those parts of the human environment affected by their decisions, and to give the environmental agencies more power to cope with the effects of unsustainable development.

Source: World Commission on Environment and Development, 1987: 10.

Box 10.13 Problems of not focusing on gender

A focus on women is recognized as legitimate in its own right and the basis of one of the most important political movements of the century. However, when translated into professional practice over the last 15 years, it has resulted in a sector which is marginalized from mainstream development policies, programmes and projects, with little impact on overall development processes and economic, social and political relations in many countries.

Source: Levy, 1992: 136.

---

and the strategies of this sector in terms of women, not gender". Levy's analysis highlights that, as in most situations, choices exist among imperfect options. Which alternative do you believe is most likely to result in the most effective treatment of gender issues?

10.8 Implications

The Earth Summit in Rio de Janeiro (1992) and the Fourth World Conference on Women in Beijing (1995) helped to draw attention to the issues of *women* and *gender* in resource and environmental management. Gender is a more appropriate focus, since it is often social relationships and customary practices that need to be shifted if women are to have the same opportunities as men.

In seeking to improve opportunities for women, it is important to recognize various roles – reproductive, productive, community managing – which women often hold in their households, communities and region. Too often, initiatives to improve the conditions of women have concentrated only upon the productive role. A gender approach reminds us that all three roles deserve attention and action. A gender approach also reminds us that women are not a homogenous

Guest Statement

Turning the compass: on re-orienting "gender/ development" studies North and West[1]

Maureen Reed, Canada

In 1988, during a graduate seminar in geography, I participated in heated discussions about how a focus on gender might reorient a research agenda concerned with environment and land use in the so-called "Third World".[2] But when it came to discussing a gender perspective on gender/development issues here in Canada, the room fell silent. Some students argued that women in Canada had no connection to the non-human environment because they did not have to use their immediate environment to meet their subsistence or market needs. Neither were women connected through institutions of power and governance such as corporate, legal and government decision making. What, then, could be said about gender and development in places of the "North" or "West"?

Today, more than a decade later, scholars continue to undertake research about uneven, inequitable or unsustainable development in developed countries in a way that often erases issues of gender relations. It is true, most women in developed countries do not use their biophysical environment to meet subsistence needs. Yet, if we consider development processes as uneven physically, economically and socially, we can easily consider how gender relates to these processes, even in the West. For example, much work in gender/development draws on concepts within the research of political economy/ecology (Walker, 1998), development theory (Marchand and Parpart, 1995) and feminist theories of gender relations (Rocheleau et al., 1996). Researchers in these fields ask questions about the reciprocal relations between social institutions and the non-human environment across places and over time that are equally pertinent to Western forms of development. Second, gender/development perspectives are pertinent to research undertaken in developed countries relating to rural studies, labour studies in resource communities, social movements and environmental justice, as well as environmental planning. Gender relations in productive, reproductive and community management activities have been considered in relation to agriculture, mining, fishing and forestry communities. There is much scope for extending these findings by comparing different development pathways across different resource sectors.

Beyond specific resource sectors, gender/development perspectives might consider how development practices are influenced by gender relations as well as specific ethnic or cultural groups. The burgeoning literature in environmental justice has also begun to challenge environment and development debates in North America and offers new frameworks for studying gender/development (Pulido,

---

[1] I use the term "gender/development studies" to avoid restricting my own discussion to the particular approach of "gender and development".

[2] I use the terminology of "Third World", "developing countries" "northern" and "western" countries following conventions that are well understood. However, I do not suggest consensus about what these terms mean.

1996). Yet more work is needed regarding how gendered expectations in society and local communities affect the framing of environmental issues and their resolutions. How, and with whose participation, are decisions made about the natural environment? How do our assumptions about productive work related to resource extraction and development affect planning processes established to manage the non-human environment?

These research efforts and ongoing questions suggest that there are important insights to be gained by extending gender/development studies to research in developed countries. As developing countries rightly demand that developed countries deal with their own environment and development problems before dictating to others how to live sustainably, researchers and practitioners of gender, sustainability and development would do well to turn their compasses North and West.

References

Marchand M H and J L Parpart (eds) 1995 *Feminism/Postmodernism/Development*. London and New York, Routledge

Pulido L 1996 *Environmentalism and Economic Justice: Two Chicano Struggles in the Southwest*. Tucson, University of Arizona Press

Rocheleau D, B Thomas-Slayter and E Wangari (eds) 1996 *Feminist Political Ecology: Global Issues and Local Experiences*. London, Routledge

Walker P 1998 Politics of Nature: an overview of political ecology. *Capitalism, Nature, Socialism* 9: 131–44

Maureen Reed is an Associate Professor in the Department of Geography at the University of Saskatchewan, Saskatoon, Canada. Her research focuses on sustainability of resource-dependent communities that are experiencing economic, social and environmental policy changes. Her current research examines the perspectives of women living in forestry communities on the west coast of Canada and who support conventional forestry.

group, and that therefore gender should not be considered in isolation from other variables such as financial and social status.

Experiences over the past two decades are both encouraging and discouraging. The well-known Chipko movement in India illustrates how women can change gender relationships related to resource-use practices by directly challenging conventional attitudes and practices. And yet, in the same country some of the joint forest management and village initiatives continue to marginalize women. Experiences with irrigation agriculture in Peru and sharecropping in Madagascar illustrate that often profound changes in societal attitudes and values toward women are needed if customary gender roles are to be changed.

The examples in this chapter have focused on experiences in developing countries, as situations there are usually more stark in gender relations. However, gender-defined roles also exist in developed countries, and there often are gender differences in approaches and attitudes to environmental issues. As the guest statement by Maureen Reed indicates, however, conflicting findings and interpretations suggest that much more work remains to be done in this field.

References and further reading

Agarwal B 1997 Gender, environment, and poverty interlinks: regional variations and temporal shifts in rural India, 1971–91. *World Development* 25: 23–52

Ahmed M R 1992 Unseen workers: a sociocultural profile of women in Bangladesh agriculture. *Society and Natural Resources* 5: 375–90

Armitage D R and B Hyma 1997 Sustainable community-based forestry development: a policy and program framework to enhance women's participation. *Singapore Journal of Tropical Geography* 18: 1–19

Awumbila M and J H Momsen 1995 Gender and the environment: women's time use as a measure of environmental change. *Global Environmental Change* 5: 337–46

Bee A 2000 Globalization, grapes and gender: women's work in traditional agro-export production in northern Chile. *Geographical Journal* 166: 255–65

Blocker T J and D L Eckberg 1989 Environmental issues as women's issues: general concerns and local hazards. *Social Science Quarterly* 70: 586–93

Breton M J 1998 *Women Pioneers for the Environment*. Boston, Northeastern University

Chen L C, W M Fitzgerald and L Bates 1995 Women, politics, and global management. *Environment* 37: 4–9, 31–3

Cuomo C J 1992 Unravelling the problems in ecofeminism. *Environmental Ethics* 14: 351–63

Cutter S L 1995 The forgotten casualties: women, children and environmental change. *Global Environmental Change* 5: 181–94

Dankelman I and J Davisdon 1988 *Women and Environment in the Third World: Alliance for the Future*. London, Earthscan Publications with IUCN

Davis D 2000 Gendered cultures of conflict and discontent: living "the crisis" in a Newfoundland community. *Women's Studies International Forum* 23: 343–53

Davis D L and J Nadel-Klein 1992 Gender, culture, and the sea: contemporary theoretical approaches. *Society and Natural Resources* 5: 135–47

Elmhust T 1998 Reconciling feminist theory and gendered resource management in Indonesia. *Area* 30: 225–35

Fratkin E and K Smith 1995 Women's changing economic roles with pastoral sedentarization: varying strategies in alternate Rendille communities. *Human Ecology* 23: 433–54

Gerrard S 2000 The gender dimension of local festivals: the fishery crisis and women's and men's political actions in north Norwegian communities. *Women's Studies International Forum* 23: 299–309

Gray L 1993 The effect of drought and economic decline on rural women in Western Sudan. *Geoforum* 24: 89–98

Greed C H 1994 *Women and Planning: Creating Gendered Realities*. London, Routledge

Jacobson J L 1992 *Gender Bias: Roadblock to Sustainable Development*. Worldwatch Paper 110, Washington, DC, Worldwatch Institute

Jarosz L 1991 Women as rice sharecroppers in Madagascar. *Society and Natural Resources* 4: 53–63

Kabeer N 1999 Resources, agency, achievements: reflections on the measurement of women's empowerment. *Development and Change* 30: 435–64

King R J H 1991 Caring about nature: feminist ethics and the environment. *Hypatia* 6: 75–89

Kinnaird V and D Hall (eds) 1994 *Tourism: A Gender Analysis*. Chichester, Wiley

Leach M 1992 Gender and the environment: traps and opportunities. *Development in Practice* 2: 12–22

Lenny J 1999 Deconstructing gendered power relations in participatory planning: towards an empowering feminist framework of participation and action. *Women's Studies International Forum* 22: 97–112

Levy C 1992 Gender and environment: the challenge of cross-cutting issues in development policy and planning. *Environment and Urbanization* 4: 134–49

Levy D E and P B Lerch 1991 Tourism as a factor in development: implications for gender and work in Barbados. *Gender and Society* 5: 67–85

Little J 1994 *Gender, Planning and the Policy Process*. Oxford, Pergamon

Locke C 1999 Constructing a gender policy for joint forest management in India. *Development and Change* 30: 265–85

Lynch B D 1991 Women and irrigation in highland Peru. *Society and Natural Resources* 4: 37–52

Mackenzie F 1990 Gender and land rights in Marang'a District, Kenya. *Journal of Peasant Studies* 17: 609–43

Mayoux L 1995 Beyond naivety: women, gender inequality and participatory development. *Development and Change* 26: 235–58

McStay J R and R E Dunlap 1983 Male–female differences in concern for environmental quality. *International Journal of Women's Studies* 6: 291–301

Meer S (ed.) 1997 *Women, Land and Authority: Perspectives from South Africa*. Cape Town, David Philip Publishers

Meinzen-Dick R S, L R Brown, H Sims Feldstein and A R Quisumbing 1997 Gender, property rights and natural resources. *World Development* 25: 1303–15

Merchant C 1989 *Ecological Revolutions: Nature, Gender, and Science in New England*. Chapel Hill, University of North Carolina Press

Merchant C 1995 *Earthcare: Women and the Environment*. London, Routledge

Mohai P 1992 Men, women and the environment: an examination of the gender gap in environmental concern and activism. *Society and Natural Resources* 5: 1–19

Moser C O N 1989 Gender planning in the Third World: meeting practical and strategic gender needs. *World Development* 17: 1799–825

Moser C O N 1993 *Gender Planning and Development: Theory, Practice and Training*. London, Routledge

Mosse D 1994 Authority, gender and knowledge: theoretical reflections on the practice of participatory rural appraisal. *Development and Change* 25: 497–526

Mosse J C 1993 *Half the World, Half a Chance: An Introduction to Gender and Development*. Oxford, Oxfam

Nesmith C and P Wright 1995 Gender, resources, and environmental management. In B Mitchell (ed.) *Resource and Environmental Management in Canada: Addressing Conflict and Uncertainty*. Toronto, Oxford University Press, pp. 80–98

Ngwa N E 1995 The role of women in environmental management: an overview of the rural Cameroonian situation. *GeoJournal* 35: 515–20

Norr J L and K F Norr 1992 Women's status in peasant-level fishing. *Society and Natural Resources* 5: 149–63

Onkoko E 1999 Women and environmental change in the Niger delta, Nigeria: evidence from Ibeno. *Gender, Place and Culture* 6: 373–78, and commentaries, 379–400

Peluso N L (ed.) 1991 Special issue: women and natural resources in developing countries. *Society and Natural Resources* 4: 90 pp

Plumwood V 1994 *Feminism and the Mastery of Nature*. London, Routledge

Preston V, D Rose, G Norcliff and J Holmes 2000 Shiftwork, childcare and domestic work: divisions of labour in Canadian paper mill communities. *Gender, Place and Culture* 7: 5–29

Prindeville D-M and J G Bretting 1998 Indigenous women activists and political participation: the case of environmental justice. *Women and Politics* 19: 39–58

Rauf F 1998 Environmental transformation and conflicts: women's perceptions from rural Punjab, Pakistan. *Development* 41(3): 91–6

Razavi S 1999 Gendered poverty and well-being: introduction. *Development and Change* 30: 409–33

Reed M G 2000 Taking stands: a feminist perspective on 'other' women's activism in forestry communities of northern Vancouver Island. *Gender, Place and Culture* 31: 363–87

Riley G 1999 *Women and Nature: Saving the "Wild" West*. Lincoln, University of Nebraska Press

Sarin M 1995 Community forestry management: where are the women? *Hindu Survey of the Environment* 25: 27–9

Scheyvens R 2000 Promoting women's empowerment through involvement in ecotourism: experiences from the Third World. *Journal of Sustainable Tourism* 8: 232–49

Scheyvens R and L Lagisa 1998 Women, disempowerment and resistance: an analysis of logging and mining activities in the Pacific. *Singapore Journal of Tropical Geography* 19: 51–70

Seager J 1993 *Earth Follies: Coming to Feminist Terms with the Global Environmental Crisis*. London, Routledge

Shrestha (Vaidya) P L 1998 Conservation and management of watershed regions by Nepalese women leading to enhancement of water potential. *International Journal of Water Resources Development* 14: 513–25

Sinclair J and L Ham 2000 Household adaptive strategies: shaping livelihood security in the Western Himalaya. *Canadian Journal of Development Studies* 21: 89–112

Stern P C, T Dietz and L Kalof 1993 Value orientations, gender and environmental concern. *Environment and Behavior* 25: 322–48

Stubbs J 2000 Gender in development: a long haul – but we're getting there! *Development in Practice* 10: 535–42

Swain M 1995 Gender in tourism. *Annals of Tourism Research* 22: 247–66

Thomas-Slayter B 1995 *Gender, Environment, and Development in Kenya: A Grassroots Perspective*. Boulder, Colorado and London, Lynne Rienner

Thomas-Slayter B and N Bhatt 1994 Land, livestock and livelihoods: changing dynamics of gender, caste and ethnicity in a Nepalese village. *Human Ecology* 22: 467–94

Thomas-Slayter B, A L Esser and M D Shields (eds) 1993 *Tools for Gender Analysis: A Guide to Field Methods for Bringing Gender into Sustainable Resource Management*. ECOGEN Research Project, International Development Program, Worcester, Mass., Clark University

Townsend J G 1995 *Women's Voices from the Rainforest*. London, Routledge

Ulluwishewa R 1996 Gender, development and ecology of emerging food insecurity in the dry zone of Sri Lanka. *Malaysian Journal of Tropical Geography* 27: 51–9

United Nations, Food and Agriculture Organization 1989 *Women in Community Forestry: A Field Guide for Project Design and Implementation*. Rome, Food and Agriculture Organization.

Van Wijk C, E de Lange and D Saunders 1996 Gender aspects in management of water. *Natural Resources Forum* 20: 91–103

Warren K J (ed.) 1995 *Ecological Feminism*. London, Routledge

Warren K (ed.) 1996 *Ecological Feminism Philosophies*. Bloomington, Indiana University Press

White G F, D J Bradley and A U White 1972 *Drawers of Water: Domestic Water Use in East Africa*. Chicago, University of Chicago Press

Wickramasinghe A 1997 Women and minority groups in environmental management. *Sustainable Development* 5: 11–20

Williams S, J Seed and A Mwau 1994 *The Oxfam Gender Training Manual*. London, England and Dublin Ireland, Oxfam

World Commission on Environment and Development 1987 *Our Common Future*. Oxford, Oxford University Press

Zweifel H 1997 Biodiversity and the appropriation of women's knowledge. *Indigenous Knowledge and Development Monitor* 5(1). Online at http://www.nufficcs.nl.ciran/ikdm

Chapter 11

Alternative Dispute Resolution

11.1 Introduction

"Conflict is a clash of interests, values, actions or directions, and has been a part of life since time began" (Johnson and Duinker, 1993: 17; Box 11.2). Thus, conflicts are inescapable, but they can be positive as well as negative. Positive aspects occur when conflict helps to identify a process for resource and environmental management which is not working effectively, to highlight poorly

---

Box 11.1 Resolving conflicts

Conflict resolution, or alternative dispute resolution (ADR), techniques are intended to facilitate consensus decision making by disputing parties, thereby avoiding legal or administrative proceedings to resolve disputes. Some characteristics of this group of techniques include: (1) focusing on the underlying interests of the disputing parties, rather than on their bargaining positions; (2) using creative thinking to dovetail unlike interests, preferences, capabilities, and risk tolerances and change disputes from zero-sum games to situations with the potential for joint gains; (3) appealing to jointly-accepted objective standards for apportioning gains; and (4) requiring consensus among parties to a decision, rather than majority rule. An independent mediator is often used to direct the process of dispute resolution.

Source: Maguire and Boiney, 1994: 33.

Box 11.2 Conflict is common

Environmental assessment is often characterized by conflict and controversy. . . . This is an inevitable consequence of the differences in values and interests that exist in a pluralistic society with respect to the use and management of land, water and other natural resources. Dispute settlement is usually difficult to achieve for two inter-related reasons: first, the benefits and costs of development are unevenly distributed and include intangibles that are hard to evaluate and compare; and, second, many affected and interested parties with diverse views and interpretations are often involved.

Source: Sadler and Armour, 1987: 1.

developed ideas or inadequate or misleading information, and to draw attention to misunderstandings. Conflict also can be helpful when, by questioning the status quo, it leads to new creative approaches. In contrast, conflict can be negative if it is ignored or consciously set aside. "An unresolved conflict breeds mis-information, misunderstanding, mistrust and biases. A conflict is bad when it allows higher and stronger barriers to be built up between the involved parties" (Johnson and Duinker, 1993: 19).

In this chapter, alternative approaches for resolving disputes are examined, with particular attention to what has become known as *alternative dispute resolution* (ADR). Section 11.2 describes the characteristics of various ways of dealing with conflicts, and Section 11.3 reviews different ADR approaches. Then, the discussion considers the conditions or factors necessary or desirable for effective use of ADR. That review is followed by some examples of the application of ADR approaches.

11.2 Different approaches to dispute resolution

When conflicts arise over resource allocation or different interests regarding the environment, at least four approaches can be used to deal with them: (1) political, (2) administrative, (3) judicial and (4) alternative dispute resolution. These approaches are not necessarily mutually exclusive; some can be used together.

Political approaches involve elected decision makers considering the range of competing values and interests, and then making a decision. In this approach, the decision makers normally are not specialists in resource and environmental management, but they do receive advice from technical experts in the public service. In addition, through various participatory mechanisms they can seek to involve the public, and to hear directly from the public about needs, aspirations and preferences. In a democracy, the decision makers are accountable to all of their constituents. However, not all constituents are equal, due to different access to financial and other resources. As a result, some constituents may have a disproportionate influence on decision makers. Where corruption is prevalent, elected decision makers may not try to balance all interests, but instead focus on the interests of a select few. Furthermore, the decision makers are usually distant from the place and the people most affected by their decisions, and may not always be aware of, or sensitive to, specific local conditions. This latter point is of less concern, of course, when the decision makers are elected at a local level.

Administrative approaches are built into resource and environmental management organizations, and allow bureaucrats to take decisions regarding some kinds of disputes. Thus, a district or regional manager may be empowered to bring conflicting groups together, listen to their views, consider information provided by technical experts, and then reach a decision. In some situations, such as the co-management discussed in Chapter 9, power can be shared with or delegated to people who will be affected by the decisions. Generally, however, administrative approaches are best suited to what might be called *routine* as opposed to *strategic* types of decisions. If the people affected by the decisions are unhappy with the outcome, there is often provision for an appeal to an administrator at a different level in the management system, or to elected officials. As with the political approaches, if corruption exists, then decisions will not necessarily reflect consideration of all interests in the system.

Judicial approaches involve litigation and the courts. This approach is well suited for situations in which parties in dispute are so entrenched in their positions, or so angry at other participants, that they will not voluntarily meet with the other parties to try and reach a resolution. The judicial approach has the power (police) to ensure that people participate at hearings, and once a decision is taken, has the power to impose and enforce sanctions (fines, prison sentences). The judicial approach is based on procedures and guidelines which have evolved over centuries. Emphasis is placed on *facts*, *precedents*, *procedures* and *argument*. Accountability is normally high, as provision exists for appeals to a higher court.

Notwithstanding the many advantages of the judicial approach, there also are some disadvantages. The main weaknesses are the *adversarial*, *time-consuming* and *expensive* aspects of the judicial approach. The adversarial nature means that opposing sides do not try to work with each other to solve a problem, but instead present only information which supports their interests, and discredit information or views supporting the interests of their opponents. The process can be time consuming and expensive, making it difficult for some parties to participate if they do not have funds to employ legal experts. Even if they can hire legal advisors, they may not be able to match the team of legal and other technical advisors that another group may be able to assemble.

For many people, another disadvantage of the judicial approach is that it usually results in *winners* and *losers*. That is, the outcome in a court decision normally is that one party wins and the other loses. While the judicial system has enough sanctions (fines, imprisonment) to make sure that the decision in the court is upheld, often the decision generates considerable ill will, and makes future cooperation unlikely. Thus, while it is important to appreciate that the judicial approach has many strengths and will always be required as a means for resolving disputes, like every approach it also has some distinct weaknesses.

Alternative dispute resolution (ADR) approaches have emerged in response to the perceived weaknesses of the judicial approach, and also in response to the growing expectations in many societies for more participation and local empowerment in resource and environmental management. Alternative dispute resolution approaches try to avoid the adversarial and winner–loser characteristics of the judicial approach. The dominant characteristics of ADR include: (1) attention to interests and needs over positions and precedents, (2) persuasion rather than coercion, (3) commitment to joint agreement rather than imposed settlement, (4) constructive communication and improved understanding instead of negative criticism and preoccupation with justifying or defending interests, (5) achievement of settlements that will be long lasting because of shared commitment, (6) effective sharing and use of information, and (7) greater flexibility.

These characteristics of ADR represent ideals which cannot always be achieved, and sometimes require satisfaction of preconditions which cannot be met

---

Box 11.3 But what if conflicts appear irreconcilable?

Conflicting interests that are perceived to be mutually exclusive present a special problem. The wilderness versus industrial development issues . . . are examples of this phenomenon. Wilderness advocates define wilderness in terms of a lack of industrial development: an area that is partly developed is not wilderness. From this perspective, the dispute becomes an all-or-nothing, win–lose contest. In other words, the integrity of the contested wilderness area, from the wilderness advocates' perspective, is a non-negotiable issue. Negotiation, from this perspective, will likely be perceived as offering nothing and may be perceived as exposing the wilderness advocate to mollification and manipulation.

Source: Wood, 1989: 45.

Box 11.4 Contrast between judicial and mediation approaches

The internal dynamics of environmental mediation are completely different than the courtroom context. Participants in mediation often develop bonds of trust, understanding, and even affection, toward their opponents. The climate of understanding and progress in working toward mutually satisfactory solutions creates subtle pressures to be reasonable and conciliatory. These dynamics may undermine the determination of unsophisticated parties to stand their ground on issues. . . . The typical low key atmosphere, and press exclusion, of the proceedings protects the parties from the scrutiny of their constituents, and shields them from the awareness that they might be sacrificing constituent concerns in the interests of achieving a settlement. The parties with less experience and sophistication may walk away with an agreement which favors their perspective much less than would have been possible in a more public, adversarial context.

The context of litigation is not conducive to intimacy and trust between contending parties. Adversarial relationships and the development of competing evidence heighten the differences between opponents. The use of expert witnesses leads to the development of elaborate information and contrasting interpretations of the same data to support different positions. The public nature of the courtroom spurs lawyers and experts to make their utmost efforts to enhance their reputations in light of future opportunities, and helps to attract the resources to involve highly skilled professionals. The public context assures widespread awareness of the proceedings, and protects litigants from the temptation to sacrifice the interests of their constituents in the desire to achieve a settlement. The dynamics of starkly competing perspectives, the context of legal precedent, and the emphasis upon proper procedure help to assure that each side ends up with the maximum benefit which is justified within the law. The internal dynamics of litigation protect the interests of weaker parties much better than environmental mediation.

Source: Blackburn, 1988: 569–70.

(Box 11.3). For example, people with different and conflicting interests may not be prepared to meet to share information, and to try and reach a long-lasting settlement. As a result, ADR is not inevitably a better approach for resolving conflicts relative to the judicial approach (Box 11.4). In this chapter, the interest is in examining the strengths and weaknesses of ADR for dispute resolution, and to review some situations in which it has been applied. Before doing that, however, it is important to appreciate that ADR is not a single approach. Different types of ADR exist, and are considered below.

11.3 Types of alternative dispute resolution

Four types of ADR exist: (1) public consultation, (2) negotiation, (3) mediation and (4) arbitration.

11.3.1 Public consultation

In Chapter 8, various aspects of partnerships and participatory approaches were examined. Basic motivations for public consultation are to allow more sharing of experience and information, to ensure that many perspectives are considered, to open up management processes so that they can be seen to be both efficient and fair, and thereby to ensure that more people will be satisfied with decisions and plans. If all of these characteristics are achieved through public participation, then many issues which might trigger conflicts can be dealt with before they emerge as full-scale disputes.

As described in the above paragraph, public consultation is a means to resolve conflict, and is an alternative to the judicial, administrative and political approaches. However, public consultation also could readily become a component of administrative and political approaches, when appointed or elected decision makers seek to consult with the public before decisions are taken, or to allow some decisions to be made by the public. Because public consultation concepts and mechanisms were discussed in detail in Chapter 8, they will not be examined further here.

11.3.2 Negotiation

Negotiation is one of three approaches normally considered to comprise ADR. Negotiation involves situations in which two or more groups meet voluntarily in order to explore jointly an issue causing conflict between them. The purpose is to reach a mutually acceptable agreement by *consensus*. No external person or group provides assistance, and the parties in dispute have to be willing to meet with the other side to examine the issue.

11.3.3 Mediation

Mediation has all of the characteristics of negotiation, plus the involvement of a neutral third party (a mediator). The third party has no power to develop or

> ### Box 11.5 Role of a mediator
>
> . . . as a guardian of the process, a mediator can intervene to correct miscommunications, to clarify ambiguous messages, and to challenge deceptive communications. Also, a mediator can point out when differences in interpretations have arisen . . .
>
> *Source*: Ozawa and Susskind, 1985: 35.

impose an agreement, but functions as conciliator, facilitator and fact finder, in order to help the parties in conflict reach an agreement (Box 11.5). A mediator may be used when the parties in conflict are prepared to meet to discuss their problem, but also when feelings may be so strong that it is unlikely face-to-face meetings would be constructive. In such a situation, the mediator might separate the groups, help them to identify the main points of contention, and then serve as a messenger to facilitate dialogue between the parties.

11.3.4 Arbitration

When arbitration is chosen, a third party is involved. Unlike mediation, however, the person serving as the arbitrator has power to make a decision, which may or not be binding. If it is binding, then the parties in dispute have agreed before the arbitration process begins to abide by the settlement developed by the arbitrator. The prospect of binding arbitration in many cases is sufficient to make parties work diligently during a mediation process, in order to avoid a situation in which a third party imposes an agreement upon them. Usually the participants in the dispute are directly involved in the selection of the arbitrator, which is one of the key differences between arbitration and judicial approaches. Normally, in a judicial situation the disputing sides have no role in determining which judge or magistrate will preside over their case.

11.3.5 Summary

The four ADR approaches represent a continuum from public consultation to arbitration, in which the process becomes increasingly more structured and the participants relinquish more and more control of the process. Which one is appropriate depends upon the history of the relationships among the groups in conflict, and particularly upon their willingness to come together voluntarily to try and reach a solution which will be long lasting and beneficial for all interests. These different approaches are often used sequentially, for example with arbitration following a failed mediation process.

11.4 Conditions for effective alternative dispute resolution

ADR is not a guaranteed recipe for effective resolution of conflicts. Numerous conditions ideally should be met before ADR is used. All of these conditions reflect beliefs that: (1) the individuals or parties in a dispute may be in the best position to identify and settle the issues causing the conflict, (2) direct face-to-face discussions can be productive, (3) voluntary commitment exists for joint problem solving, and (4) a genuine desire is present to work towards consensus and reach a mutually agreeable settlement. If these beliefs are not realistic, then ADR is unlikely to be effective. If these beliefs are valid, however, then attention must turn to some other considerations.

11.4.1 Acknowledgement of a dispute

It may seem too obvious to mention, but a key aspect is that all parties recognize the existence of a dispute, and are able to agree upon the components or dimensions of the problem. Situations can occur, however, in which one group feels its interests are being damaged by the activities of another party. However, if the latter does not recognize or acknowledge the problem which is bothering the first party, the prospects for mutual problem solving are slim to non-existent.

11.4.2 Motivation to find a joint solution through ADR

For ADR to be effective, all parties must conclude that meeting together to search for a mutually acceptable solution is preferable to any other option they could pursue (Box 11.6). For example, if one party to a dispute decides to "play hard ball", and to ignore the concerns of others being affected by its decisions or activities, there is little benefit for the others who feel their interests are being damaged to come together through some form of ADR. The motivation or incentive to use ADR normally is to avoid the time, expense and adversarial nature of a judicial approach. Thus, there can be some very compelling reasons for parties to want to work together. However, if any party concludes that it

---

Box 11.6 Incentive to use ADR

A key element that determines whether a dispute may be mediated is whether or not there are sufficient incentives for the conflicting parties to enter into negotiations. Initially, however, each party must decide whether or not to recognise various groups and their interests. Mediation, and negotiations as a whole, are processes revolving around a desire, to varying degrees, to accommodate the opposing party. Implicit within this process is the recognition of the legitimacy of the opposition's demand and the opposition's right to represent those interests. The decision to mediate a dispute must be based upon a shared perception that it will provide the least-costly method of resolution and will provide the highest joint benefit. This type of cost benefit analysis is extremely difficult to gauge, given the unpredictability of negotiations and the potential for the variation of goals and objectives during negotiations.

Source: Rankin, 1989: 14.

---

has a better alternative (ignore the concerns of others; go to court), then ADR will not be effective.

11.4.3 Representation of interests

If a long-lasting, mutually agreeable solution is to be found, all significant interests should be represented in the ADR process. Achieving such representation can be challenging for various reasons. Governments can usually send representatives, since for the public servants such involvement is part of their job. Large corporations also can normally provide representatives, as they have staff whose jobs include such activity. But representatives from small businesses, labour groups or many small non-government organizations often find it difficult to send representatives for a sustained period because time spent at the ADR meetings is time taken away from jobs and involves loss of income. Thus, *financial* considerations can constrain an appropriate mix of representation.

The *scope* of an issue can also create problems for representation. In a conflict over deciding about development of a remote area for mining or timber extraction, or for designation as a national park or protected area, the interests to be represented regarding the heritage aspects could be those of citizens who live far away from the area. Or, if the site is judged to be of global significance, then the question becomes how to ensure that representatives relative to its global value are incorporated into the process.

Another challenge is to achieve *intergenerational* representation. In Chapter 4, it was noted that intergenerational equity is a basic component of sustainable development. How to achieve representation for "future generations" is clearly problematical, other than encouraging or expecting participants to take a temporal perspective that is longer than their own generation.

11.4.4 Involvement in design of the ADR process

A well-accepted tenet of ADR is that the process to be used needs to be agreed upon before substantive aspects of the dispute are addressed. Normally, the various parties in the dispute participate in the design of the process. A number of matters require attention.

Since ADR usually strives to reach agreements on the basis of *consensus*, it is important to have a common interpretation about the meaning of "consensus". Consensus normally implies a "general agreement" as opposed to an agreement based on a majority or a unilateral decision by an individual in a position of authority. More than any other basis for reaching decisions, consensus requires the greatest amount of trust, good will and mutual respect. Consensus also treats everyone as equals, since no-one need fear that he or she will be overwhelmed by a majority vote if they hold a minority opinion. In order to reach consensus, people usually are less concerned about the number of votes that will be given to a particular option than they are to identify aspects for accommodation and innovative resolution. Consensus is not a perfect basis on which

---

Box 11.7 A search for consensus can be problematical

The ideal of harmony, so prevalent in the West today . . . can be used to suppress criticism. It locks out those who continue to protest and discourages disagreement among those who have accepted a seat at the negotiation table. Confrontations can destroy the trust and assumption of shared goals which negotiation requires, jeopardizing the whole process of conflict resolution. Aiming for consensus, the process reinforces the status quo; radical change is unlikely to emerge when any change has to be agreed to by all parties.

Sources: Bedir, 1994: 236.

---

to make decisions, however, as it can lead to pressure for conformity and can suppress innovative ideas which are not generally supported (Box 11.7).

There is no single definition for what constitutes a "consensus" – the key concern is that the parties in a dispute agree in advance what the criteria should be to identify a consensual agreement. For example, 100% agreement, or unaniminity may represent an ideal, but another interpretation could be lack of dissent (with silence meaning acceptance), or agreement by a "vast majority" (only a few parties dissenting). Provision should also usually be made for a "fallback" position, such as agreeing to focus initially on aspects for which near unanimous agreement can be reached, and then moving on to more difficult matters. Other procedural issues need to be resolved. These include arrangements for different parties to communicate with their constituents, understanding about whether or not there is to be confidentiality of discussions, arrangements about the sharing of information, and creation of deadlines or targets. The key is that such matters should be sorted out before discussions begin on the substance of the conflict.

11.4.5 Acceptance of need for challenging constructively

The notion of "challenging constructively" could be included as an component of the process (Section 11.4.4 above), but is significant enough to deserve separate discussion. If trust, good will and respect are to be built up, it is important that participants seek to be constructive in their dialogues, rather than engage in destructive challenging that often occurs in an adversarial situation. Some key considerations in this regard include striving to avoid a negative and combative manner, seeking to bring people to the discussions who have relevant experience and interests, trying to gain a joint understanding of issues through systematic consideration of assumptions, data and logic, and viewing the group as "problem solvers" rather than as "argument destroyers".

11.4.6 Scope for compromise

"Compromise" often has a negative connotation, in that it can imply giving up or sacrificing important values or principles. If ADR is to work, however, there

---

Box 11.8 Willingness to compromise: coastal erosion example

The erosion of coastal barrier islands presents a classic environmental policy problem characterized by both uncertainty and conflict.

Stakeholders with an interest in shoreline property cling to the conventional characterization of the problem. Owners of shoreline property view erosion as a threat to their economic interests. They insist on the right to protect their property and strongly support government programs that subsidize protective measures such as flood insurance and publicly funded erosion control projects. Real estate and development business, along with local governments that rely on oceanfront property for a substantial portion of their tax revenues, have similar interests and concerns.

Environmentalists, scientists, and people with an interest in the public value of the beach and its environment challenge the right to protect private property at the expense of the natural system. Shoreline protection structures such as groins and seawalls are criticized for intruding visually and physically onto the public trust beach and for interfering with natural geologic and ecologic processes. Publicly-funded beach protection and restoration projects, and emergency response initiatives following major storms, are viewed as inequitable subsidies for the elite who own beachfront property.

Source: Deyle, 1994: 461.

---

has to be a willingness to accept the validity of another party to hold a different perspective; to try to understand, if not always agree, with those other perspectives; and to search for solutions that accommodate diverse interests (Box 11.8). If a party comes to an ADR meeting with the view that it will not be flexible, and will accept solutions that satisfy only its interests, then the likelihood of ADR being effective is very low.

On the other hand, at some point a party may indicate an item associated with the dispute involves a matter of such principle that compromise is not possible. It is also possible that a solution could have such a heavy and onerous impact on one or more parties that they are unwilling to accept the decision, even if most of the other parties would benefit significantly. If a participant can clearly demonstrate that a decision is problematical because of a principle or an unreasonably negative impact for her or his group, and therefore is unable to compromise, then "it becomes incumbent upon the rest of the group to make an explicit effort to address those concerns. In many instances, simply identifying the point of dissension and having it accepted by the rest of the group is half the battle in resolving impediments to a general agreement" (British Columbia Round Table on the Environment and the Economy, 1991a: 5).

11.4.7 Acceptance of a "principled approach"

Fisher and Ury (1981) developed a "principled approach" for dispute resolution, which they contrasted to a "positional approach". In the positional approach, parties arrive for negotiations having already decided on a desirable solution, and

---

Box 11.9 Characteristics of a principled approach

(1) separate the people from the problem
(2) focus on interests, not positions
(3) invent options for mutual gain
(4) insist on explicit, "objective" criteria to guide decisions

Source: After Fisher and Ury, 1981.

---

attempt to persuade or coerce others to accept their terms and solution. Thus, they arrive with a "position", and their goal is to achieve it. Such an approach tends to constrain flexibility, and a willingness to be open-minded about alternative solutions. A positional approach also tends to create a win–lose situation, as some parties attain their positions and others do not achieve theirs.

In contrast, a principled position emphasizes avoiding the acceptance of a position from the outset, but instead stresses working with other parties to develop a creative solution which will meet most people's needs. The attributes of the principled approach are shown in Box 11.9.

(1) *Separate the people from the problem.* Fisher and Ury argue that it is important to be "soft" on the people but "hard" on the problem. In other words, the goal should be to look beyond the idiosyncrasies of the people involved, and instead concentrate on the problems of common concern. Clearly, at times, people are part of the problem, especially if they are bloody-minded, vindictive or Machiavellian. However, other than in such obvious instances in which the people are indeed a key part of the problem, the idea is that people should overlook differences in personality and style, and focus on the substantive problems. This approach does not require parties to view one another as friends, but the argument is that they should view themselves as joint problem solvers, rather than as adversaries or argument destroyers.

(2) *Focus on interests rather than positions.* In their work on dispute resolution, Fisher and Ury have found that groups with different publicly declared *positions* often have similar or shared *interests*. Thus, for example, two opposing parties may find that they share a common interest in avoiding the pollution of an estuary. However, one may have a public position that a new factory is essential for providing jobs in an area with a high unemployment rate, and the other may have a public position opposing the factory. More productive discussions would occur if they focused on their common interest in protecting the water quality in the estuary, and then explored alternative ways of accommodating the needs of the factory and the quality of the estuary.

(3) *Invent options for mutual gain.* Rather than each party considering only those solutions that provide benefits to itself, the belief is that it is more constructive to search for solutions that generate benefits for all parties. This view does not mean that every party will always get everything it wants. However, if some groups get most or all of what they want, and

others get little or nothing, a decision is likely to face many challenges over a long period of time.

(4) *Insist on explicit, "objective" criteria.* If options for mutual gain are to be developed, then it is important that the parties in a conflict agree, at the outset of their negotiations, about the criteria against which possible solutions will be assessed. If it can be established at the outset which criteria will be used, then the expectation is that options for mutual gain will more likely be considered on the basis of "principled" rather than "positional" criteria. The use of such criteria should allow the parties in the conflict both to use reason and to be open to reason concerning identification and evaluation of options. The intent then would become to identify one or more solutions that provided the greatest benefits relative to the agreed upon criteria, rather than the positions of any one group.

The use of these four elements of a "principled" approach does not guarantee a mutually acceptable solution will be generated, and conflicts will be resolved. However, they do provide a systematic structure or framework to help guide groups who agree to use some form of ADR to address their dispute.

11.4.8 Capacity for implementation

Unlike the judicial approach, which has the power of the courts and law enforcement systems to uphold decisions, the voluntary, mutual-agreement approach of ADR does not have a built-in mechanism for implementing agreements. As a result, it is important to specify what arrangements will be made to implement decisions resulting from ADR. Furthermore, to provide credibility for the process, it is also useful to design some form of *monitoring* and *reporting* so that accountability is provided (see also Chapters 12 and 13). Without such accountability, the legitimation or credibility of the ADR agreement may fall in doubt if people do not have evidence of action following the agreements.

11.5 Applications of alternative dispute resolution

Kartez and Bowman's (1993) analysis of the experience with resource developments in Colorado and Texas illustrates effectively the factors which help and hinder ADR approaches in resolving conflicts. Their basic message is that "quick deals" often are no better than "no deal" in environmentally based conflicts, if a long-term resolution is desired.

11.5.1 Colorado and Homestake Mining

In 1980, the Homestake Mining Company negotiated an agreement with Colorado environmental groups and affected residents regarding an open-pit uranium mine. This experience is often cited as one of the earliest examples of successful use of voluntary dispute resolution procedures as an alternative to what could have been lengthy and costly litigation. Prior to the use of voluntary dispute resolution, Homestake had experienced four years of conflict in obtaining permits for the mine, and anticipated up to five more years of costly legal processes. In 18 months of mediated direct negotiation, an agreement was reached which led to granting of the permits.

The conflict

The proposed open-pit uranium mine was to be sited in the high country of a national forest, triggering vocal protests from a coalition involving local and statewide conservation groups, as well as from local residents concerned about environmental degradation and safety. Other concerns reflected ideological views about the appropriateness, or otherwize, of nuclear power. The mining company had completed a required environmental impact statement (EIS) as part of the permitting process, but the conservation groups immediately criticized the EIS and threatened litigation due to what they considered to be inadequacies in the statement. As Kartez and Bowman (1993: 321) remarked, "Although the opponents 'got the company's attention', they realized they would be unable to 'get the assurances they really wanted' on mine reclamation, backfilling, long-term water quality maintenance, and revegetation. Litigation over procedural issues is notoriously inept at solving complex environmental management problems that really require creative solutions."

For its part, while Homestake believed that the adequacy of the EIS and reclamation plan would be verified, they wanted assurance that they could indeed resolve the environmental problems, rather than simply winning on legal or technical grounds. The counsel for Homestake noted that his client wanted to avoid a situation in which opponents would demand that every possible technical issue, however remote, had to be answered definitively before the mine was approved. He noted that reliance on administrative appeals and litigation almost guaranteed such a tactic would be used by opponents. Thus, all sides (the mining company, the conservation groups, the local residents) had reached a stage in which alternatives to a "win–lose" adversarial process became appealing. In the jargon of ADR, the issue had become *ripe*. The parties had arrived at a point where they recognized the presence of a conflict, and when voluntary searching for a mutually agreeable solution appeared to be more attractive than any other method to resolve the dispute.

The ADR process

Kartez and Bowman commented on four problems that the parties had to overcome if ADR was to be used. The first issue was whether there was scope for compromise through negotiations. The *position* of the opponents that the mine project should be halted appeared to leave little scope for consideration of alternatives. However, informal debate prior to the actual mediation discussions revealed that the basic *interests* of the parties in dispute were more similar than the publicly declared positions: both sides shared an interest that the mine

---

Box 11.10 Consensus and compromise

Rarely do negotiation and consensus processes result in "all win" solutions. This approach searches for a middle ground and areas of accommodation or compromise. For many, however, the term "compromise" conjures up visions of sacrificing "good" solutions for expediency, and long-term solutions for "quick fixes".

In every dispute, there may be interests that cannot and should not be compromised: for example, those that define the fundamental identify of an individual or organization. It is important that all parties in negotiations understand where those essential interests lie. In addition, compromise does not mean sacrificing principles or fundamental values. Consensus-building can recognize fundamental values and still reach accommodations on such things as "how" rather than "whether" these values will be satisfied.

Source: British Columbia Round Table on the Environment and the Economy 1991a: 7.

---

property should be reclaimed properly. As a result, discussion shifted from whether or not the mine site should be reclaimed, to how it might be reclaimed (Box 11.10). This change reflected the similarities in interest. The conservation groups' position that the mine operation should be disallowed had been its way of trying to avoid degradation and safety problems. Those groups were willing to consider other solutions that would meet their concerns. As a result, "the significance of this step was that the players defined a problem that might be solved, based on a common interest" (Kartez and Bowman, 1993: 321).

The second issue involved *representation* of interests for the mediation discussions. An "anti-nuke" group refused to negotiate with Homestake on any basis. As a result, the mining company, the conservation groups and the residents agreed to search for a voluntary settlement through mediation, recognizing that any agreement would be susceptible to attack from the "anti-nuke" group which would not be at the table.

The third problem to be resolved was *data, access to data* and the *role of experts*. The parties in the dispute agreed that technical experts would not be the main negotiators, but would serve as resource people. Furthermore, the mining company agreed to provide answers to a detailed set of technical questions. The outcome was that rather than information being used strategically by each side to bolster its position and interests, information became a shared resource to help in identifying new solutions. This decision moved the discussions away from the common challenge and counter challenge regarding assumptions and facts, and focused attention upon solving shared concerns.

The fourth issue involved the relationship between the negotiating parties and the *government regulatory agencies*. The mediators emphasized that it was important for all the parties involved in the negotiations to communicate early and continuously with the regulatory agencies. The advantage of this approach was that the regulatory agencies would be aware of the issues and possible solutions being considered, and could provide advice about their legal, administrative

> ## Box 11.11 The settlement
>
> For the opponents, the settlements led Homestake to accept higher standards for managing mine reclamation and environmental impacts and for ongoing rather than one-shot mitigation. Homestake received assurance that the uncertainty of legal tactics-of-delay would be avoided.
>
> *Source*: Kartez and Bowman, 1993: 323.

and political feasibility. In this manner, time was not spent on options which were "non-starters", and in addition the regulatory agencies did not have to worry that their roles and statutory authority were being undermined by the voluntary, negotiation process.

Implications

The dispute over the open-pit uranium mining operation met many of the conditions which support the effective use of ADR. The parties in dispute believed that ADR was likely to be more helpful than a judicial approach, and there was willingness by each party to recognize the needs and concerns of the other parties in the conflict. There was willingness to ensure full representation of interests, to the extent that was practical, to share information and expert understanding, and to work with the regulatory agencies.

The outcome was a resolution containing three key elements. A *Statement of Understanding* specified the mining company's commitment to manage water quality, restore the mine pit, and monitor erosion and revegetation. These provisions also were to be included in the permits required from the regulatory agencies. A *Mediation Agreement* stipulated that the company would make specified data about the reclamation work available to the conservation groups, and also that it would fund research related to high-altitude revegetation. Finally, a *Covenant-Not-To-Sue* was signed by each member of the environmental coalition, committing them to refrain from legal action as long as Homestake abided by its commitments outlined in the other two documents (Box 11.11).

11.5.2 Texas, Mitsubishi Metals Corporation and Texas Copper Company

During early 1989, the Mitsubishi Metals Corporation (MMC) announced that it intended to construct a quarter-billion dollar smelter in Texas City which would be operated by a subsidiary, Texas Copper Company (TCC). By the end of the year, the proposal was under vigorous criticism from environmental groups concerned about the possible impact from the discharge of heavy metals into Galveston Bay. In 1991, MMC signed an agreement with some of its critics, hoping to avoid protracted legal proceedings. However, as the comments in Box 11.12 show, the outcome was not as satisfactory as the experience in Colorado.

Box 11.12 The Texas Copper conflict

Although an agreement was signed in one evening, there was continued acrimony and threats of legal action by other opponents of the facility, new requirements by the state regulatory agency after a change of administration (requirements which the project proponents rejected), eventual withdrawal of the proposed project by the company, and charges of bad faith from most quarters.

Source: Kartez and Bowman, 1993: 320.

The conflict

The conflict divided groups into those who saw the smelter as an important addition to the local economy, and those who viewed it as a threat to the ecology of Galveston Bay. After reviewing an initial environmental impact statement, the US Environmental Protection Agency and the Corps of Engineers agreed that the smelter posed no significant threat of pollution, which meant a full impact statement was not required. However, during the early part of 1990, critics appeared before the Texas Water Commission to express concern about possible pollution in the bay. The critics included representatives of several conservation groups, an upper-class community sited on a bay island, and a broadly based coalition opposing the smelter due to concern about its impact on health of nearby residents. The protest quickly became highly visible, as the critics initiated petitions, released press statements and began public demonstrations. The critics also complained that Texas Copper "was withholding key information, being deceptive, and letting a Japanese corporation exploit Galveston Bay" (Kartez and Bowman, 1993: 323). The critics demanded a full environmental impact assessment, and also that Texas Copper should be required to use state-of-the-art wastewater technology if the smelter were to be built.

In June 1990, the Texas Water Commission did grant a discharge permit to Texas Copper, but its own legal department challenged that decision. When the decision to grant the permit was upheld, opponents filed for a hearing, and late in the year the Water Commission reopened hearings which were to focus on technical aspects in dispute. By the spring of 1991, the company had been given a three-year permit with specific directions regarding how to handle discharges. Opponents of the smelter immediately filed for another hearing.

A dispute settlement process

By the middle of 1991, the positions of the opposing parties had hardened and polarized. Concern of the opponents about technical issues had become less significant than their lack of trust in the president of Texas Copper. As a result, substantive issues became subsidiary to issues related to personalities and previous decisions. Thus, the parties in dispute were increasingly focusing on people rather than the problem, and on positions rather than on interests. They also had not established mutually acceptable criteria for evaluating options, and were not searching for a solution with benefits for all parties. The dispute was increasingly adversarial.

Against this background, the Japanese chief operating officer contacted the lawyer for the Galveston Bay Foundation, one of the opponents, and arranged to have a dinner. The result of the dinner meeting was an agreement between Texas Copper, the Galveston Bay Foundation, and the Galveston Bay Conservation and Preservation Association. The agreement was very similar to that in Colorado with Homestake Mining. The two opponents agreed not to pursue legal action regarding the water permit. The company agreed to install additional wastewater treatment capacity which would be subject to approval by the two groups, as well as by the regulatory agencies. The director of the Galveston Bay Foundation indicated that her group had never been opposed to the smelter *per se*, but to unnecessary contamination of the bay. Thus, they were pleased to have reached a mutually developed solution to a shared concern – how to avoid pollution of Galveston Bay.

Thus, the parties at the dinner meeting concentrated upon substantive problems rather than the people, upon interests rather than positions, and on finding a mutually beneficial solution. However, the effectiveness of such an agreement is "also influenced by who is or is not involved, and the conditions under which settlement occurs" (Kartez and Bowman, 1993: 325). The agreement reached in one evening had excluded participation of the two most vociferous opponents: the bay island community, and the grass-roots group concerned about health impacts.

The bay island community and grass-roots group were shocked at the agreement, and felt left out and betrayed by the Galveston Bay Foundation and the Galveston Bay Conservation and Preservation Association. They announced that their opposition would continue, and that they would pursue their concerns through the courts. Furthermore, the agreement about appropriate technology was not approved by the chair of the regulatory agency, who simply said that Texas Copper would have to meet a zero discharge target within one year of operation. This decision meant that a cooperative and joint evaluation of the treatment technologies was unlikely to be feasible. Within a few months of the dinner agreement, the Japanese parent corporation reduced the staff at Texas Copper and announced cancellation of the smelter project, referring to unreasonable local opposition and environmental standards.

Implications

The dispute in Texas generated by the proposed smelter and associated concerns about pollution of Galveston Bay had many of the same characteristics of the Colorado experience with the open-pit uranium operation. Both situations involved concerns about what were acceptable environmental impacts. Each included a mix of opponents with differing interests. Both involved concern about technical options, cumulative impacts and capacity for mitigation measures. Both included the need for obtaining multiple permits. All of these

> ## Box 11.13 Key elements for successful ADR
>
> . . . deciding the boundaries of who should be included as legitimate parties is one of the most important elements of effective pre-negotiation protocols or preparations. . . . A large part of the durability of outcomes from voluntary negotiations is "procedural satisfaction." That is the parties' belief that they have been better served by using a voluntary process than by fighting on, and that they would use such a process again.
>
> . . . No competent mediator would have failed to tell the Texas Copper combatants that their exclusion of parties, rushing of negotiations, and avoidance of coordination with regulators was a recipe for useless negotiations.
>
> . . . One clear lesson is that environmental issues by their nature require an open process. The minute the process is closed for tactical reasons, any opponent often has a case to dispute the outcome by indicting the process, including the technical adequacy of any scientific analysis which by its nature requires open argument.
>
> *Source*: Kartez and Bowman, 1993: 327.

common characteristics normally provide a strong incentive for use of ADR, to avoid what could be prolonged and potential intractable battles.

The main reason that ADR was successful in Colorado and was not successful in Texas can be linked to a key consideration essential for ADR: "appropriate representation and willing participation" (Kartez and Bowman, 1993: 326; Box 11.13). These two aspects were not incorporated into the design of the dispute settlement process used by Texas Copper and its opponents. As a result, a quickly developed solution could not be implemented.

11.6 Implications

It is has been stated before in this book that we often do not manage resources and the environment, but that we manage human interaction with resources and the environment. When that is indeed the situation, then very often planners and managers have to decide how best to address and resolve conflicts. The impacts of resource development or use of the environment normally generate benefits and costs which have different implications for the interests of various groups. Almost inevitably, one or more groups will feel disadvantaged, and will protest a proposed policy or development related to use of resources or the environment. As a result, an important skill is conflict or dispute resolution.

Societies have developed various ways to address conflicts. The judicial approach is the most frequently used. The use of courts and their well-established processes and procedures is often the most appropriate way to deal with a dispute. However, increasing concerns about the time, costs and adversarial nature of the judicial approach have led to increasing attention being given to alternative dispute resolution (ADR). It has never been argued that ADR should replace the judicial approach, but that some disputes, handled through ADR, might be resolved at less cost and reduced time.

Guest Statement

From ADR to AG: potentials and challenges

Anthony (Tony) H J Dorcey, Canada

Over the last decade there have been major innovations in approaches to conflict resolution in environment and natural resources management. In part, this has been a response to the increasing demands, complexity and uncertainty that have together generated increasing conflict. But it has also been associated with the remarkable shift to seeing these conflicts in the more comprehensive context of sustainability governance. In the first half of the 1990s Canada was at the forefront of these innovations in alternative governance (AG), stimulated by the Brundtland Report's imperative of sustainable development and the development in the USA of alternative dispute resolution (ADR) techniques and processes. By the end of the decade, Canada had lost the initiative as its governments retreated from overly ambitious and seemingly threatening experiments and the cutting-edge of innovation shifted back to the USA and was taken up internationally by both developed and developing countries.

Today's AG perspective on the resolution of conflicts in environmental and natural resources management is significantly different from the earlier, much narrower perspective of ADR. Whereas ADR had its origins in using negotiation and mediation to avoid the costs and uncertainties of resolving conflicts through the courts, AG focuses much more broadly on their use in the overall governance system, considering how they might be utilized to advantage in every aspect of state, civil society and business activity. In addition, AG focuses on how to reach agreements proactively, and not just reactively resolve disputes.

The innovations in AG approaches are exemplified by the multi-stakeholder round-table processes created for drafting new policies, legislation and regulations; developing watershed and regional resource management plans; and implementing stewardship initiatives. These processes may be initiated by either government or civil society or business. They use a wide variety of negotiation-based techniques and employ various types of facilitators and mediators throughout the process. Such decision-making processes are often described as being "negotiatory" or "shared" or "consensus-based". Particular applications have been given distinct labels such as "round tables", "policy dialogues", "reg-neg (regulatory negotiation)", "consensus conferences", "charettes" and "co-management". But negotiation and mediation techniques and processes have also been increasingly used to transform the operation of traditional mechanisms such as public hearings, advisory committees and community boards.

For some people, AG represents a radical new model of democracy; for others it implies at least a shift from the dominant managerialist to more pluralist and populist forms of governance. In addressing the more fundamental issues of power and inequity that underlay environmental and natural resource management conflicts, it is threatening to those who hold power. It is thus not surprising that innovations such as those in Canada in the early 1990s induce a backlash. But it also has to be recognized that many of the innovations were hugely ambitious experiments that inevitably failed to deliver on all that they promised. Unfortunately, insufficient attention has been given to evaluating these AG approaches and claims as to their relative merits are not well substantiated by empirical studies.

In the future, attention should focus on adaptive strategies which build evaluation into AG innovations. These strategies should draw on the emerging understanding of best management practices by practitioners and academics, focusing in particular on the critical role of process facilitators, explicit mandates from sponsors of the process, the appropriate design of the process for the context, and the fundamental issues of empowerment. These adaptive strategies should employ not only participant observation but also action research methods. Done this way, the inevitable next wave of innovation will overcome many of the challenges that frustrated the earlier experiments and more fully meet the huge promise of AG.

Anthony (Tony) H J Dorcey is Director of and Professor in the School of Community and Regional Planning at the University of British Columbia, Vancouver, Canada. His teaching, research and professional practice focus on the use of negotiation, facilitation and mediation in sustainability governance. During the last decade he has been involved in the experimental development of AG as a member of the BC Round Table on Environment and Economy, inaugural Chair of the Fraser Basin Management Board, and facilitator of international multi-stakeholder processes such as the one that resulted in the establishment of the World Commission on Dams.

There are different types of ADR, such as negotiation, mediation and arbitration. However, each shares core ideas that voluntary searches for solutions which represent mutual gain for parties in a dispute can be feasible. For ADR to be used, however, it is critical to recognize that some preconditions must be satisfied. Some of the most critical include general recognition of the presence of a dispute, motivation to consider ADR for joint problem solving, willingness to compromise, preparedness to use a "principled approach", and determination to facilitate implementation of an agreement. If these preconditions cannot be satisfied, ADR is unlikely to be effective.

It can be anticipated that ADR will be used increasingly in the future, especially in countries with a tradition of democratic governance. Even if ADR is not directly applicable, awareness of and sensitivity to some of the key ideas associated with that approach are likely to make planners and managers more effective in resolving disputes related to resource and environmental issues. Furthermore, as the guest statement by Tony Dorcey reveals, ADR may provide the stimulus for alternative types of governance, which may represent more profound shifts in approach to resource and environmental management.

References and further reading

Allor D J 1993 Alternative forums for citizen participation: formal mediation of urban land use disputes. *International Journal of Conflict Management* 4: 167–80

Amy D 1987 *The Politics of Environmental Mediation.* New York, Columbia University Press

Bedir S 1994 Consensus or conflict? *Ecologist* 24: 236–7

Bingham G 1986 *Resolving Environmental Disputes: A Decade of Experience.* Washington, DC, The Conservation Foundation

Blackburn J W 1988 Environmental mediation as an alternative to mitigation. *Policy Studies Journal* 16: 562–74

British Columbia Round Table on the Environment and the Economy. Dispute Resolution Core Group 1991a *Consensus Processes in British Columbia.* Volume 1, Victoria, British Columbia Round Table on the Environment and the Economy

British Columbia Round Table on the Environment and the Economy. Dispute Resolution Core Group 1991b *Implementing Consensus Processes in British Columbia.* Volume 2, Victoria, British Columbia Round Table on the Environment and the Economy

Buckle L G and S Thomas-Buckle 1986 Placing environmental mediation in context: lessons from failed mediations. *Environmental Impact Assessment Review* 6: 55–70

Burkardt N, B L Lamb and J G Taylor 1998 Desire to bargain and negotiation success: lessons about the need to negotiate from six hydropower disputes. *Environmental Management* 22: 877–86

Busenberg G J 1999 Collaborative and adversarial analysis in environmental policy. *Policy Sciences* 32: 1–11

Carpenter S L and W J D Kennedy 1988 *Managing Public Disputes: A Practical Guide to Handling Conflict and Reaching Agreements.* San Francisco, Jossey-Bass

Chenoweth J L 1998 Conflict in water use in Victoria, Australia: Botte's Divide. *Australian Geographical Studies* 36: 248–61

Cormick G W 1980 The "theory" and practice of environmental mediation. *Environmental Professional* 2: 24–33

Cormick G W 1989 Strategic issues in structuring multi-party public policy negotiations. *Negotiation Journal* 5: 125–32

Cumbler J T 2000 Conflict, accommodation and compromise: Connecticut's attempt to control industrial wastes in the Progressive Era. *Environmental History* 5: 314–35

Deyle R E 1994 Conflict, uncertainty, and the role of planning and analysis in public policy innovation. *Policy Studies Journal* 22: 457–73

Dorcey A H J 1986 *Bargaining in the Governance of Pacific Coastal Resources: Research and Reform.* Vancouver, BC, University of British Columbia, Westwater Research Centre

Dorcey A H J and C L Riek 1987 Negotiation-based approaches to the settlement of environmental disputes in Canada. In Canadian Environmental Assessment Research Council *The Place of Negotiation in Environmental Assessment.* Ottawa, Canadian Environmental Assessment Research Council, pp. 7–36

Doremus H 1999 Nature, knowledge and profit: the Yellowstone bioprospecting controversy and the core purposes of America's national parks. *Ecology Law Quarterly* 26: 401–88

Fisher R and W Ury 1981 *Getting to Yes: Negotiating without Giving In.* Boston, Houghton Mifflin

Gadgil M and R Guhn 1994 Ecological conflicts and the environmental movement in India. *Development and Change* 25: 101–36

Glavovic B 1996 Resolving people–park conflicts through negotiation: reflections on the Richtersveld experience. *Journal of Environmental Planning and Management* 39: 483–506

Grandy M 1995 Political conflict over waste-to-energy schemes: the case of incineration in New York. *Land Use Policy* 12: 29–36

Haftendorn H 2000 Water and international conflict. *Third World Quarterly* 21: 51–68

Harashina S 1995 Environmental dispute resolution: process and information exchange. *Environmental Impact Assessment Review* 15: 69–80

Hollander G M 1995 Agroenvironmental conflict and world food system theory: sugarcane in the Everglades Agricultural Area. *Journal of Rural Studies* 11: 309–18

Hurwitz D 1995 Fishing for compromises through NAFTA and environmental dispute-settlement: the tuna–dolphin controversy. *Natural Resources Journal* 35: 501–40

Innes J E 1996 Planning through consensus building: a new perspective on the comprehensive planning ideal. *Journal of the American Planning Association* 62: 460–72

Johnson P J and P N Duinker 1993 *Beyond Dispute: Collaborative Approaches to Resolving Natural Resource and Environmental Conflicts.* Thunder Bay, Ontario, Lakehead University, School of Forestry

Kartez J D and P Bowman 1993 Quick deals and raw deals: a perspective on abuses of public ADR principles in Texas resource conflicts. *Environmental Impact Assessment Review* 13: 319–30

Kelly R A and D K Alper 1995 *Transforming British Columbia's War in the Woods: An Assessment of Vancouver Island Regional Negotiation Process of the Commission on Resources and Environment.* Victoria, University of Victoria, Institute for Dispute Resolution

Kliot N 1993 *Water Resources and Conflict in the Middle East.* London, Routledge

Knopman D S, M M Susman and M K Landy 1999 Civic environmentalism: tackling tough land-use problems with innovative governance. *Environment* 41(10): 24–32

Krakover S 1999 Urban settlement program and land dispute resolution: the State of Israel versus the Negev Bedouin. *GeoJournal* 47: 551–61

Maguire L A and L G Boiney 1994 Resolving environmental disputes: a framework incorporating decision analysis and dispute resolution techniques. *Journal of Environmental Management* 42: 31–48

Margairo A V, S Laska, J Mason and C Forsyth 1993 Captives of conflict: the TEDS case. *Society and Natural Resources* 6: 273–90

McDorman T L 2000 Global ocean governance and international adjudicative dispute resolution. *Ocean and Coastal Management* 43: 255–75

McManus E C 2000 Mutual trust and mutual gains: a recent Crown land use planning exercise in Ontario. *Forestry Chronicle* 76: 425–27

McMullin S L and L A Nielson 1991 Resolution of natural resource allocation conflicts through effective public participation. *Policy Studies Journal* 19: 553–9

Mercer D 1999 Closing the gap: Australian–Indonesian relations, the "perilous moment" and the maritime boundary zone. *Tijdschrift voor Economische en Sociale Geografie* 90: 61–79

Montrie C 2000 Expedient environmentalism: opposition to coal surface mining in Appalachia and the United Mine Workers of America, 1945–1977. *Environmental History* 5: 75–98

Moore C and M Santosa 1995 Developing appropriate environmental conflict management procedures in Indonesia: integrating traditional and new approaches. *Cultural Survival Quarterly* 19: 23–9

Nepal S K and K E Weber 1995 The quandary of local people–park relations in Nepal's Royal Chitwan National Park. *Environmental Management* 19: 853–66

Osborne P L 1995 Biological and cultural diversity in Papua New Guinea: conservation, conflicts, constraints and compromise. *Ambio* 24: 231–7

Ozawa C P and L Susskind 1985 Mediating science-intensive policy disputes. *Journal of Policy Analysis and Management* 5: 23–39

Rankin M 1989 The Wilderness Advisory Committee of British Columbia: new directions in environmental dispute resolution? *Environmental and Planning Law Journal* March: 5–17

Rogers P 1993 The value of cooperation in resolving international river basin disputes. *Natural Resources Forum* 17: 117–32

Sadler B and A Armour 1987 Common ground: on the relationship of environmental assessment and negotiation. In *The Place of Negotiation in Environmental Assessment*. Ottawa, Canadian Environmental Assessment Research Council: 1–6

Schmidtz D 2000 Natural enemies: an anatomy of environmental conflict. *Environmental Ethics* 22: 397–408

Shaftoe D (ed.) 1993 *Responding to Changing Times: Environmental Mediation in Canada*. Waterloo, Ontario, Conrad Grebel College, The Network: Interaction for Conflict Resolution

Stern A J 1991 Using environmental impacts for dispute resolution. *Environmental Impact Assessment Review* 11: 81–7

Susskind L E and J Cruikshank 1987 *Breaking the Impasse: Consensual Approaches to Resolving Public Disputes*. New York, Basic Books

Talbot A R 1983 *Settling Things: Six Case Studies in Environmental Mediation*. Washington, DC, The Conservation Foundation

Tarock A 1999 The politics of the pipeline: the Iran and Afghanistan conflict. *Third World Quarterly* 20: 801–20

Weed T J 1994 Central America's "peace parks" and regional conflict resolution. *International Environmental Affairs* 6: 175–90

West L 1987 Mediated settlement of environmental disputes: Grassy Narrows and White Dog revisited. *Environmental Law* 18: 131–50

Wiesmann U, F N Gichuki, B P Kiteme and H Liniger 2000 Mitigating conflicts over scarce water resources in the highland–lowland system of Mount Kenya. *Mountain Research and Development* 20: 10–15

Wood P M 1989 Resolving wilderness land-use conflicts by using principled negotiation. *Forest Planning Canada* 8: 42–7

Chapter 12

Implementation

12.1 Introduction

Implementation is usually interpreted to mean taking action, or taking something such as a promise or statement of intent and translating it into specific activity. In resource and environmental management, a challenge often is to move from *normative* planning (what should be done) to *operational* planning (what will be done) (also discussed in Chapter 8). An often heard criticism is that the world is littered with good intentions, policies and plans, but little follow up action. As the statement in Box 12.1 indicates, expressions of intent without associated activity normally have little value.

In this chapter, attention focuses upon some of the factors which hinder and facilitate effective implementation of policies and programmes. Ideas to improve our capacity for implementation are explored, and experiences with implementation are reviewed.

Box 12.1

Policies, by themselves, have very little value. Without the development of implementation strategies and the will to carry those policies into actual practice, all that is left are hollow words.

Broad statements of policy create expectations which must be met. A failure to meet expectations creates significant credibility gaps which, at a minimum, hamper further action. Moreover, a failure to implement good policy is also a failure to address significant problems in a meaningful way.

... policy makers must be held accountable not only to enunciate policy, but also to insure that the means exist to carry that policy out. We cannot be satisfied with the creation of policy alone. We must force the policy makers to address what is necessary to bridge the policy-practice gap. ... The key, ..., is to understand that most policy is not self-implementing and requires a conscious effort toward implementation before it will be actually realized in practice.

Source: Somach, 1993: 19, 20, 22.

12.2 Implementation issues

Weale (1992: 43) remarked that "implementation failure is like original sin: it is everywhere and it seems eradicable". If we are to eradicate, or at least reduce, such failure, it is important to have a clear understanding about what implementation implies, what some of the most important obstacles are, and what type of an implementation framework might help to structure our approach. Each is considered below.

12.2.1 Dimensions of implementation

Weale (1992) suggested attention needs to be given to two matters regarding implementation failure:

- policy outcomes not complying with policy objectives or expectations; and
- government or other organizations failing to recognize a problem or to take decisions about it – in other words, there is a lack of policy, or the intentions are not sufficient relative to the need.

Table 12.1 shows the relationship between a focus on policy or problems, with regard to both *outputs* (laws, regulations and agencies created to deal with a policy problem) and *outcomes* (changes in environmental degradation or resource use). Each of the four cells in this table draws attention to different issues and questions. For example:

Cell 1. Attention here is concerned with the extent to which governments or any other organizations have been able to take their declared intentions and change them into tangible *outputs*. Interpretation of the significance of outputs is not always straight forward, however. For example, some environmental legislation may have been created, so an output has been generated. However, if the government is slow or reluctant to use and enforce the act, the *outcome* may be negligible.

Cell 2. Here the interest is not just on the creation of an output, but also on the extent to which environmental degradation has been stopped or reversed, or

Table 12.1 Classification of implementation problems; for description of cells (1)–(4) see text

	Focus of analysis	
Orientation to problem	Output	Outcomes
Orientation to policy intentions	(1)	(2)
Orientation to problem	(3)	(4)

Source: Weale, 1992: 45.

Box 12.2 What is "implementation"?

Implementation, to us, means just what Webster and Roget say it does: to carry out, accomplish, fulfill, produce, complete. But what is it that is being implemented? A policy, naturally. There must be something out there prior to implementation; otherwise there would be nothing to move toward in the process of implementation.

Source: Pressman and Wildavsky, 1973: xiii.

---

resource patterns have become more efficient or equitable. The major challenge is to determine the influence of the implemented policy on the *outcome*, given that there could well be other variables contributing to it. Thus, air pollution may improve as a result of a Clean Air Act, but other contributing factors could include changes in relative prices of fuel contributing to greater use of a less polluting fuel, or development of more efficient heating units from homes, leading to reduced fuel consumption and therefore reduced emissions.

Cells 3 and 4. While in Cells 1 and 2 the focus is upon the matching of outputs or outcomes with intent, in Cells 3 and 4 attention is upon whether the outputs and outcomes are the most appropriate relative to a problem. An analogy can be made with a book review. The approach indicated by Cells 1 and 2 would be to review the book relative to the objectives stated by the author. The approach in Cell 3 would be to review the book relative to some other objectives or criteria deemed to be more pertinent or significant. The key here is to determine the range of options considered in developing what was to be implemented, and then to judge whether the chosen alternative was the most appropriate. From the perspective of Cell 4, the interest would be to judge whether the impact (outcome) would have been more significant if some other measure had been implemented. The viewpoint taken in Cells 3 and 4 diverges from that reflected in the comments by Pressman and Wildavsky contained in Box 12.2. They consider that judgements about implementation should only consider actions which follow policy. In other words, they position their view of implementation in Cells 1 and 2.

The value of Table 12.1 is not so much to determine into which cell or category implementation initiatives fit, or whether they fit neatly into one or another cell. More significant is the idea that we should differentiate between *outputs* and *outcomes* when considering the (in)effectiveness of implementation. Furthermore, attention to policy intention and problem orientation broadens our outlook, and should encourage us to start any review of implementation by considering the "bigger picture".

12.2.2 Important obstacles for implementation

From the research dealing with implementation, it is possible to identify several major factors which can affect implementation. These include: (1) tractability

Box 12.3 Considerations which increase intractability

During the period since the heady days of the 1960s many of the poorer LDCs [lesser developed countries] have been subject to intense pressure from the world system: commodity prices fell, energy prices rose astronomically, interest payments on debts became a drain on capital resources and the perpetual balance of payments crises, reschedulings and conditionalities focused attention in two areas. The first of these was running faster to stand still (the "White Queen" phenomenon) and the other was the inability to look much beyond tomorrow. Neither of these is conducive to the longer-term perspective of environmental management. In a situation of cutback the conservation budget is often the first to go in an LDC, simply because it does not give a rapid and visible return to the balance of payments.

Source: Baker, 1989: 33–4.

of the problem, (2) lack of clarity of goals, (3) commitment of those responsible for implementation, (4) resources (means) available to achieve goals (ends), (5) inadequate access to information, (6) inappropriate assumptions about cause–effect relationships, (7) dynamics of enforcement, (8) conditions specific to developing countries, and (9) different styles due to cultural variations.

(1) Tractability

As already noted in Chapter 2, some problems are more complex than others, leading to their being characterized as "messes" or as "wicked". Thus, the effectiveness of implementation will be influenced by the tractability, or resolvability, of the problem to which action is addressed (Box 12.3). It may be relatively easy to implement a Greenways trail system in a community in a developed country. It may be more difficult to reverse soil erosion and other environmental degradation in a developing country as a result of people who, below a subsistence level, are over-harvesting a resource. The latter problem is a symptom of structural poverty and inequity in a society, and increased extension services and information programmes focused on land-use practices will not by themselves remove the fundamental causes of the environmental degradation.

Tractability is also influenced by the diversity of behaviour to be modified. The greater the diversity, the more difficult it will be to develop responses applicable to all situations, and therefore the less likely that objectives will be realized. For example, water pollution regulations ideally have to match with the circumstances of the thousands of point (industrial factories, municipal sewage treatment plants) and non-point (farms, road surfaces) sources. The same dilemma is encountered for air pollution, when both point (factory and household emissions) and non-point (automobiles, trucks, trains, aeroplanes, ships) sources exist. Variation among the kinds, magnitudes, timing, duration and significance of so many sources "makes the writing of precise overall regulations essentially impossible" (Sabatier and Mazmanian, 1981: 8).

Other factors contribute to increased intractability. The larger the proportion of the population for which behavioural change is needed, or the greater the extent of behavioural change required, the more intractable will be the problem. All of these "tractability factors" contribute to a uneven playing field when comparisons are made regarding the effectiveness of different implementation initiatives. Some problems are fairly straight forward and resolvable. Others are more like a Gordian knot, and may defy all but extraordinary or ingenious solutions.

(2) Lack of clarity of goals

It is much easier to determine implementation "success" if a well-established vision and clearly defined objectives exist. As we will see in Section 12.3, some people believe that sharply defined objectives are most likely to lead to their realization because discretionary interpretation is less likely to distort what was intended. This perspective is called a *programmed approach* in Section 12.3.1. However, having well-defined goals is not sufficient for successful implementation. It is also important for the priority of objectives regarding resource and environmental management to be established relative to objectives of other policies. Such priority setting is often difficult, since at any given time societies normally pursue multiple objectives (protect biodiversity; create jobs; increase exports). Some of these will conflict with each other, and often there is expectation that some objectives will always prevail over others.

However, we will also see in Section 12.3 that not everyone agrees with the merits of having clear and unambiguous objectives. Those who support an *adaptive approach* argue that some deliberate vagueness and ambiguity are highly desirable, as they allow resource and environmental managers discretion to ensure that policies are implemented with regard to local conditions and needs (Section 12.3.2). Nevertheless, if there is no consensus about what the goals and objectives are, such an approach makes it more difficult to track or monitor the success in implementing policies or programmes. This problem will be considered further in Chapter 13.

(3) Lack of commitment

Highly committed and enthusiastic resource and environmental managers are often capable of implementing even poorly crafted or designed policies so that they will provide benefits (Box 12.4). In contrast, unmotivated or incompetent people may be unable to implement the most sophisticated and carefully designed policy. Sustained commitment and interest are critical for effective implementation.

Box 12.4 Commitment and leadership

... the variable most directly affecting the policy outputs of implementing agencies, namely, the commitment of agency officials to the realization of statutory objectives.

Source: Sabatier and Mazmanian, 1981: 20–1.

However, as Weale (1992) indicated, there can be many reasons for the leaders of an implementing agency to move forward in a reluctant or tentative manner. For example, the governments in both the USA and Britain revealed great reluctance to curtail the contribution from their countries to acid precipitation. One of the main reasons for such reluctance was that the costs of curtailment would be borne in their own countries, while many of the benefits to the environment would accrue to neighbouring countries (Canada in the case of the USA; the Scandinavian countries in the case of Britain). Another reason for hesitation can be conflict with other, higher priorities. A government may be sincerely concerned about environmental degradation, but it could have a greater concern about economic development which could contribute to creation of jobs, regional stability or national security. Or, again, it could be genuinely concerned about the environment, but give higher priority to reducing a national debt or deficit, and therefore be unprepared to allocate the necessary funds to reduce waste emissions, and to begin to rehabilitate degraded areas.

(4) Lack of means

Even if a strong commitment exists to deal with environmental problems, governments often find that they do not have the necessary means or tools to implement the most desirable package of activities (Box 12.5). Inadequate means can involve much more than shortages of money. For example, constitutional constraints may limit what can be implemented. In many countries, responsibility for resources and the environment is shared between central and state governments. The result is that joint initiatives are often required. However, if the other level of government with a shared responsibility does not give the same priority to an environmental initiative, it is likely that unilateral implementation will be only partially effective. Thus, for example, the German central government has concurrent power for air pollution with the *Wassergenossenschaften* and the *Länder* levels of government, but not the same shared power for water pollution because of unwillingness of the other levels to give up traditional authority. As a result, more progress has been achieved regarding air pollution than for water pollution.

Box 12.5 Sustaining commitment

In general, the commitment of agency officials to statutory objectives, and the consequent probability of their successful implementation, will be highest in a new agency with high visibility that was created after an intense political campaign. After the initial period, however, the degree of commitment will probably decline over time as the most committed people become burned out and disillusioned with bureaucratic routine, to be replaced by officials much more interested in security than in taking risks to attain policy goals.

Source: Sabatier and Mazmanian, 1981: 20.

Box 12.6 Consequences of administrative fragmentation

..., governments can find that one arm of government is unaware of what another arm of government is doing. Just at the time that the UK's environment department was grasping the importance of fossil fuel combustion and global environmental change, the transport department was publishing plans that proposed greatly to expand the motorway system with only the most cursory acknowledgement of their damaging environmental potential. Just at the time that the Dutch environment ministry ... was beginning to warn of the adverse environmental implications of increased NO_x emissions from increases in vehicle mileages, so the transport ministry raised the speed limit on motorways.

Source: Weale, 1992: 52.

Political constraints can also be limiting, in that both governments and its public servants have to be sensitive to public expectations and values. Thus, Weale (1992) explained that in Britain the Director-General of the Health and Safety Executive noted his organization had allocated more resources to regulating the nuclear power industry than the mining and construction industries – notwithstanding that fatality rates are much higher for the last two industries. However, the public has a higher-level concern about nuclear power, and it would be politically unwise for the Health and Safety Executive not to be seen to allocate substantial resources to the nuclear industry.

Another constraint can be the administrative or institutional structure for implementation. As already discussed in Chapter 5, the design of government agencies only occasionally reflects or matches the holistic character of ecosystems. As in most organizations, government departments are divided and subdivided into parts with specialized functions (Box 12.6). This is done to provide focus and to achieve efficiency. However, this division and subdivision of organizations and management functions makes it difficult to maintain a holistic approach. To illustrate, Weale (1992) explained that in Holland, nature conservation matters were handled in the ministry responsible for agriculture, while pollution control was allocated to the environmental ministry. One consequence was a difficulty in generating strategic thinking which crossed over nature conservation and pollution issues. As another example, in most countries the regulation of automobiles and other vehicles is done by a transportation agency. While road traffic is a primary non-point source of pollution, however, pollution usually is not the concern of a transportation agency. The outcome of such administrative arrangements can be difficulty in coordination and communication, considerable duplication and overall loss of efficiency – all of which are obstacles to implementation.

(5) Access to information

A major problem can occur when different participants do not have access to information. This dilemma often occurs with regard to environmental policy.

Officials in regulatory agencies usually do not know the full implications of adopting various pollution-reducing technologies for a specific industry or firm, nor do they always understand the implications of new costs to firms as a result of changed environmental standards. Industries or individual firms often know much more about their waste streams and costs compared with the regulator. And, the general public usually does not know what type of lobbying occurs behind closed doors between industry and government. As a result, there is not an even playing field for information. Some participants have greater access than others to information, and to the extent that "knowledge is power", may be able to facilitate or frustrate implementation activities. This takes on more significance in the context of Weale's (1992: 55) comment that "implementation is always a process of interpretation in which new information is added or required".

(6) Assumptions about cause–effect relationships

If a policy or programme is to have its anticipated impact, there should be understanding of the causal linkages between stated objectives and activities. In many situations, our understanding of biophysical and human systems is inadequate for us to be confident that we understand such cause-and-effect relationships, and therefore to be confident that interventions will produce anticipated effects.

However, even if our understanding of ecological and social systems were good, there is still scope for the outputs or outcomes of implementation to be different than expected. One explanation is the concept of *perverse effects*. A good example of this is the effort to implement a coastal protection policy in California. Weale (1992: 56) explained that the policy was effective in protecting visual amenities, but was very inadequate in ensuring access to the resource. The faulty assumption was that planning controls by themselves would be adequate to ensure public access to coastal areas. However, as the comments in Box 12.7 show, if an adaptive approach is used and we learn from such an error, then gradually management can be improved.

Despite the attractiveness of an ideal of learning from mistakes, such self-correction is not always easy to incorporate. First, there needs to be willingness within an organization to test assumptions, which requires an approach that encourages self-evaluation and accepts external criticisms. Such conditions are difficult to establish, since most organizations are hesitant to expose themselves to ongoing evaluation by their own officers, their clients or the public at large.

Box 12.7 Learning from mistakes

. . . opportunities are needed within a policy system to learn from mistakes and to correct policy in the light of experience. In the absence of these opportunities, implementation failures remain unremedied.

Source: Weale, 1992: 57.

---

(7) Dynamics of enforcement

Despite all of the concepts and theories that can be brought to bear regarding implementation, most implementation is done by people working at the field level. Such officials rarely can implement policies or procedures in a mechanical way, but instead have to make interpretations and exert judgement. For example, "when dischargers are in breach of limits the inspector has to make a series of judgements as to culpability, intentionality, likelihood of recurrence and so on" (Weale, 1992: 57).

Another complication is that field officers or inspectors often are more oriented towards negotiating to achieve compliance with environmental regulations, rather than with only enforcing rules. It is usual for some latitude to be given to an offender, in the expectation that such consideration will result in future compliance with regulations. An inspector has to decide whether strict enforcement will lead to improved environmental conditions, or may so antagonize business and industry that in the long run it will be more difficult to achieve environmental improvements. Thus, once again, judgement and discretion are involved, so outcomes from implementation may be uneven across a jurisdiction. This situation also indicates that we need to judge the effectiveness of implementation efforts in both the short and long term. Something which may appear to be ineffective in the short term could well be effective if a longer time perspective were taken. The opposite is also possible. That is, implementation that appears to be very effective in the short term may not turn out to be sustainable.

(8) Factors in developing countries

The seven factors identified above address implementation issues primarily in the context of developed nations. As Smith (1985: 135) has commented, however, while Third World countries have many impressive policies and objectives, "immense slippage" often occurs in translating the policies into programmes and projects. In his words, "many policies remain only symbolic statements by political leaders or laws on statute books, while others that are implemented achieve little of what was originally intended".

Third World countries encounter all of the implementation obstacles of their developed country counterparts. In certain respects, however, they also deal with some other, more formidable challenges. By almost any standards of measurement, the problems to be addressed, whether ranging from meeting basic needs, alleviating poverty or reversing environmental degradation, are usually of a magnitude and intensity well beyond those of developed countries. As a result, the *tractability* of many problems is almost overwhelming. Using the language of Chapter 2, the problems are wicked, or messes.

Beyond the *tractability* discussed by Sabatier and Mazmanian, Smith (1985) noted the following problems regarding implementation in developing countries: (1) government bureaucracies which are ineffective, inefficient and not task oriented; (2) managerial leadership skills which are poor; and (3) corruption which is well entrenched. While Smith (1985: 135) recognized that all governments have policies and laws to stop corruption, nevertheless, "corruption in some form exists in all political and administrative systems". Numerous reasons can be identified for the presence of corruption, ranging from low salaries for public officials who are almost forced to seek bribes in order to support their families, to some well-connected senior officials who have opportunities to earn significant incomes through payoffs. These comments do not imply that corruption does not and cannot occur in developed countries. However, the pressures or incentives for corruption are often greater in lesser developed countries. The outcome is that some initiatives may not get implemented, or will do so only in a token manner, if no supporters provide payoffs to the public officials responsible for their implementation.

(9) Cultural differences

This factor becomes particularly critical in situations involving multilateral approaches to resource and environmental management, in which two or more countries must work together. As Rayner (1991) has commented, capitalist-oriented Kenya and socialist Tanzania may have difficulty in agreeing on measures to implement joint land-use controls to protect biodiversity in cross-boundary regions. Implementation may founder or suffer because the two countries have different approaches to resource and environmental management, each of which reflects distinct cultural traditions and practices, or differing ideologies.

Rayner (1991: 96–7) argued that there are three managerial or institutional cultures: market, hierarchal and collective. Which one dominates in a country will to a large extent reflect cultural norms. To illustrate, he suggested that market-oriented cultures will "favor implementation policies that maximize the discretion of individual decision makers and firms. We may therefore suppose that they will favor carrots rather than sticks". In contrast, societies that emphasize the collective welfare are more likely to emphasize sticks over carrots. They are likely to prefer "command and control" approaches to implementation procedures which result in the uniform application of policy with minimal scope for discretion by individual managers of firms because such discretion in principle violates principles of equity. Rayner recognized that cultural preference by itself "will not determine which implementation instrument will be selected". Nevertheless, culturally based preferences may have a strong influence in choice of ends and means, which can cause major difficulties when several countries have to reach agreement on ends and then develop compatible means for implementation. In Chapter 7, cultural differences were shown to have a significant impact on the implementation of environmental impact assessment procedures.

Summary

The nine obstacles noted above help us to understand why implementation of policies and programmes often does not go smoothly, and to why unexpected outputs and outcomes occur. Beyond these considerations, it is also necessary to appreciate that judgements will be made about what should and can be

---

Box 12.8 Expectations can vary

It may make sense to choose a convention or agreement that can be most fully and flexibly implemented, rather than a more stringent agreement that, because it is essentially unenforceable, remains of symbolic rather than instrumental importance. In other circumstances, however, it may be perfectly appropriate to conclude symbolic agreements that cannot be implemented by states, with the intention of encouraging action by non-governmental institutions (such as industry or voluntary organizations).

Source: Rayner, 1991: 99.

---

accomplished, and policies and implementation procedures may be crafted accordingly, as indicated in the comments in Box 12.8. In his guest statement, Fred van Zyl shares his conclusions about important aspects requiring attention for effective implementation, based on his experience in South Africa.

12.3 Adaptive versus programmed approaches

Berman (1980) has explained that two schools of thought have arisen regarding the best way to achieve implementation. One school, called *programmed implementation*, reflects the view that implementation problems can be overcome by systematic and explicit pre-programming of procedures for implementation. The other school, *adaptive implementation*, reflects a conclusion that implementation will be facilitated through an approach that allows adjustments to changing circumstances, events and decisions. Thus, the proponents of these two schools have diagnosed the problems of implementation to come from different sources, and offer quite different solutions. Such perspectives are not new, as it is possible to recognize elements of *synoptic* or *comprehensive rational planning* in the programmed approach, and aspects of *incrementalism* in the adaptive approach (Chapter 2). And, as Berman's comment in Box 12.9 reminds us, it often is not helpful to think in "either/or" terms. Both approaches have merit, and the challenge for analysts and planners is to determine which approach will be most useful in given situations.

Box 12.9 Approaches to implementation

There is no universally best way to implement policy. Either programmed or adaptive implementation can be effective if applied to the appropriate policy situation, but a mismatch between approach and situation aggravates the very implementation problems these approaches seek to overcome.

Source: Berman, 1980: 206.

Guest Statement

Implementation in South Africa

Fred van Zyl, South Africa

South Africa has recently gone through a major transformation process. This applies not only to the political reform, but also to the way in which the environment is being managed. This change is reflected in new legislation, such as the Constitution of South Africa (1996), the New Water Act (Act 36 of 1998), the Water Services Act (Act 108 of 1997), the Forestry Act (Act 84 of 1998), the Environmental Management Act (1998), the Environmental Conservation Act (1999), as well as the Local Government and Municipal Acts.

The new way of thinking promotes a change from traditional engineering approaches to more holistic and integrated management practices. This includes proactive planning, integrated management, participatory and cooperative governance, partnerships with stakeholders, business principles, focus on sustainability and viability, transformation from information systems to integrated knowledge bases and very importantly an elevated status for the environment.

Integrated strategic planning and management have become a core element of the business. The Cabinet of South Africa is driving a major incentive to develop integrated rural development and economic development strategies. All the main Government Departments are actively participating in this initiative. It focuses on addressing poverty and economic development by creating a healthy, sustainable and prosperous environment. This is supported by integrated spatial development initiatives that coordinate and facilitate focused node and corridor development. Various legislation requires integrated development planning on a local government level. These plans include water services management, services delivery, environmental health, economic development as well as sustainable water resources management. Other aspects included are effective water use, environmental protection, infrastructure delivery, environmental and social management, as well as institutional and financial aspects.

Due to the scarcity of water in South Africa, as well as economic constraints, much emphasis is now placed on effective water use, demand and conservation management, environmental protection and sustainable water resource management. This requires development and implementation of appropriate technologies, awareness and best practices.

Nineteen Catchment Management Authorities are being establish throughout South Africa with the core objective to ensure integrated and participatory management. They form part of a national strategic framework and are based on the principles of partnerships and shared responsibilities. In all of this, the environment has received a prominent status. The National Water Act has declared basic water use and the environment as preferential water uses, referred to as the reserve. All rivers are being classified in terms of their environmental importance, which in future will dictate all other water allocations and use. This is supported by the Environmental Act, which requires proper environmental impact assessments, the development of environmental strategic plans by all major institutions and departments, as well as regular state of the environment reports.

The approach to services delivery has changed from the implementation of once-off *ad-hoc* projects to the development and implementation of on-going services businesses. All these approaches require a common understanding of the business, the development of common value systems, setting of common goals and team efforts. Focus is also placed on empowerment, capacity building and development of a planning culture. New standards, norms and regulations are being developed to support effective water management and services delivery. Major development is taking place in enhancing information systems to knowledge bases.

Tremendous vision, lateral thinking, leadership and coordination are needed to achieve these goals.

Fred van Zyl, born in Pretoria, South Africa, and qualified as a civil engineer at the University of Pretoria, has 30 years' experience in water resources management. His experience varies from design and construction of large dams and schemes, water rights, allocations, systems and schemes management, hydrology, water resources planning, water quality management, catchment management as well as water services planning and management. Some of his main achievements are his leadership in the establishment of catchment management in South Africa, the initiation of water for the environment programmes, the transformation from a pollution control approach to total water quality management as well as the development of unique water quality guidelines for South Africa. He is also well known for his competency in strategic planning, information systems, and his role and leadership in the transformation of water management from a purely technical field to a strategic business concept. He is presently the Director of Macro Planning and Information Systems in the Department of Water Affairs and Forestry, South Africa.

12.3.1 Programmed implementation

The school associated with programmed implementation believes that implementation difficulties arise from several sources:

Source 1. Ambiguity or vagueness in policy goals, caused by or leading to misunderstandings, confusion or conflicts of values. Such ambiguity is thought to leave implementers without sufficient direction or guidance. The solution is for officials to provide specific, detailed and consistent objectives to be followed by those given the task of implementation.

Source 2. Involvement of too many participants with overlapping or conflicting responsibilities. The outcome is that no one person or agency is willing or able to take final authority or responsibility. The consequences are that implementers are able to do what they wish, with discretionary decisions thwarting realization of stated goals and objectives. The solution is to establish clear lines of authority, to minimize the number of participants and to limit the scope for discretion.

Source 3. Resistance, ineffectiveness or inefficiency from those charged with doing the implementation. It is believed that lower-level implementers are more comfortable with a well-established routine, and that any policy initiatives that modify such routines will usually be met with resistance, inefficiency and ineffectiveness. The solution is to constrain the amount of discretion by providing explicit guidelines for operating procedures, monitoring behaviour so that implementers can be held accountable for what is (or is not) done, and providing incentives (salary bonuses, status awards) for desired actions.

Thus, the characteristics of a programmed approach include a well-specified plan with clearly defined objectives, unambiguous lines of responsibility, limited participation and minimal discretion.

12.3.2 Adaptive implementation

A different diagnosis and solution are offered by the adaptive approach. From this perspective, implementation difficulties occur because of over-specification and rigidity of goals, failure to involve a wide enough mix of people in decision making, and undue control over implementers. The solution is to create processes which allow a policy to be adjusted and revised (that is, adapted) as a result of changing circumstances. Outcomes are not assumed to be automatic or guaranteed. The issues are related to a number of sources.

Source 1. Over-specification and rigidity of goals. The solution is to use general and even vague goals. Or, if goals cannot be agreed upon, then agreement on means is acceptable (as with the incrementalists in Chapter 2, or as shown by Figures 1.6 and 1.7). Vagueness or ambiguity provides scope to custom design implementation arrangements to suit differing conditions prevailing within a jurisdiction. The disadvantage is that some people will complain when they see people in other areas being treated differently from them, especially if they feel that those people are getting what is viewed to be preferential treatment.

Source 2. Not enough groups or interests are involved. The adaptive approach seeks active participation of relevant participants. This is done in the belief that more participants will bring more information and perspectives to help define issues and develop solutions. Another reason for this approach is the conviction that if people are involved in creation of a policy or a solution, then they will be more likely to be motivated to make it succeed.

Source 3. Not enough discretion. It is believed that too much direction and control stifle creativity and enthusiasm, and also lead to standardized approaches that may not fit local conditions. The solution is to give local implementers discretion to make modifications relative to local needs and conditions. It is also thought that this approach will allow people to "learn by doing", rather than mechanically following a set of guidelines which may not make a lot of sense for their situation. Monitoring has a different role. For programmed implementation, monitoring is used to determine whether objectives are being realized, and if expected outcomes and impacts are occurring. In adaptive implementation, monitoring is used to determine if there is a need to make

Table 12.2   Policy situations

| | Situation type | |
Characteristics	Structured	Unstructured
Scope of Change	Incremental	Major
Certainty of technology or theory	Certain within risk	Uncertain
Conflict over policy's goals and means	Low conflict	High conflict
Structure of institutional setting	Tightly coupled	Loosely coupled
Stability of environment	Stable	Unstable

Source: Berman, 1980: 214.

modifications to objectives or processes in the light of accumulated experience with implementation.

Berman (1980) concluded that different policy situations exist. The trick in selecting the most appropriate implementation strategy (adaptive or programmed) is to be able to read a situation and determine which conditions are dominant. Table 12.2 shows different characteristics of policy situations. If a situation were characterized by all the attributes in the "structured" column, a programmed approach would likely be most appropriate. An adaptive approach would most likely be the best choice if the situation had the attributes shown in the "unstructured" column.

The main message from the discussion of programmed and adaptive implementation strategies is that different policy situations do exist, and that each strategy offers advantages. Traditionally, the programmed approach has been favoured, but such an approach assumes need for only moderate changes, considerable certainty about means to be used, low conflict regarding goals, agencies that work collaboratively, and relative stability. However, we are aware from Chapter 2 that complexity, uncertainty and turbulence are often encountered, and in such situations an adaptive approach will likely be more effective. And, it is also possible that, over the life time of a policy or programme, characteristics of both structured and unstructured conditions will occur. Thus, it could be that aspects of both approaches may be useful at different times. It should not be assumed, however, that it always and inevitably is a structured situation which will occur, and that a programmed approach automatically should be used.

12.4 Implementation: collaborative and stakeholder approaches

Based on analysis of experience with collaborative approaches related to ecosystem management in Australia and the USA, Margerum (1999a) identified both weaknesses in and opportunities for implementing stakeholder and collaborative approaches. His main conclusion was that, through collaboration, stakeholders often create new levels of understanding and also reach consensus, referred to as "shared capital", but they also often struggle to translate such capital into action.

Three phases of collaboration were identified: (1) *problem setting*, in which stakeholders come together, obtain commitment, and agree on processes and mechanisms to achieve collaboration; (2) *direction setting*, in which the stakeholders identify issues, exchange information, address conflicts, agree upon common goals, seek consensus and identify post-plan actions; and (3) *implementation*, during which stakeholders determine actions, roles, tasks and responsibilities, design an approach to implementation, take action and monitor results.

Stakeholders often make considerable progress in the *planning phases*, and are able to achieve significant understanding, identify common goals and deal with conflicting interests. The outcome becomes "shared capital", which includes trust, agreed norms and networks; "intellectual capital", in the form of agreed definitions and facts as well as mutual understanding; and "political capital", in the form of alliances and agreements. Many groups also are often effective in creating management strategies, information brochures as well as general information and education programmes, fairs, restoration projects and "cleanup days". However, in Margerum's view, many stakeholders often do not make progress regarding the goals established during the planning process. In other words, they make progress with respect to creating shared, intellectual and political capital, collectively labelled as social capital, but are not successful in effecting substantive action.

Margerum concluded that stakeholders are often unsuccessful in implementing a collaborative approach because of: (1) poor communication, (2) problems in resolving conflicts, (3) personality differences, (4) extremely difficult (i.e. intractable) problems, (5) long histories of antagonism, and (6) inadequate funding to support implementation. These reasons confirm many of the key factors identified in Section 12.2.2. Some other generic problems were also identified, and are discussed in turn below.

(1) Structural factors

Disparity in power and resources among stakeholders creates disadvantages for the weaker groups, which can lead to reduced support by them for implementation. Furthermore, rather than using a collaborative approach, some groups may opt to deal with issues through other means, such as the courts or legislatures. When this occurs, collaborative efforts can be undermined or overtaken.

(2) Lack of strategic direction

The most frequently identified weakness associated with collaborative initiatives is failure to establish priorities and identify specific actions. Instead of prioritizing, many stakeholder groups combine all ideas and possible actions. The result is a "wish list" rather than a well-focused set of strategic objectives and related actions. When commonly identified goals are agreed upon, they are often too broad and too vague to set a strategic direction. Such vagueness, comprehensiveness and lack of prioritization usually occur due to stakeholder groups coalescing objectives and actions into a comprehensive list instead of being selective.

(3) Lack of community involvement during implementation

Each stakeholder group usually includes a mix of individuals with a range of interests. Such diversity can be a strength for collaboration, but it also can lead to complacency about community involvement. Stakeholder groups often view themselves as *representatives* of their community (which usually occurs through democratic elections) instead of being *representative* of the community (which occurs when groups with many interests are involved). Stakeholder representatives are chosen because they reflect a set of interests within the community, unlike elected officials who are expected to be concerned with all interests.

A further challenge is that during the process of collaboration, stakeholder representatives often identify and agree upon common goals, which may or may not be supported by their different constituent groups. The outcome is that while the representatives of stakeholder groups may reach agreement among themselves, the same level of agreement is not achieved by the general membership of the various groups which they represent. This dilemma reinforces the importance of stakeholder representatives communicating regularly with the group which they represent.

(4) Lack of stakeholder commitment to implementation

Enthusiasm and commitment from key agencies with responsibility and authority for the resource and environmental issues are normally needed if a collaborative approach is to lead to effective implementation. Margerum found that state and local government agencies often do not show strong commitment to a collaborative approach. Such agencies send a delegate to meetings and share information, but often do not modify their policies and programmes with regard to the strategies generated by the stakeholder group.

Lack of commitment can occur due to policy constraints which limit their discretion, which is common if a programmed approach is favoured by senior managers or the elected official to whom they are accountable. In addition, they may have legally specified responsibilities which they cannot disregard or ignore, even if those go against the vision or preferred actions of the stake-holder group.

To overcome the above barriers to implementation, Margerum suggested stakeholders can use various approaches, alone or in combination. He argued that these should include using a common information set, a cooperative plan or policy, and joint decision making.

(1) Common information set

Agreeing upon, and creating, a common information set can remove disagreements about what are facts, and what assumptions underlie them. By sharing information, stakeholders can also become sensitive to each others' perspectives and analyses, and also gain an enhanced understanding of the ecosystem within which issues are embedded. These accomplishments can result in better, or at least more informed, decision making. Nevertheless, the benefits of a common information set are limited for several reasons. First, if the ecosystem is changing rapidly, agreed information may become quickly outdated, and may not remain relevant in planning for the future. Second, such an approach assumes that mutually agreed upon information will be incorporated into decisions by the management agencies, and this may or may not occur. And third, it assumes that having common information, each group will know how to use it effectively. As with the second point, this may or may not be the case.

(2) Cooperative plans and policies

Stakeholders normally seek to use a cooperative plan or policy to guide implementation initiatives. Based on a common information set, the stakeholders may develop objectives and actions, identify responsibilities and determine tasks, and agree on a procedure to monitor progress. Such an approach, when successful, can result in significant changes in management and behaviour among a range of decision makers.

However, cooperative plans and policies also have limitations. First, they remain applicable only as long as the information and analysis on which they are based are relevant. Second, they assume that actions can be allocated among stakeholder groups, and that the actions will be implemented in a coordinated manner. Such coordinated implementation can be a challenge, as each stakeholder group has to deal with the actions allocated to it in the context of other responsibilities and priorities it has, as well as with regard to available financial and human resources. If one stakeholder cannot move forward with its implementation responsibilities, that may undermine the rationale for a planned sequence of initiatives.

(3) Joint decision making

Joint decision making has significant potential to facilitate implementation of actions developed through collaborative processes, but is infrequently used. Joint decision making is often difficult to achieve because of legally defined roles and responsibilities of some stakeholders and their reluctance to give up some autonomy or share decisions; the difficulty to agree on which matters will be handled jointly and which ones by individual stakeholder groups; and the extra time required for consultation throughout the implementation period due to the need for groups with different responsibilities to talk to one another prior to meeting to make joint decisions.

The above points emphasize that, in addition to obstacles, there are opportunities to achieve effective implementation based on a collaborative process, as discussed earlier in Chapters 8 and 9. However, as with most aspects of life, there are no perfect answers, reminding us of the importance to consider a mix of approaches and methods, rather than ever only relying upon one.

12.5 Implementation: forest management in the USA

Brown and Harris (1992) conducted a study of the USA Forest Service regarding its evolution during the 1980s. A special interest was the capacity for *policy implementation*, or those activities which followed policy directives. In their view, implementation issues required more attention because "well-intentioned policy formulation by no means ensures realization of desired policy outcomes" (Brown and Harris, 1992: 459).

Among other matters, they considered how the values and attitudes of field officers influenced the achievement of the Forest Service goals. Values and attitudes were examined because they believed that policy makers and implementers often search for, and select, data which reflect their own values, and facilitate implementation of activities consistent with their values. Also, where implementers are given discretion to make choices, or have to make choices because of contradictory directives, it is likely that decisions will reflect basic values and attitudes. Brown and Harris believed that it was normal rather than unusual for policy implementers to have to deal with a "smorgasbord" of goals, a situation in which they had to interpret directives, assign priorities and resolve conflicts.

The introduction of the National Forest Management Act (Box 12.10) in the mid 1970s was greeted as a landmark statute that would help to resolve many conflicts being encountered in forest management. However, the act also illustrates a directive which contained multiple and inconsistent signals, forcing implementers to exercise discretion, judgement and choice.

The comments of Brown and Harris highlight the challenges faced by those given responsibility for implementation of the act. Furthermore, if their conclusion that the most controversial and pressing national forest management policy outcomes will be determined by choices made during the implementation process, and that those choices are made by field managers using their discretion, then the values and attitudes of the forest managers become very significant.

During the 1980s, they found that the values of the field forest managers were increasingly challenging the traditionally dominant principle of *timber primacy* in the Forest Service. While timber management remained a central and important goal for the Forest Service, more managers were accepting the legitimacy of other goals and interests. More interest was being given to maintenance

---

Box 12.10 Implementing the National Forest Management Act

But the official goals established by the NFMA were multiple, conflicting, difficult to put into operation, and difficult to achieve. As a result, the Forest Service has to choose which NFMA goals to implement and pursue, which goals to satisfy, and which goals to displace or substantially ignore; these decisions are reflected in activities that actually get carried out on the ground.

Source: Brown and Harris, 1992: 459.

of healthy ecosystems, and to non-consumptive uses of forests. While this increasingly diverse set of values and mind sets of forestry professionals bodes well for a more balanced approach to management, it will also enhance the conflicts among some of the basic choices that foresters have to make. In the short term, such conflicts may make the implementation process less smooth. In the longer term, however, it should result in the goals, objectives and activities being implemented to reflect the overall mix of interests in later twentieth century society.

12.6 Implementation: integrated catchment management in Australia

During the mid 1980s, the state government of New South Wales introduced the concept of *total catchment management*, which quickly became known as "TCM". The intent was to manage soil, land, water, vegetation and other environmental components together in order to achieve sustainable use and production. Particular emphasis was to be placed on coordination of the interests, activities and uses of all land users to avoid degradation, and/or to facilitate rehabilitation of degraded landscapes. TCM was consistent with an ecosystem approach, as outlined in Chapter 5. A more detailed outline of TCM is given in Box 12.11.

In May 1988, the Australian Water Resources Council co-sponsored a National Workshop on integrated catchment management. Representatives from all the states and territorial governments went to that workshop with proposed policies and strategies for *integrated catchment management*, or ICM, which had become the favoured term for what earlier had been called TCM in New South Wales.

Box 12.11 Total catchment management, New South Wales, Australia, mid 1980s

Total Catchment Management involves the co-ordinated use and management of land, water, vegetation and other physical resources and activities within a catchment, to ensure minimal degradation and erosion of soils and minimal impact on water yield and quality and on other features of the environment.

Specifically, Total Catchment Management aims to –

- encourage effective co-ordination of policies and activities of relevant departments, authorities, companies and individuals which impinge on the conservation, sustainable use and management of the State's catchments, including soil, water and vegetation;
- ensure the continuing stability and productivity of the soils, a satisfactory yield of water of high quality and the maintenance of an appropriate protective and productive vegetative cover; and
- ensure that land within the State's catchment is used within its capability in a manner which retains as far as possible, options for future use.

Source: Cunningham, 1986: 4.

At the conclusion of the workshop, delegates voted to indicate which state or territory had developed the best approach. They were unanimous that the team from Western Australia had developed the best proposal for ICM. However, their ideas still had to be implemented. In the remainder of this section, the experience in making the transition from concept to implementation in Western Australia is considered (Mitchell and Hollick, 1993).

The government of Western Australia made some key decisions regarding implementation of ICM. First, implementation would occur through coordination of the policies and activities of existing agencies. Second, a small secretariat would be created to facilitate the desired coordination. Third, priorities would be established.

Thus, no new legislation or new agencies would be created, based on the belief that institutional arrangements in Western Australia were not unduly complex, and using existing capacity was preferable to creating more bureaucracy. The coordinating mechanism became the Integrated Catchment Management Coordinating Group. This group was later served by a small secretariat called the Office of Catchment Management (OCM). OCM had very few staff, and a modest budget. Finally, Community Catchment Groups were also established, to allow local people to become involved in identifying problems and in developing solutions. Technical Advisory Groups, consisting of individuals with technical expertise, were established to work in parallel with the Community Catchment Groups.

A review of the experience with implementation was conducted using the six-part framework shown in Figure 12.1 (Mitchell, 1990). The parts of the framework include:

- **Context.** Attention focuses upon historical, cultural, economic and institutional dimensions, as well as the state of the biophysical environment.
- **Legitimation or credibility.** If a policy or programme is to be implemented, it receives legitimation through political commitment, statute, financial support or administrative support. Legitimation is strengthened when more than one of these elements is present.
- **Functions.** Decisions must be taken regarding which management functions are to be integrated relative to a vision. Functions can be *generic* (policy development, data collection, planning, development, regulation, enforcement) or *substantive* (water supply, pollution control, flood plain management, wetland management). It is not automatic that all functions should be integrated, and indeed there can be good arguments to separate some of them, such as development and enforcement.
- **Structures.** These involve the number and type of organizations responsible for management functions. They can be few in number and centralized, or more in number and decentralized. However they are crafted, there always will be *edge* or *boundary* problems among agencies, which involve areas of overlapping interest and responsibility.
- **Processes and mechanisms.** No matter how well functions and structures are designed, there will be mismatches, or areas of overlap and underlap.

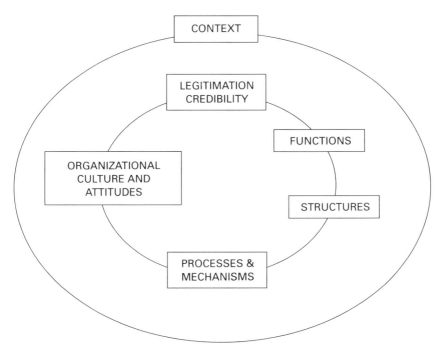

Figure 12.1 Framework for integrated resource management (from Mitchell's work)

Thus, there is a need for processes and mechanisms, such as inter-departmental working groups or public participation, to address *edge* or *boundary* issues.

- **Organizational culture and attitudes.** The effectiveness of implementation will usually be strongly influenced by the people who have the responsibility for that, as well as the *culture* of an organization (does it promote competition or cooperation?) and the *attitudes* of participants (do they strive to share and cooperate?).

These points were used to consider the effectiveness of implementation of ICM in Western Australia.

Regarding *legitimation*, ICM was supported by a state policy, and by creation of an inter-agency coordinating group and a small secretariat (Office of Catchment Management) to support that group. Also, modest resources were provided to help form Community Catchment Groups. However, as with most jurisdictions, the mandates of different line agencies (agriculture, forestry, wildlife, water) often conflicted, and neither the Office of Catchment Management nor the coordinating group had any real power to force competing agencies to resolve their differences. This meant that the OCM and the coordinating group had to rely on persuasion and exhortation, which in many instances fell on inattentive ears.

The *functions* to be integrated were never clearly identified. As a result, while the coordinating group served as a forum to coordinate policies and activities, and had some notable successes, there was no systematic consideration at a state level regarding which functions needed to be integrated, or how that would be done. *Structures* also served as an impediment, as well-established line agencies with readily identifiable constituencies (agriculture, water) seemed more comfortable in protecting long-standing mandates and activities than in actively sharing through joint programmes. And, as already noted, because of no strong political commitment, enabling legislation or significant financial support, a good strategy for those opposed to ICM was simply to go slowly and outwait it. This approach, used by some line agencies, was effective as in late 1995 the OCM was terminated, and its responsibilities were assigned to various other agencies.

The inter-agency working group was the principal *mechanism* used to search for coordination and cooperation. A critical ingredient for the success of such committees is that senior representatives of participating agencies attend the meetings on a regular basis. A good way to undermine or undercut such a working group is to send junior or intermediate people who do not have the authority to make commitments on behalf of their organizations, and continuously have to take matters back to their agencies for consultation. When this happens, or when people do not regularly attend meetings, the time taken to make decisions becomes much longer. And, because decisions are not being taken and action is not occurring, this situation will and can be used by participating agencies to suggest that the working group is not effective and become justification for sending even more junior people. Thus, once this pattern starts, it can become self-perpetuating and self-fulfilling.

As important as any of the variables was the *organizational cultures* and *attitudes.* Some senior officers in line agencies were sceptical or antagonistic to the concept of ICM, or to the individuals appointed to the OCM. As a result, where cooperation might have been forthcoming, in many instances it was not (Box 12.12). In addition, the agriculture ministry had established Land Care Community Groups in many parts of the state, and was more inclined to devote resources to them than to the Community Catchment Groups. Thus, considerable effort at developing functions, structures and processes is unlikely to be effective unless there is enthusiasm and willingness to make an integrated approach work.

Based on experience with implementing policies and programmes, a checklist of critical factors needing attention to facilitate action after the preparation of a policy or a plan, especially if a community-based approach is to be used, can be established (Stout, 1998: 6). These are complementary to the checklist in

---

Box 12.12 Need for human cooperation

Too often the emphasis is on "amalgamating and restructuring agencies in an attempt to foster cooperation. Not surprisingly, this does not work. The crucial thing is to get people at all levels working together, and you cannot make this happen simply by amalgamating agencies – you still have to do the hard work of fostering cooperation and a team spirit."

Source: Robinson, 1992: 4.

Chapter 5 regarding key factors related to application of an ecosystem approach, and also reflect many of the principles identified in Chapters 3, 5, 8 and 11.

Considerations for community-based implementation of policies or plans

1. Patience. Patience. Patience. We didn't get to where we are today overnight, and we won't get to where we are going tomorrow. When we set a lofty goal, break it down in smaller steps. Before you know it, you will have reached your goal.

2. Conflict can be healthy – if managed positively. Conflicting views or ideas often become a third view or idea that can be near healthy for the group's efforts and the watershed's health.

3. Ask not, "do you like it?" but ask, "can you live with it?". Remember that you probably will propose many ideas before the group reaches a common point of agreement. What is important in reaching a consensus is that everyone can agree to live with a decision.

4. Celebrate your successes. Regardless of how small, celebrate progress. Whether your groups measure progress by the number of canoe trips, kilometres of buffer strips, or hectares of no-till farming, reaching benchmarks is important. One more bonus tip: Be kind to each other; you may need that person to agree with you later.

5. Seek common interests, not positions. By working to find the common interest of all stakeholders, you will establish a strong foundation for an effective watershed management plan. One way to do this is to get past opposing positions by asking why stakeholders have taken a particular position. Keep asking "why" again and again. It usually takes seven layers of "whys" to uncover the interest that is common to other stakeholders.

6. Encourage teaching. Allow stakeholders to teach each other. No idea is too simple to be discussed. For example, a farmer can teach the basics of watering, fertilizer application and pest management to homeowners.

7. Ask for free advice and in-kind services. For example, if you need a video, ask the local television station for script and production assistance. If you need monitoring or assistance, work with your local government department and your local school system. And don't forget that saying thank you in public will go a long way toward getting additional help the next time. One bonus tip: No-one gives money to a group without a plan for how to use it. Financial assistance can come from unusual places and innovative sources once the group has a solid plan.

8. Great leaders plant seeds and nurture them. They facilitate the group to reach consensus, plant new and different ideas when necessary, and assist the group in nurturing those ideas. Effective leaders are great communicators, they listen and expand on others' ideas, and make sure every idea is explored and that all stakeholders are heard.

9. Bring everyone to the table. Successful efforts include everyone who has a stake. This enables the group to build a consensus on what needs to be

done and how to do it. Leaving a critical stakeholder out of the process at any step may cause unnecessary problems later.

10. Think small. The smaller the area, the easier the partners can relate or connect to it. In addition, the smaller the area, the faster it will react to changes in management practices.

12.7 Implications

In Chapter 3, it was argued that effective planning and management requires a *vision* (an appreciation of which ends are wanted), a *process* (agreement of means to use in reaching the ends) and a *product* (a strategy or plan to guide the use of means). However, this chapter has emphasized that a fourth essential component is the capacity for *implementation*. If there is not the will and ability to implement, then all the visions, processes and plans are unlikely to achieve desired changes.

Many obstacles can thwart implementation. Indeed, the examples in this chapter indicate the challenges which can be encountered during implementation of policies or programmes. If implementation failure is not to be like "original sin" (occurring everywhere and being ineradicable), then resource and environmental managers need to be thinking about implementation in parallel to thinking about visions and planning processes. Too often it seems to be assumed that implementation will logically follow a well-crafted policy or plan. However, many people and societies are resistant to change, and new policies or policy changes usually imply an intent to change the status quo. If existing interests and institutional inertia are to be overcome, then considerable thought and time must be devoted to the implementation component of resource and environmental management. If such effort is not made, then the likelihood of implementation failure is likely to be very high.

References and further reading

Baker R 1989 Institutional innovation, development and environmental management: an "administrative trap" revisited, Part I. *Public Administration and Development* 9: 29–47

Bellamy J A and A K L Johnson 2000 Integrated resource management: moving from rhetoric to practice in Australian agriculture. *Environmental Management* 25: 265–80

Berman P 1980 Thinking about programmed and adaptive implementation: matching strategies to situations. In H M Ingram and D E Mann (eds) *Why Policies Succeed or Fail.* Beverly Hills, California, Sage, pp. 205–27

Boehmer-Christiansen S 1994 Policy and environmental management. *Journal of Environmental Planning and Management* 37: 69–85

Brekke J 1987 The model-guided method for monitoring program implementation. *Evaluation Review* 11: 281–99

Brown G and C C Harris 1992 The United States Forest Service: changing of the guard. *Natural Resources Journal* 32: 449–66

Cook B J, J L Emel and R E Kasperson 1991/2 A problem of politics or technique? Insights from waste-management strategies in Sweden and France. *Policy Studies Review* 10: 103–13

Courtner H J 1976 A case analysis of policy implementation: the National Environmental Policy Act of 1969. *Natural Resources Journal* 16: 323–30

Cowell R 2000 Environmental compensation and the mediation of environmental change: making capital out of Cardiff Bay. *Journal of Environmental Planning and Management* 43: 689–710

Cunningham G M 1986 Total catchment management – resource management for the future. *Journal of Soil Conservation, New South Wales* 42: 4–5

Eckerberg K and B Forsberg 1998 Implementing Agenda 21 in local government: the Swedish experience. *Local Environment* 3: 333–47

Emmett B 2000 Managing for sustainable development: 99% perspiration. *Forestry Chronicle* 76: 93–7

Gardner A 1990 Legislative implementation of integrated catchment management in Western Australia. *Environmental and Planning Law* 7: 199–208

Gilg A W and M P Kelly 1997 The delivery of planning policy in Great Britain: explaining the implementation gap: new evidence from a case study in rural England. *Environment and Planning C* 15: 19–36

Gow D D and E R Morss 1988 The notorious nine: critical problems in project implementation. *World Development* 16: 1399–418

Guerin T F 2000 Overcoming the constraints to the adoption of sustainable land management in Australia. *Technological Forecasting and Social Change* 65: 205–37

Harvey L D D 1995 Creating a global warming implementation regime. *Global Environmental Change* 5: 415–32

Hull A 1995 New models for implementation theory: striking a consensus on windfarms. *Journal of Environmental Planning and Management* 38: 285–306

Hull R B and P H Gobster 2000 Restoring forest ecosystems: the human dimension. *Journal of Forestry* 98(8): 32–6

Jacobi P, D Baena Segura and M Kjellén 1999 Governmental responses to air pollution: summary of a study of the implementation of *rodizio* in Sao Paulo. *Environment and Urbanization* 11: 79–88

Johnson A K L, D Shrubsole and M Merrin 1996 Integrated catchment management in northern Australia: from concept to implementation. *Land Use Policy* 13: 303–16

Jordan A 1998 The ozone endgame: the implementation of the Montreal Protocol in the United Kingdom. *Environmental Politics* 7: 23–52

Jordan A 1999 The implementation of EU environmental policy: a policy problem without a political solution? *Environment and Planning C* 17: 69–90

Keynan Z 1999 Implementing marine environmental protection law in China: progress, problems and prospects. *Marine Policy* 23: 207–25

Ledoux L, S Crooks, A Jordan and R K Turner 2000 Implementing EU biodiversity policy: UK experiences. *Land Use Policy* 17: 257–68

Leu W-S, W P Williams and A W Bark 1995 An evaluation of the implementation of environmental assessment by UK local authorities. *Project Appraisal* 10: 90–102

Loske R and S Oberthur 1994 Joint implementation under the Climate Change Convention. *International Environmental Affairs* 6: 45–58

Margerum R D 1999a Getting past yes: from capital creation to action. *Journal of the American Planning Association* 65: 181–92

Margerum R D 1999b Implementing integrated planning and management: a typology of approaches. *Australian Planner* 36: 155–61

Mazmanian D A and P A Sabatier (eds) 1981 *Effective Policy Implementation*. Lexington, Massachusetts, D C Heath and Co.

Mitchell B 1990 Integrated water management. In B Mitchell (ed.) *Integrated Water Management: International Experiences and Perspectives*. London, Belhaven, pp. 1–21

Mitchell B and M Hollick 1993 Integrated catchment management in Western Australia: transition from concept to implementation. *Environmental Management* 17: 735–43

Morah E U 1996 Obstacles to optimal policy implementation in developing countries. *Third World Planning Review* 18: 137–46

Morgan R K 1995 Progress with implementing the environmental assessment requirements of the Resource Management Act in New Zealand. *Journal of Environmental Planning and Management* 38: 333–48

Noble J H, J S Banta and J S Rosenburg (eds) 1977 *Groping Through the Maze*. Washington, DC, The Conservation Foundation

Padgitt S and P Lasley 1993 Implementing conservation compliance: perspectives from Iowa farmers. *Journal of Soil and Water Conservation* 48: 393–400

Parikh J K 1995 "Joint" implementation and North–South cooperation for climate change. *International Environmental Affairs* 7: 22–41

Pickvance C G 2000 Local-level influences on environmental policy implementation in Eastern Europe: a theoretical framework and a Hungarian case study. *Environment and Planning C* 18: 469–85

Pressman J L and A B Wildavsky 1973 *Implementation*. Berkeley, University of California Press

Rayner S 1991 A cultural perspective on the structure and implementation of global environmental agreements. *Evaluation Review* 15: 75–102

Rayner S 1999 Mapping institutional diversity for implementing the Lisbon principles. *Ecological Economics* 31: 259–74

Robertson-Snape F 1999 Corruption, collusion and nepotism in Indonesia. *Third World Quarterly* 30: 589–602

Robinson S 1992 Horses for courses are galloping along in WA. *Catchment Matters* 5, September: 4–6

Robinson S J and R Humphries 1997 Towards best practice: observation on Western Australian legal and institutional arrangements for ICE 1987–1997. In *Advancing Integrated Resource Management: Processes and Policies*. 2nd National Workshop on Integrated Catchment Management, Waverly, Victoria, River Basin Management Society, October

Rosenthal I, P J McNulty and L D Helsing 1998 The role of the community in the implementation of the EPA's rule on risk management programs for chemical accidental release prevention. *Risk Analysis* 18: 171–9

Sabatier P A and D A Mazmanian 1981 The implementation of public policy: a framework of analysis. In D A Mazmanian and P A Sabatier (eds) *Effective Policy Implementation*. Lexington, Massachusetts, D C Heath, pp. 3–36

Schwarze R 2000 Activities implemented jointly: another look at the facts. *Ecological Economics* 32: 255–67

Sheirer M and E Rezmovic 1983 Measuring the degree of program implementation: a methodological review. *Evaluation Review* 7: 599–633

Silva E 1994 Thinking politically about sustainable development in the tropical forests of Latin America. *Development and Change* 25: 697–721

Smith T B 1985 Evaluating development policies and programmes in the Third World. *Public Administration and Development* 5: 129–44

Somach S L 1993 Closing the policy-practice gap in water resources planning. *Water Resources Update* 90, Winter: 19–22

Stout G E 1998 Sustainable development requires the full cooperation of users. *Water International* 23: 3–7

Talen E 1996 Do plans get implemented? A review of evaluation in planning. *Journal of Planning Literature* 10: 24,859

Thompson D B 2000 Political obstacles to the implementation of emissions markets: lessons from RECLAIM. *Natural Resources Journal* 40: 645–97

Victor D G and E B Skolnikoff 1999 Translating intent into action: implementing environmental commitments. *Environment* 41(2): 16–20, 39–44

Wallis R L and S J Robinson 1991 Integrated catchment management: the Western Australian experience. *Environment* 33: 231–40

Weale A 1992 Implementation failure: a suitable case for review? In E Lykke (ed.) *Achieving Environmental Goals: The Concept and Practice of Environmental Performance Review*. London, Belhaven, pp. 43–63

Wescott G 2000 The development and initial implementation of Australia's "integrated and comprehensive" Oceans Policy. *Ocean and Coastal Management* 43: 853–78

Wood A 1993 The multilateral fund for the implementation of the Montreal Protocol. *International Environmental Affairs* 5: 335–54

Wynne B 1993 Implementation of greenhouse gas reductions in the European Community: institutional and cultural factors. *Global Environmental Change* 3: 101–28

Chapter 13

Monitoring and Evaluation

13.1 Introduction

Monitoring can be done in many ways, by members of non-governmental organizations observing and recording changing conditions in their local area, to government-based programmes, to remotely sensed images generated from orbiting satellites (Box 13.1). However monitoring is conducted, it is usually done for one or more of the following reasons: (1) to document general environmental conditions; (2) to establish environmental baselines, trends and cumulative effects; (3) to document environmental loading, sources and sinks; (4) to test environmental models and verify research; (5) to educate the public about environmental conditions; and (6) to provide information for decision making.

Box 13.1 Monitoring can be done in novel ways: Skywatch surveillance by the Ninety-Nines

Since 1978 in Ontario, Canada, a group of female pilots have donated their time to fly surveillance missions to document violation of environmental court orders by prohibited practices which are continuing. The photographs from the air reveal shapes or patterns that from the ground would be difficult to notice, such as a buried tanker truck which is slowly leaking toxic materials, or illegal dumping pits which have been covered. Air pollution plumes also can be identified.

The women are members of the "Ninety-Nines", a worldwide organization for women pilots formed in 1929 with Amelia Earhart as the founding president. About twenty members of the Toronto chapter fly in the Skywatch surveillance program.

The provincial government rents a Cessna air plane for the pilots, and a surveillance officer from its Ministry of Environment acts as navigator and photographer on the Ninety-Nine flights. On average, a Skywatch team is in the air somewhere over Ontario during half the working days in summer and a quarter of the working days in winter.

The results of this monitoring have been significant. In a court case, the introduction of photographs taken from the air is usually more effective than a verbal description of alleged polluting activity. The outcome has been more successful prosecutions and higher fines.

Source: Based on a report by Cameron Smith, 1996: D6.

Box 13.2 Monitoring and evaluation

Monitoring and evaluation are different but related, and both are needed. *Monitoring* is a tool for obtaining data and provides essential information related to the changing conditions. *Evaluation*, a companion activity, is a tool for translating data into useful information. Together, they are tools to assist managers determine whether programs are effective.

Source: Based on Shindler, Cheek and Stankey, 1999: 3.

Monitoring focuses upon *describing* changing conditions, and *explaining* cause–effect relationships. When assessments of the effectiveness, efficiency or equity of public and private sector initiatives related to changing conditions are included, then an *evaluative* component is added (Box 13.2). Evaluation is not always included, as elected and senior government managers are often hesitant to subject themselves to such scrutiny. Guidelines about monitoring and evaluation are outlined by Adrian McDonald in his guest statement.

In the remainder of this chapter, attention focuses on monitoring and evaluation, with regard to *state of environment reporting*, *public involvement in adaptive management* and *environmental auditing*.

13.2 State of the environment reports

State of the environment (SOE) *reporting* emerged during the late 1980s, following recognition that monitoring or tracking of progress was needed if sustainable development was to be achieved (Box 13.3). In other words, without monitoring it would be difficult to know whether policies or actions were moving a society toward or away from characteristics consistent with sustainability. As a result, national, provincial or state, and municipal governments began to initiate state of the environment reporting, often as a follow up to the preparation of a conservation or sustainable development strategy (Nelson, 1995).

Box 13.3 State of the environment reporting defined

The purpose of SOE reporting is to provide timely, accurate and accessible information on ecosystem conditions and trends, their significance and societal responses, emphasizing the use of indicators. This information should increase public understanding and education, and inform priority setting and decision-making about matters related to the environment by providing objective and scientifically valid information. The information should also establish linkages between environmental conditions and socio-economic factors, reflecting the holistic and integrative nature of the relationship that should exist between humans and the environment.

Source: Dovetail Consulting, 1995: 8.

Guest Statement

Monitoring and evaluating environmental conditions

Adrian McDonald, UK

Monitoring for a purpose: Monitoring is an expensive, long-term commitment. Thus, monitoring must have a clear purpose if it is to be cost justifiable. Monitoring is often related to risk – searching for values above a significant threshold. Alternatively, monitoring may be related to performance – to identify change, improvements and progress toward targets. Evaluation must always be present to give a purpose and justification to monitoring.

Hindsight helps: An environmental incident often raises the claim that the rare 'X' has been damaged. Sadly, we seldom have information on the pre-existing conditions. Leeds University accidentally discharged 12,000 litres of oil to a high quality urban stream populated by the English crayfish, a rare species threatened by invading American crayfish. The monitoring body, the Environment Agency, will not close the incident file until the river returns to its original conditions. But the "original condition" is vague. Evaluating baseline conditions and understanding the transferability of baseline estimates are vital.

Sophisticated enough to suffice, simple enough to be practical: Concern over *Cryptosporidia*, a protozoan parasite that infects people through water supplies, has required all UK water companies to evaluate catchment risk. Sophisticated, GIS-based information systems can map risk – from farming, road spills, industry, etc., but such clever systems can fail because they: (1) address aspects outside the responsibility of an individual manager, (2) are inherently complex, and (3) rely on data that ages. Alternatively, checklists that "blackmark" a catchment with a road (a road spill risk exists) or plantation forestry (a felling impact risk exists) are unsophisticated, equating land erosion with industrial discharge. Monitoring must be fit for purpose.

Ignorance can be bliss: Many people swim in the sea. Water quality needs to be monitored to ensure safe swimming. But what is safe and who wants to know? For years, recreational water was monitored using *Escherichia coli* against an arbitrary threshold. Controlled cohort epidemiology in the 1980–1990s established the relations between water quality and health. We know the bacterial concentration at which people become ill. But does a politician want to know this? Tasked with choosing a value, s/he may find it awkward that a scientist can now say "Minster, if you choose that value then of the 'x'00, 000 swimming, 'y' will be ill".

A process not a task: One of the most successful approaches to monitoring the sustainable development of a city has the acronym, PICABUE. It is not prescriptive or reliant on a suite of indicators, or composite index. Instead, it provides a process whereby the long-established wisdom of a variety of stakeholders aids an inclusive recognition of issues, and the selection of mechanisms for choosing appropriate monitoring and evaluation strategies.

Perhaps the best guide to monitoring environmental conditions lies in the words of Aristotle:

> It is the mark of an instructed mind to rest satisfied with the degree of precision that the nature of the subject permits and not to seek an exactness where only an approximation of truth is possible.

Adrian McDonald is Professor of Environmental Management and Dean of the Earth and Environment Faculty at the University of Leeds, UK. A Scot, his first degree was from the University of Edinburgh in Resource Management. His doctorate, also from Edinburgh, concerned flood risk assessment and management. He has worked with many water companies in the UK – on demand estimation and forecasting, leakage estimation, river and catchment management, resource modelling and renewable energies from biomass. A fundamental interest has been to have the results of sound technical research inform policy and management. Adrian was a technical advisor to the UK Inter-Ministerial Conference on Water Security for the 21st Century.

SOE reporting was usually introduced for one or more of the following reasons: (1) to provide early warning signals to decision makers about changing environmental conditions to facilitate policy or institutional changes; (2) to encourage, and ensure, accountability of public agencies for their decisions and initiatives; (3) to identify inadequate knowledge, and therefore assist in prioritizing research needs; and (4) to sensitize the public about the implications of decisions and actions.

13.2.1 Types of questions addressed in state of the environment reporting

SOE reporting usually includes some mix of four questions, and occasionally will include another three (Dovetail Consulting, 1995: 5, 9). Each is considered below.

1. What is happening in the environment?
2. Why is it happening?
3. Why is it significant?
4. What is being done about it?

The third question requires consideration of threats to integrity of ecosystems, human health, and human values and cultures. At this stage, the relationships among environmental, economic and social systems are usually addressed. The fourth question necessitates understanding of ongoing resource and environmental policies, practices and initiatives, and about the information decision makers require and can use. It is a given that, when dealing with this question, social and economic considerations are also considered.

5. What has happened since the previous SOE report?
6. What further action should be taken?
7. What conclusions can be reached about the performance of resource and environmental organizations?

The fifth question seeks to identify changes as a result of ongoing degradation, as well as from remediation and rehabilitation efforts. The intent is to document patterns or trends which have resulted from (in)action by planners and managers. In the sixth question, the intent is *normative* or *prescriptive*, in that the thrust is to recommend interventions to stop or reverse undesirable change, or to enhance desirable change. Consideration of the connections between the state of the environment, and social and economic aspects, is essential here. The sixth question, leading to possible recommendations, begins to take SOE reporting into the "political" realm. As a result, government-generated SOE reports usually do not address this question. Nevertheless, they may strive to provide enough information for individuals or members of non-government organizations to use the SOE material to reach their own conclusions about what needs to be done.

The seventh question is even more "political" than the sixth, and it is only relatively recently that it has begun to be, or advocated to be, included in SOE reports. For example, the International Union for the Conservation of Nature and Natural Resources (1991: 75) argued that governments should monitor "performance of policy, laws and other institutional arrangements". When it has been included, evaluation often is presented in the form of a "report card", an example of which will be considered later in Section 13.2.4.

A final comment about orientation of SOE is that in some countries the practice has evolved from SOE (state of the environment) to SOS (state of sustainability) reporting. This shift partially reflects the agreement at the Earth Summit during 1992 that all countries endorsing *Agenda 21* were obliged to provide regular reports about progress in achieving sustainable development. SOE reporting emphasizes the natural environment, and considers social and economic systems only to the extent that they have direct implications for the environment. In contrast, SOS reporting allocates equal attention to social, economic and environmental systems, and to the relationships among them.

13.2.2 Types, spatial focus and target audiences of SOE reports

The *type* of SOE reporting can vary, with at least four different types being produced. They include reports which are: (1) comprehensive (national, state, city); (2) sectoral (forests, agriculture, water, fisheries); (3) issue-based (global change, wastes); and (4) indicator-based (air quality, water quality). The most common SOE reports have a sectoral or issue orientation. The comprehensive type has been the least often produced, likely because of the many challenges of integrating all of the components. Another way of characterizing SOE reporting is by the type of analytical model used. A *stress–response* model is often used. This model directs attention to indicators of either human or environmental stress, and the way in which people and societies respond to such stresses (or to opportunities).

The *spatial focus* also can vary. SOE reports can be based on either a political or administrative region, or on a natural area or bioregion. The appeal of using political or administrative regions is that if the SOE reporting is to provide information for decision making, such decisions are normally taken by elected people who are accountable to constituents who live in politically or administratively defined areas. On the other hand, many resource and environmental issues do not respect human-made boundaries, and in such instances use of other spatial units may be more sensible, as long as there is decision-making capacity related to such natural units. There appears to be increasing use of an ecosystem approach in SOE reporting, as that often facilitates consideration of environmental processes, patterns and relationships.

The choice of *target audience* is an important decision, since that should influence the type of information assembled and the manner in which it is reported. Most SOE reports have been aimed at what might be called the "informed public" and "decision makers". However, many SOE reports are designed with a more general educational purpose in mind, and then decisions have to be taken regarding what age group and educational level is the primary target.

13.2.3 Indicators

Developing a set of environmental indicators for a region, a sector or an issue normally requires a partnership approach in which the experience and knowledge of the public and private sectors, and the local people, can be combined. Many believe that the success of state of the environment reporting is determined to a considerable extent by the choice of indicators (Moldan, Billharz and Mattravers, 1997). Indicators must be both meaningful and understandable (Box 13.4). Sophisticated indicators which cannot be understood or interpreted by users are unlikely to result in effective monitoring of environmental conditions and trends.

Using a modification of the ideas of Gélinas' (1990: 3), a set of indicators should be able to:

- translate and synthesize complex scientific or experiential data into understandable information that can be communicated effectively to users;
- enhance appreciation and understanding of how and why an environment is changing;

Box 13.4 SOE indicators

State of the Environment (SOE) reporting provides continuous assessments of environmental conditions and trends . . . Because the environment is a complex system of inter-related components, there is no easy way to assess or measure the state of the environment. There are currently thousands of environmental data parameters, captured as part of baseline inventory and survey programs by environmental, natural resource and socio-economic agencies. Consequently, it is difficult to assimilate and interpret the masses of environmental data available. **SOE indicators are key measures** that must represent the state of the environment and that collectively provide a comprehensive profile of environmental quality, natural resource assets, and agents of environmental change.

Source: Gélinas, 1990: 3 (emphasis in the original).

- influence decisions taken by elected officials, technical experts, the media and the public;
- provide a measure of quality of life for an area or people, and serve as a measure of progress toward sustainable development or other objectives;
- become a departure point to evaluate the effectiveness of policies, programmes, projects or other activities and initiatives.

13.2.4 Examples of state of environment reporting

Acid rain   Environment Canada (1991) has developed a state of the environment reporting programme, in which environmental indicators are used to provide information about significant trends regarding the environment, natural resource sustainability and related human activities. For each issue or region, a framework illustrates the links between resource and environmental systems, and human systems. Figure 13.1 shows the framework used for acid rain, caused by pollutants such as sulphur dioxide (SO_2) and nitrogen oxides (NO_x) which exist in the atmosphere but become chemically transformed into sulphuric acid and nitric acid, respectively (Environment Canada, 1999). The acids fall to earth in diluted form as wet deposition (rain, drizzle, hail, freezing rain, snow) or dry deposition (acid gas or dust). While normal rain is slightly acidic, acid rain can be up to 100 times more acidic. Once the acid rain becomes part of terrestrial and aquatic systems, it can change the chemical composition of lakes and rivers, negatively affecting fish and other organisms, can place stress on vegetation, and can damage human-built structures, ranging from statues and monuments to buildings. More than 90% of the SO_2 and NO_x emissions in North America are caused by human activity. The main Canadian

What are the links?

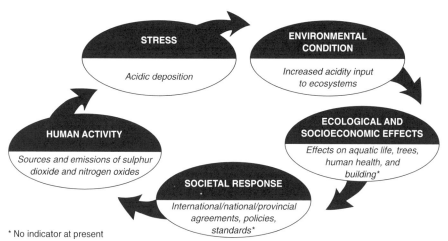

Figure 13.1   Framework for acid rain indicators in Canada (Environment Canada, 1999)

sources for SO_2 are smelting or refining of sulphur-based metal ores and combustion of fossil fuels for energy. Transportation (automobiles, trucks, trains, etc.) count for more than 50% of NO_x emissions; power generation contributes another 10%; and combustion processes from industrial, commercial and residential sources contribute 30%. Figure 13.2 presents information about two indicators related to acid rain: (1) emissions of sulphur dioxide and (2) wet sulphate deposition. Such information is used to track the progress of acid rain reduction initiatives.

Water The Moreton Bay catchment includes an area of some 19,700 km^2 and is the base for one of the fastest growing metropolitan areas in Australia – Brisbane (Figure 13.3). About 2 million people, or 60% of the population of Queensland, live in the Morton Bay catchment, which includes the estuary and bay, as well as the Brisbane River and other river catchments. Over the past decades, serious environmental degradation has occurred, leading to scientific study of the ecosystem (Dennison and Abal, 1999), and the development of a "Healthy Waterways" strategy.

To help citizens understand the seriousness of the degradation, a report card was developed to portray the ecological health and water quality of the catchment

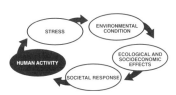

Acid Rain

Indicator: Emissions of sulphur dioxide

▶ Eastern Canadian emissions of SO_2 remained relatively constant from 1994 to 1997. In 1997, emissions of SO_2 in the seven easternmost provinces remained 24% below the target of 2.3 million tonnes.

▶ Smelting of metal ores, the largest emission source in Canada, accounted for 49% of total eastern Canadian SO_2 emissions in 1997. Power generation and other sources contributed 21% and 30%, respectively.

▶ Canada met its goal of limiting annual national SO_2 emissions to 3.2 million tonnes in 1992. By 1995, its SO_2 emissions were down to approximately 2.7 million tonnes, a 42% reduction from the 1980 level of 4.6 million tonnes.

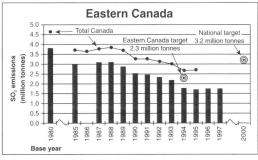

Notes:
i) Eastern Canadian emissions data for 1986 are unavailable.
ii) Canadian emission estimates are based on Canada's 1998 report to the United Nation Economic Commission for Europe (UNECE). They may differ from previously released estimates because of changes in methodology.

Canadian sources:
Eastern Canadian emissions: Environment Canada. 1990–97. *Annual Report on the Federal–Provincial Agreements for the Eastern Canada Acid Rain Program.*
Total Canadian emissions: Pollution Data Branch, Environmental Protection Service. Environment Canada.

(a)

Figure 13.2a Indications for acid rain: emissions of sulphur dioxide (Environment Canada, 1999)

Indicator: Wet sulphate deposition

▶ The area in eastern Canada receiving 20 kg/ha per year or more of wet sulphate deposition declined by 61% between the two five-year periods, 1980–84 and 1991–95, reflecting the reduction in SO_2 emissions in both Canada and the United States.

▶ The 20 kg/ha per year wet sulphate deposition target was established as an interim objective in the 1980s. Recent research confirms that this target is too high, as 20 kg/ha per year exceeds the buffering capacity of many acid-sensitive lakes in eastern Canada.

▶ Emission changes combined with variations in precipitation and weather patterns cause changes in the shape and size of the yearly deposition patterns. Thus, emission reductions may not be immediately translated into deposition reductions.

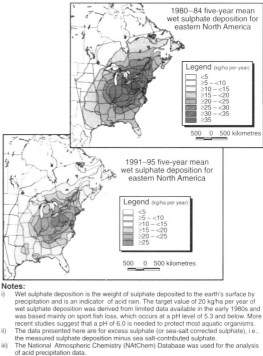

Notes:
i) Wet sulphate deposition is the weight of sulphate deposited to the earth's surface by precipitation and is an indicator of acid rain. The target value of 20 kg/ha per year of wet sulphate deposition was derived from limited data available in the early 1980s and was based mainly on sport fish loss, which occurs at a pH level of 5.3 and below. More recent studies suggest that a pH of 6.0 is needed to protect most aquatic organisms.
ii) The data presented here are for excess sulphate (or sea-salt corrected sulphate), i.e., the measured sulphate deposition minus sea salt-contributed sulphate.
iii) The National Atmospheric Chemistry (NAtChem) Database was used for the analysis of acid precipitation data.
iv) National figures on dry sulphate deposition are not available.

Source:
R. Vet, C.-U. Ro, and D. Ord, National Atmospheric Chemistry Database and Analysis Facility, Atmospheric Environment Service, Environment Canada, Downsview, Ontario.

Figure 13.2b Indications for acid rain: wet sulphate deposition (Environment Canada, 1999)

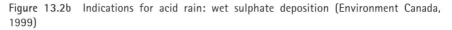

(Moreton Bay Catchment Water Quality Management Strategy Team, 1998). It was explained that a report card is valuable because it allows a mark to be given for existing conditions, which can then serve as a baseline in future years to determine what difference remediation initiatives have had. For the Moreton Bay system, a decision was taken to develop grades for eight different sites, with the intent to reassess the grades on a regular basis.

In the Healthy Waterways report of 1998, the grading system ranges from "A" to "F". To obtain an "A", an area would represent a productive, stable, resilient and balanced ecological system. Such a high quality system would

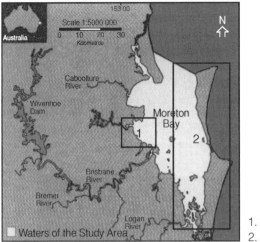

1. Bramble Bay
2. Eastern Moreton Bay

Figure 13.3   Moreton Bay Catchment, Queensland, Australia

have the capacity to bounce back after being exposed to a threat or stress, whether from human or biophysical sources. Furthermore, to obtain an "A", a place would have to exhibit biodiversity, interpreted as having a large number of different kinds of living species. At the other extreme, a "F" grade indicates an area whose natural system has not only declined, but has actually failed, in that it does not function well and has no biodiversity. It would be a place which was out of balance, and not ecologically healthy. Such an area could be improved, but it would need serious and sustained attention. Figure 13.4 provides the report card marks and comments for two of the sites in Moreton Bay: Bramble Bay and eastern Moreton Bay.

Implications   The examples related to indicators for acid rain in Canada and water management in Australia illustrate different ways in which monitoring can be used. The acid rain indicators are oriented to a more technical or professional audience. In the Moreton Bay catchment, a technical report presented detailed information for specialists, but a report card was also provided to popularize the results. The Moreton Bay report also illustrates a comprehensive kind of monitoring, whereas the acid rain information is an example of a sectoral report. The Moreton Bay report covered all seven of the questions identified earlier, whereas the acid rain indicators focused on the first four. These different characteristics indicate that there is no one, correct model for monitoring or state of environment reporting. Choices exist, and resource and environmental managers have to decide which mix is most appropriate for their needs and situation.

13.3 Public involvement in adaptive management

In Chapter 6, the concept of adaptive environmental management (AEM) was examined, and in Chapter 8 partnerships and participation were considered.

327

Report card mark for Bramble Bay

F

Human impact on the natural system in Bramble Bay has been such that the system has failed and continues to fail. It can't cope with the pollutants that pour into it every day. It has lost its natural ability to function effectively.

Reasons for this failure include:

• the massive amounts of mud/sediment arriving in the bay;

• the constant re-suspension of that sediment by tide and wind that keeps the water 'muddy' and makes it impossible for sunlight to reach the sea bottom;

• the total absence of seagrass;

• the added sewage nutrients that drive the massive 'bloom-bust' phytoplankton growth cycles that can turn the water green overnight; and

• very poor tidal flushing because Bramble Bay is a long way from the sea, so what's in the water tends to stay there for months.

We need to understand that Bramble Bay is always going to do it tough because of its geographical position. It sits at 'the end of the pipe' and it has little or no water exchange – what comes there, tends to stay there.

Encouraging factors

There is not much good news. As you'll read later, lots of activity is underway to reduce the stress on the natural system and encourage it to function better. As a whole catchment, we have to face the fact that Bramble Bay is a failed system and that it will be tough to lift its grade in the years ahead.

This is not a local problem. This failure affects everyone in the catchment. The stormwater and sediments and sewage nutrients that end up in Bramble Bay don't just come from the local area, they come from all over the catchment. That's why it will take a whole catchment effort to make any substantial improvement in the quality of water.

Future science

A major investigation is planned into the mud/sediment in Bramble Bay.

Before we can improve the situation, we need to reduce the flow of sediments and nutrients into the bay. Before we can do that, we need to know exactly where the sediment is coming from. That identification task is scheduled for the 1999-2001 scientific study in the catchment.

Bramble Bay

Figure 13.4   Report card for Bramble Bay and Eastern Moreton Bay, Queensland, Australia (Moreton Bay Water Quality Management Strategy Team 1998 31, 65)

Report card mark for eastern Moreton Bay — A

This is the benchmark for the rest of the system. That is not to say that Bramble or Deception Bay were ever crowded with the same amount of jostling marine life but the fact is they were once as healthy as this in their own right.

Dugong and calf, eastern Moreton Bay

DEPARTMENT OF ENVIRONMENT AND HERITAGE

Encouraging factors

Eastern Moreton Bay is an example of what Healthy Waterways is all about. Here we have an active vital system that is showing little, if any, sign of degradation. It is now protected to a large degree because some areas have been proclaimed national and/or international wildlife parks and reserves.

Future science

Ongoing study of this area is vital. It is our insurance policy for the eastern bay.

By understanding what is happening here, we have an example of a healthy system in the bay as a guide for any future restoration activity in the western bay.

Monitoring and evaluation are key attributes of adaptive environmental management, but they pose challenges, as noted by Shindler, Cheek and Stankey (1999), regarding public involvement. While evaluation usually emphasizes "outcomes", in AEM it is also appropriate for several reasons to consider the *process* and *context* related to public involvement. First, *process* focuses upon how tasks are handled, including the willingness of people to work together, to set objectives and to create organizational structures which facilitate collaboration. Small process initiatives which encourage cooperation can make major contributions to the achievement of long-term substantive objectives. Second, the *context* can shape both processes and outcomes. Therefore, it is necessary to recognize the situational characteristics of management issues, ranging from unique characteristics of a local setting to the particular interests of participants. Indeed, Shindler, Cheek and Stankey (1999: 3–4) suggested that the context is the most critical factor for citizens, and often is the factor which leads to their deciding to become involved.

Shindler, Cheek and Stankey (1999: 4) also argued that monitoring and evaluation should be designed to allow opportunity to learn from surprises (unintended outcomes). For such learning to occur, they noted that intended outcomes should be clearly identified and documented. If that is not done, then it becomes difficult to determine the differences between what was expected and what actually happens.

Finally, they argued that because success can be interpreted differently by various participants, it is essential to clarify what constitutes "success". Usually, success is judged relative to "tangible outcomes", such as completing an important project on time and on budget, or reducing the number of appeals and litigation. However, success can also be judged with reference to less tangible or obvious outcomes. For example, when mutual learning occurs and trust is created, better relationships among stakeholders may be established which will lead to greater willingness to work together and to seek solutions meeting the needs of all stakeholders. In such cases, the judgement of a success is not focused on solving "the problem", but on creating the capacity for stakeholders to work together constructively on an ongoing basis to deal with changing problems.

In terms of monitoring and evaluating public involvement in resource and environmental management, Shindler, Cheek and Stankey (1999: 8–12) identify six attributes against which to judge success:

(1) **Inclusive**. Public involvement is successful if the processes include all affected parties and achieve broad representation. Using a variety of mechanisms and forums helps to reach different segments of the population. However, a key issue is not so much whether all groups are represented, bur rather whether those affected by or interested in the outcome have opportunity to participate in decision making. As a result, the concept of inclusiveness is related more to quality than quantity of participation.

(2) **Interactive and sincere leadership**. Successful public participation occurs when one or two key agency staff members are committed to public participation, have high-level interpersonal skills and become actively involved in the participatory process. Personal and interactive forms of involvement are more effective than impersonal types of communication, such as submission of briefs, questionnaire surveys or public meetings intended for one-way flow of information.

(3) **Innovative and flexible.** Use of new and different methods for participation often leads to successful participatory processes. Furthermore, flexibility in the design of processes, giving special regard to local situations and needs, as well as to specific characteristics of a project, also contribute to effectiveness.

(4) **Early plus continuous equal enduring.** If the public is consulted sporadically, or only late in the planning process, the message provided to many is that the lead agency has already taken the key decisions before engaging the public in consultation. A key to success is to incorporate the public early and as frequently as practical. This approach also requires the agency to have thought through very clearly what the role of the participatory process is to be, and what outcomes are expected from it.

(5) **Designed with fundamental organizational strategies.** Success usually is achieved if basic organizational planning strategies are used in the design of the participatory process. The most fundamental need is to set clear objectives – both for the overall project, and for the participatory

---

Box 13.5 Achieving trust and credibility

Concerns about trust and credibility will continue to be a primary focus for managers and their publics. Interpersonal trust – characterized in part by a person being able to believe what another says and to think that the other person is interested in joint gains, not just individual gains – is important to the functioning of groups, to decision making, and to support for agencies. . . . But building trust is difficult; it takes time and requires many opportunities for parties to interact, to get to know one another, and for participants' voices to be heard. Trust is also fragile, requiring maintenance of contact between parties and diligent and ongoing attention.

Source: Shindler, Cheek and Stankey, 1999: 16.

---

component. Other basic organizational considerations include attention to detail, commitment to participants and good leadership.

(6) **Result in action.** While citizens increasingly value the opportunity to participate, they also expect a tangible difference as a result of their participation. Thus, participatory processes are successful when individuals and groups can observe on-the-ground progress due to their involvement. Trust and commitment will be created when people see that their efforts are contributing to real action (Box 13.5).

To be able to determine the success of a participatory process through monitoring and evaluation, special care needs to be given to contextual factors. Specifically, attention should be given to understanding the management setting and community characteristics, the agencies and institutions, as well as the attitudes and skills of, and relations among, the stakeholders. All these characteristics will provide insight into what "history" may colour people's views about other participants, and their willingness to work in a cooperative and collaborative manner.

To monitor and evaluate particpatory processes, Shindler, Cheek and Stankey (1999: 25–9) propose four discussion points to assist in determining progress toward goals and objectives. These are:

(1) **Plan.** What is being planned? Does it incorporate learning from earlier interactions and evaluations? Attention is given to specific early actions that build the foundation to achieve longer-term goals. It also helps to understand the basic planning objectives, as well as to validate an appropriate problem(s) being addressed.

(2) **Act.** What is happening regarding programmes, projects and participation? Here the focus is upon the details of the participation programme. As with the questions related to planning, the intent is to generate feedback at the very early stages of the overall process, a time which is often ignored.

(3) **Monitor.** What are the results, or outcomes? Here it is important to consider short- and long-term outcomes, as well as tangible and intangible

outcomes. Often short-term, intangible outcomes lay the foundation to realize long-term, tangible outcomes.

(4) **Evaluate.** What do the results from planning, acting and monitoring mean? When unintended outcomes occur, what are the explanations for them? What are implications of intended and unintended outcomes for design of future participatory processes?

While they provide detailed sets of questions for each discussion point, Shindler, Cheek and Stankey (1999: 25) emphasize that because no two participatory processes will occur in identical contexts, monitors and evaluators should not expect that in every situation all four discussion points will have the same importance nor will the same questions be applicable within each discussion point. Their position reinforces the view in this book that it is important to custom-design to solve problems, and not to rely upon a standardized recipe, cookbook or model.

13.4 Environmental auditing

In order to prepare a vision and a policy for an organization, public or private, it is important to identify relevant environmental issues. Eckel, Fisher and Russell (1992: 18) suggested that two principal methods can be used for identifying such issues: consulting with stakeholders and environmental auditing. Stakeholder consultation was addressed in Chapter 8; environmental consulting is examined here (Box 13.6).

Environmental auditing emerged during the 1980s, and is increasingly being used as public agencies and private firms strive to be more efficient and cost effective. Another motivation has been to avoid *liability*, by being able to demonstrate that *due diligence* has been practised. In other words, organizations need to be able to establish that they have systematically examined the environmental implications of their activities, and have taken what would be considered to be reasonable measures to avoid negative impacts. As Dunn (1995: 1–2) has indicated, environmental audits can be conducted to achieve one or more of the following purposes: (1) assess regulatory compliance, (2) assess environmental risks, (3) assist facility managers to improve performance, (4) identify waste reduction opportunities, (5) identify cost saving potentials, (6) demonstrate due diligence, and (7) improve public image.

Box 13.6 What is environmental auditing?

A systematic process of objectively obtaining and evaluating evidence regarding a verifiable assertion about an environmental matter, to ascertain the degree of correspondence between the assertion and established standards and criteria, and then communicating the results to the client.

Source: Canadian Standards Association, 1994: 2.

Box 13.7 Time frame

Because the audit assesses performance, it is impossible to look into the future, and it is impractical and irrelevant to look far into the past. Typically, the audit will examine performance over a time period of three years or less.

Source: Dunn, 1995: 5.

Eckel, Fisher and Russell (1992) noted that environmental auditing requires a team which includes more people than only those with accounting expertise. The actual mix will vary depending upon the situation, but could easily include some combination of engineers, lawyers, natural scientists and social scientists. The audit can be completed by a team internal to the organization, or by external consultants. Normally, a mix of internal and external people is required to ensure a balance of people who understand the activities of the organization, and also have the necessary auditing expertise. Finally, the time horizon is usually very short, as noted in the comment in Box 13.7. In the following sections, some strategic aspects of environmental audits are considered, and then an example of an audit will be given.

13.4.1 Establishing scope and objectives

One of the first tasks is to determine the scope and objectives for the audit. Regarding the scope, decisions have to be taken about: (1) spatial scale (building, factory site, region, country); (2) operating unit (group, department, division, entire firm); (3) environmental considerations (water quality, air quality, solid and hazardous wastes); (4) land and vegetation; (5) environmental management systems; and (6) worker health and safety. The mix of considerations will have implications for the expertise and time required to complete the audit.

For objectives, two general aspects require attention. First, the purpose of the audit needs to be established. That is, whether a site is to be developed for residential or heavy-industry use will have implications for the standards that should be considered. However, even if the use is clearly specified, further decisions may be required. For example, if a site with contaminated soils is to be used for heavy industry, the nature of the audit could be different depending upon whether the intent is to leave the soil in its present state, or to surface the area with asphalt or concrete. If the latter choice is made, and it has been determined that the contaminants in the soil are unlikely to move off the site, then the questions to be asked and information to be collected during the audit will be different than if the soil will be excavated or "cleaned up".

Once decisions have been made about the purpose, then it is possible to decide which criteria will be used to guide the audit. Normally, more than one set of criteria can apply. As Dunn (1995: 5) indicated, the audit criteria could include one or more of: (1) government regulations, (2) permits and approvals, (3) municipal or regional bylaws, (4) government guidelines, (5) internal policies and procedures, (6) institutional guidelines (e.g. World Bank, commercial bank), (7) industry standards (e.g. ISO standards), (8) industry guidelines, or (9) generally accepted good environmental practices. The final point needs some elaboration. Many firms seek to operate well-above minimum thresholds or guidelines established by government regulations, as they do not want to risk being closed down due to a low-probability but high-impact event. Thus, a judgement is involved regarding what corporate practices will be followed with reference to external regulations or guidelines. The key point is that the audit criteria are not set in stone and are not absolute. They can vary from situation to situation, and usually involve judgement calls by planners and managers.

13.4.2 Performance measures and indicators

After the scope and objectives have been established, decisions must be taken regarding what measures and indicators will be used to determine whether or not objectives are being achieved. In conjunction with such measures and indicators, there also must be capacity to monitor performance, to ensure evaluation of the results, and to modify either the objectives or the performance in light of the monitoring and assessment.

Eckel, Fisher and Russell (1992: 20) argue that performance measures should have several characteristics. They should be able to: (1) reflect cause-and-effect relationships by being based on well-accepted and verifiable theories regarding the links between actions and outcomes; (2) be as measurable and quantifiable as possible, to minimize ambiguity in interpretation of results or recommended corrective actions; and (3) reinforce one another. The recommended procedure is to use indicators which highlight both *inputs* and *outputs*. In this way, it is usually possible to determine if difficulties in realizing objectives are because of weaknesses in the cause-and-effect models and the recommended actions, or to the manner in which the recommendations are being implemented (see Chapter 12).

Input indicators

Many indicators for inputs are available, and include the following: (1) presence of an external member of the board of directors chosen because of environmental expertise; (2) approval of capital and operating expenditures for environmental concerns; (3) creation of an environmental affairs unit or department; (4) establishment of a community advisory committee, or community outreach programmes related to environmental issues; (5) presence of recycling activities; and (6) provision of employee education programmes regarding environmental management.

Use of such indicators helps to establish the commitment of the organization to deal with environmental problems, and thereby the degree to which legitimation or credibility has been given to environmental concerns. However, inputs by themselves do not ensure desired outputs or outcomes. The fact that a member of the board of directors is selected for environmental expertise will have little effect if that member is not actively involved on the board, does not have influence or "clout" with other board members, or has not pushed for initiatives which actually affect environmental performance. Or, education programmes by themselves do not ensure that the employees learned what was taught, or were given opportunity to apply their new environmental knowledge. It is for this reason that good audits use both input and output indicators.

Output indicators

As with inputs, there are many possible indicators for outputs. Some include: (1) volume and types of materials processed by internal and waste recycling programmes; (2) volume and types of waste generated and treated; (3) measures of efficiency, energy conservation and rates of spoilage for products and production processes; (4) air emission rates for contaminants; (5) water quality concentration measures; (6) volume of pesticides or other hazardous chemicals used; (6) monetary value of damages to the natural and social environments; (7) frequency with which applicable legal or regulatory requirements are exceeded; (8) number, kind and volume of hazardous or toxic spills; (9) number and type of environmentally related complaints received from stakeholders; and (10) number and kind of environmental liabilities, lawsuits or unethical business practices.

Implications

The indicators for outputs are selected to provide evidence of whether an organization's environmental policies and objectives are being realized, and whether they satisfy regulations (Box 13.8). In that manner, when the answer is affirmative, an organization can claim that it is practising *due diligence* or *due care*, and is doing everything that could reasonably be expected. However, while some of the output indicators are relatively easy to define, others are much more difficult to put into practice. Furthermore, whether or not indicators are easy to define, obtaining information for some of them can be a major challenge. This situation, usually more common with output than with input indicators, is a main reason why audits often focus on inputs. As already noted, however, while it is often possible with input indicators to determine whether or not they are present, it is much more difficult to establish a cause-and-effect relationship with desired outcomes. For this reason, wherever possible it is good practice to use indicators for both inputs and outputs.

Box 13.8 Using indicators

Existing accounting and financial systems are particularly designed to focus on input measures, such as expenditures on environmental research and development, waste disposal, and site remediation, or disclosure of the extent of environmental liabilities and contingencies. While these systems may provide information on efficiency and productivity measures, they concentrate on financial performance rather than broader environmental performance issues.

Source: Eckel, Fisher and Russell, 1992: 21.

Box 13.9 What is an environmental audit?

The phrase "environmental audit" means different things to different people. Here it means a report on the relationship of a First Nation to its environment and resource base, that identifies current problems and needs, and that can be used as a basis for planning a sustainable future for the First Nation.

Source: Walpole Island Heritage Centre and Chreod Ltd, 1995: 1.

13.4.3 Environmental audit model for First Nations

An environmental audit has been completed for the First Nations or aboriginal people who live on Walpole Island, at the mouth of the St Clair River, where it enters Lake St Clair in Ontario. The total area of Walpole Island and several smaller islands is 350 km^2, and almost 7,000 hectares or 48% of the area is marshland, making it the largest and, likely, most significant wetland in the entire Great Lakes system. Walpole Island is also downstream of the chemical companies near Sarnia, which has become known as "chemical valley". Long-term disposal of wastes, which have gradually been curtailed, and still-occurring accidental spills, into the river have degraded the environment and posed a threat to well being of people and other living species.

Based on the audit experience, the Walpole Island Heritage Centre and Chreod Ltd developed an environmental audit model that could be used by other First Nations people (Box 13.9). That model is outlined below. It should be noted that in September 1995, the Walpole Island Heritage Centre was one of 50 organizations from around the world to receive an award from the Friends of the United Nations for its community-level work regarding sustainable development.

The Walpole Island Heritage Centre and Chreod Ltd (1995) suggested that most environmental audit reports should have three distinct sections: (1) setting and context, (2) analysis of individual environmental elements, and (3) synthesis and recommendations. The audit report should highlight distinctive features as well as potential environmental or other conflicts, provide a bibliography or list of sources on which audit statements are based so that further information can be obtained, and identify gaps in knowledge and understanding that need to be filled if a sustainable relationship between people and their environment is to be achieved. Each of the three main sections is outlined in more detail below.

(1) Setting and context

Information should be provided regarding the *physical setting* of the area regarding its landscape, climate, vegetation and water; its *population and society*, with particular regard to changes in numbers of people, age structures, settlement patterns, employment and health; and its *history and political context*.

(2) Analysis of individual environmental elements

As with many impact assessments, the model provides a checklist for possible environmental elements to be considered. It is recommended that the audit should review the adequacy and significance of information, noting where gaps are present. The checklist includes:

- *Solid and liquid wastes.* Solid waste collection and disposal systems; separation of hazardous, recyclable and residual wastes; type of landfill and other disposal sites; long-term adequacy of present systems; feasible alternatives; types of sewerage systems, proportion of population serviced by different sewerage systems, maintenance information and needs; implications for wildlife and human health, available public health data.

- *Water supply.* Systems in use (individual, collective); sources (surface or ground water); treatment; quality monitoring; sources of contaminants and prospects for improvement; long-term adequacy of present and alternative systems.

- *Air quality.* Air pollution from local and distant sources; types and adequacy of data; implications for public health, and plant and animal life (including agriculture).

- *Toxic dumps and other contaminated sites.* Location and condition of known or suspected toxic sites (including earlier or currently uncontrolled dumps and landfills, underground storage tanks); sources external to the area under audit affecting air, surface waters or ground water; potential or known health hazards; plans or need for removal or remediation.

- *Natural hazards and other dangers.* Nature and frequency of the hazard (riverine or lake flooding, forest fires, tornados); risk of accidents or other dangers from human activity (release of toxic material from train derailment or traffic accident); potential for damage to humans and local environment; opportunity for hazard avoidance or risk reduction; adequacy of emergency planning.

- *Agriculture.* Kinds and character of activity (crops, owner-operated or rented, size of units); tillage practices; use of fertilizers and pesticides; known or potential impacts on soils, watercourses and adjacent areas; evidence of soil erosion; long-term sustainability.

- *Forests.* Extent, character and ownership; existence and implementation of long-term management plans; fire protection; liability from pests and diseases; value of forests for associated flora and fauna.

- *Fish and wildlife resources.* Character of and significance for economy and society; recent and long-term trends (declining, recovering, sustainable); vulnerability to persistent and bioaccumulating contaminants; implications for human health; long-term plans and prospects.

- *Important habitats, species and sites.* Knowledge about or probability of rare or endangered species or habitats within the area; data or information about archaeological or historical sites; vulnerability of sites and species to natural erosion or development plans (roads, water and sewerage systems); long-term plans and prospects.

- *Knowledge, education and systems for sustainability.* Knowledge and documentation, based on both traditional and scientific sources, about the environment and resource base; extent to which such knowledge is accessible; level of commitment by leaders to sustainability (measured by policy statements and actions, voluntary recycling); incorporation of sustainability and environmental protection in development planning; access to professional expertise.

This checklist of environmental elements provides a systematic way for an audit team to identify potentially significant matters for a First Nations area. A similar or modified checklist could be developed for any site, region or firm.

(3) Synthesis and recommendations

An audit for an area or region could involve collection of substantial information regarding input and output indicators. The final section of an environmental audit report, therefore, should systematically identify the key linkages between the environment and the people living there. Furthermore, it usually can be expected that numerous opportunities for improvement will be identified, and thus there may be many recommendations. However, it is not sufficient to provide a list of recommendations. If that is done, a likely result is that the recipients of the report will decide that they "can tackle only a few of the recommendations, with those selected being chosen more on the basis of simplicity or available funding than because of urgency or overall importance" (Walpole Island Heritage Centre and Chreod Ltd, 1995: 19). The most useful way to present recommendations is outlined in Box 13.10.

The above discussion illustrates how important monitoring, through state of the environment or state of sustainability reports, can be for environmental audits. Environmental audits require systematic data about inputs and outputs significant for environmental management, and SOE or SOS reports often contain such data. In many situations, however, an environmental audit team will draw upon SOE or SOS data but will have to collect other information to document the extent to which an organization is achieving its environmental policies, goals and objectives, and the extent to which it is complying with any external regulations or expectations.

Box 13.10 Presentation of recommendations

The auditor, . . . , needs to offer recommendations with some indication of relative importance and priority, and the reasons for these priorities. One way is to offer a small number of formal "recommendations", covering the key needs and inadequacies, with less important matters offered as "suggestions". Of course, the First Nations may not agree with the priorities indicated by the auditor, but the main reason for conducting an environmental audit is to provide an overall view of a complex situation; the auditor's advice on what are the key elements of that situation is an essential part of the audit.

Source: Walpole Island Heritage Centre and Chreod Ltd, 1995: 19.

13.5 Implications

In this chapter, attention has focused upon state of the environment reports, public participation in adaptive management and environmental audits. Other kinds of monitoring can be and have been done in resource and environmental management. For example, a need exists to track the perceptions, attitudes and values of people and societies over time. Such tracking is not often done, and yet if there is to be an attempt to shift basic values it seems sensible that planners and managers should know current perspectives, and how they are evolving. In addition, there is a need to monitor and evaluate resource and environmental policies and programmes. There is a tradition of such research in resource and environmental management, and some of the references at the end of this chapter provide an entry point to examine this type of monitoring and evaluation.

State of environment reporting and environmental auditing also highlight conceptual and methodological issues which must be addressed. The scope and objectives need to be clarified, as they have major implications for the kind of information that will be collected. Also, we have become aware that choices need to be taken between the mix of data regarding inputs and outputs. While ultimately we are most interested in *outcomes*, it often is the case that inputs are easier than outputs to measure. As a result, indicators of inputs tend to be used most frequently in monitoring or audits. An important research need is to develop more output indicators that are conceptually and operationally sound.

Monitoring and assessment considerations are similar to implementation. Too often, resource and environmental managers do not give them enough attention, preferring to focus on the specification of ends and means. However, if we are unable to implement, or to track what is happening, then it is difficult to determine if we are on the desired path. In the future, an opportunity and challenge will be to determine how local citizens' groups and non-government organizations can become more involved in monitoring and assessment. Many governments are reducing their activities, and partnerships for monitoring and assessment offer opportunities to maintain capacity for tracking progress, even at a time of reduced financial resources.

References and further reading

Babu S C and W Reidhead 2000 Monitoring natural resources for policy interventions: a conceptual framework, issues, and challenges. *Land Use Policy* 17: 1–11

Barrett B F D 1995 From environmental auditing to integrated environmental management: local government experience in the United Kingdom and Japan. *Journal of Environmental Planning and Management* 38: 307–31

Bell S and S Morse 1999 *Sustainability Indicators: Measuring the Immeasurable?* London, Earthscan Publications Ltd

Benarie M 1988 Delphi and Delphi-like approaches with special regard to environmental standard setting. *Technological Forecasting and Social Change* 33: 149–58

Born S M, K D Genshkow, T L Filbert, N Hernandez-Mora, M L Keefer and K A White 1998 Socio-economic and institutional dimensions of dam removals: the Wisconsin experience. *Environmental Management* 22: 359–70

Boscolo M, M Powell, M Delaney, S Brown and R Faris 2000 The cost of inventorying and monitoring carbon: lessons from the Noel Kempff Climate Action Project. *Journal of Forestry* 98(9): 24–31

Brandt N, F Burström and B Frostell 1999 Local environmental monitoring in Sweden. *Local Environment* 4: 181–93

Cahill L B (ed.) 1984 *Environmental Audits*. Third edition. Rockville, MD, Government Institutes Ltd

Canadian Institute of Chartered Accountants 1992 *Environmental Auditing and the Role of the Accounting Profession*. Toronto, Canadian Institute of Chartered Accountants

Canadian Standards Association 1994 *Guidelines for Environmental Auditing: Statement of Principles and General Practices*. Publication Z75194, Rexdale (Toronto), Canadian Standards Association

Canter L W 1993 The role of environmental monitoring in responsible project management. *Environmental Professional* 15: 76–87

Chadwick A and J Glasson 1999 Auditing the socio-economic impacts of a major construction project: the case of Sizewell B nuclear power station. *Journal of Environmental Planning and Management* 42: 811–36

Cheremisinoff P N and N P Cheremisinoff 1993 *Professional Environmental Auditor's Guidebook*. Park Ridge, NJ, Noyes Publications

Chipeniuk R 1999 Vernacular bio-indicators and citizen monitoring of environmental change. In J G Nelson, R Butler and G Wall (eds) *Tourism and Sustainable Development: Monitoring, Planning, Managing, Decision Making*. Department of Geography Publication Series No. 52 and Heritage Resources Centre Joint Publication No. 2, Waterloo, Ontario, University of Waterloo, pp. 373–87

Culhane P J 1993 Post-EIS environmental auditing: a first step to making rational environmental assessment a reality. *Environmental Professional* 15: 66–75

Cuthill M 2000 An interpretive approach to developing volunteer-based coastal monitoring programmes. *Local Environment* 5: 127–37

Darmstadter J 2000 Greening the GDP: Is it desirable? Is it feasible? *Resources* 139: 11–15

Dennison W C and D G Abal 1999 *Moreton Bay Study: A Scientific Basis for the Healthy Waterways Campaign*. Brisbane, Queensland, Australia, South East Regional Water Quality Management Strategy Team, Brisbane City Council

Dickman M D 1993 Waterways walkabout: learning to see what's missing. In S Lerner (ed.) *Environmental Stewardship: Studies in Active Earthkeeping*. Department of Geography Publication Series No. 39, Waterloo, Ontario, University of Waterloo, pp. 367–73

Dorcey A H J and J R Griggs (eds) 1991 *Water in Sustainable Development: Exploring our Common Future in the Fraser River Basin*. Vancouver, BC, University of British Columbia, Westwater Research Centre

Dovetail Consulting 1995 *A Strategy for the Harmonization of State of Environment Reporting across CCME Member Jurisdictions*. Prepared for the Canadian Council of Ministers of the Environment State of Environment Reporting Task Group, Winnipeg, Manitoba

Dunn K 1995 *Fundamentals of Environmental Auditing*. Toronto: 9th Annual Toronto Environmental Conference and Trade Show

Eckel L, K Fisher and G Russell 1992 Environmental performance measurement. *CMA Magazine* March: 16–23

Edwards F N (ed.) 1992 *Environmental Auditing: the Challenge of the 1990's*. Calgary: University of Calgary Press, The Banff Centre for Management

Eisenberg N A, M P Lee, T J McCartin, K I McConnell, M Thaggard and A C Campbell 1999 Development of a performance assessment capability in the waste management programs of the US Nuclear Regulatory Commission. *Risk Analysis* 19: 847–76

Elkin T J 1990 State of the environment reports and national conservation strategies: the linkage. *Alternatives* 16: 52–61

Ellesfson P V, M A Kilgore and M J Phillips 2001 Monitoring compliance with BMPs: the experience of state forestry agencies. *Journal of Forestry* 99(1): 11–17

Environment Canada 1991 *The State of Canada's Environment*. Ottawa, Minister of Supply and Services Canada

Environment Canada 1995 *The State of Canada's Climate: Monitoring Variability and Change*. SOE Report No. 95-1, Ottawa, Minister of Public Works and Government Services Canada

Environment Canada 1999 *Acid Rain*. Overview SOE Bulletin No. 00-3, Fall 1999, Ottawa, Environment Canada

Faulkner B and C Tideswell 1997 A framework for monitoring community aspects of tourism. *Journal of Sustainable Tourism* 5: 3–28

Finlayson C M and D S Mitchell 1999 Australian wetlands: the monitoring challenge. *Wetlands Ecology and Management* 7: 105–12

Fraser Basin Management Program 1995a *Board Report Card: Assessing Progress toward Sustainability in the Fraser Basin*. Vancouver, BC, Fraser Basin Management Program

Fraser Basin Management Program 1995b *State of the Fraser Basin: Assessing Progress towards Sustainability*. Vancouver, BC, Fraser Basin Management Program

Gélinas R 1990 Towards a set of SOE indicators. *State of the Environment Reporting*. No. 5, Ottawa, Environment Canada, p. 3

Goodall B 1995 Environmental auditing: a tool for assessing the environmental performance of tourism firms. *Geographical Journal* 161: 29–37

Greeno J L, G S Hedstrom and M DiBerto 1985 *Environmental Auditing Fundamentals and Techniques*. Toronto, Wiley

Greer T, I Douglas, K Bidin, W Sinum and J Suhaimi 1995 Monitoring geomorphological disturbance and recovery in commercially logged tropical forest, Sabah, East Malaysia, and implications for management. *Singapore Journal of Tropical Geography* 16: 1–21

Hartwell M A, V Myers, T Young, A Bartaska, N Gassman, J H Gentile, C C Harwell, S Appelbaum, J Barko, B Causey, C Johnson, A McLean, R Smola, P Templet and S Tosini 1999 A framework for an ecosystem integrity report card. *BioScience* 49: 543–56

Hicks B B and T G Brydges 1994 A strategy for integrated monitoring. *Environmental Management* 18: 1–12

Howe G and T Zdan 1992 Monitoring sustainable development: an overview of concepts to link natural resource and economic accounting systems. *Canadian Water Resources Journal* 17: 373–82

International Institute for Sustainable Development 1997 *Compendium of Sustainable Development Indicator Initiatives and Publications: IUCN Monitoring and Evaluation*. Online at: http://www.iisd1.iisd.ca/measure

International Union for Conservation of Nature and Natural Resources, United Nations Environmental Programme, and World Wildlife Fund 1991 *Caring for the Earth: a Strategy for Sustainable Living*. Gland, Switzerland, IUCN

Jacobs M 1992 Environmental auditing and management in local government. *Journal of the Institute for Water and Environmental Management* 6: 583–7

Johnson V and R Nurick 1999 Towards community-based indicators for monitoring quality of life and the impact of industry in South Durban. *Environment and Urbanization* 10: 233–50

Jordan C, E Mihalyfalvy, M K Garret and R V Smith 1994 Modelling of nitrate leaching on a regional scale using a GIS. *Journal of Environmental Management* 42: 279–98

Judd G 1996 Environmental accounting and reporting practices. In B Ibbotson and J D Phyper (eds) *Environmental Management in Canada*, Toronto, McGraw-Hill Ryerson, pp. 61–84

Kaiserman M and B Kelly (eds) 1994 *Guidelines for Environmental Auditing: Statement of Principles and General Practices.* Toronto, Canadian Standards Association

Kammerbauer J, B Cordoba, R Escolán, S Flores, V Ramirez and J Zeledón 2001 Identification of development indicators in tropical mountainous regions and some implications for natural resource policy designs: an integrated community case study. *Ecological Economics* 36: 45–60

Kreutzwiser R D and M J Slaats 1994 The utility of evaluation research to land use regulations: the case of the Ontario shoreline development. *Applied Geography* 14: 169–81

Lange J H, G Kaisir and K W Thomulka 1994 Environmental site assessments and audits: building inspection requirements. *Environmental Management* 18: 151–60

Loucks D P 1997 Quantifying trends in system sustainability. *Hydrological Sciences Journal* 42: 513–30

MacDonald L H 1994 Developing a monitoring project. *Journal of Soil and Water Conservation* 49: 221–7

Marsden S 1998 Importance of context in measuring the effectiveness of strategic environmental assessment. *Impact Assessment and Project Appraisal* 16: 255–66

Maktav D, F Sunar, D Yalin and E Aslan 2000 Monitoring Loggerhead sea turtle (*Caretta caretta*) nests in Turkey using GIS. *Coastal Management* 28: 123–32

Martin V 1993 How field naturalists gear up for monitoring. In S Lerner (ed.) *Environmental Stewardship: Studies in Active Earthkeeping.* Department of Geography Publication Series No. 39, Waterloo, Ontario, University of Waterloo, pp. 375–81

Matsuto T and N Tanaka 1993 Data analysis of daily collection tonnage of residential solid waste in Japan. *Waste Management and Research* 93: 333–43

McLean R 1994 *International Trends in Environmental Auditing.* Brussels, Arthur D Little International Inc.

Moldan B, S Billharz and R Mattravers (eds) 1997 *Sustainability Indicators: Report of the Project on Indicators of Sustainable Development.* Toronto, Wiley

Moreton Bay Catchment Water Quality Management Strategy Team 1998 *The Crew Member's Guide to the Health of Our Waterways.* Brisbane, Queensland, Australia, Brisbane City Council, Moreton Bay Catchment Water Quality Management Strategy Team

Nakayma M 1998 Post-project review and environmental impact assessment for Saguling Dam for involuntary resettlement. *International Journal of Water Resource Development* 14: 217–29

Nelson J G 1995 Sustainable development, conservation strategies, and heritage. In B Mitchell (ed.) *Resource and Environmental Management in Canada: Addressing Conflict and Uncertainty.* Toronto, Oxford University Press, pp. 384–405

O'Neill R, C T Hunsaker, K B Jones, K H Riiters, J D Wickham, P M Schwartz, I A Goodman, B L Jacson and W S Baillargeon 1997 Monitoring environmental quality at the landscape scale. *BioScience* 47: 513–19

Organization for Economic Co-operation and Development 1985 *OECD State of the Environment Report*. Paris, Organization for Economic Co-operation and Development

Petts J and G Eduljee 1994 Integration of monitoring, auditing and environmental assessment: waste facility issues. *Project Appraisal* 9: 231–42

Phyper J D and B Ibbotson 1996 Environmental audits. In B Ibbotson and J D Phyper (eds) *Environmental Management in Canada*. Toronto, McGraw-Hill Ryerson, pp. 211–28

Pilgrim W and R N Hughes 1994 Lead, cadmium, arsenic and zinc in the ecosystem surrounding a lead smelter. *Environmental Monitoring and Assessment* 32: 1–20

Prochazkova D 1995 Theoretical background of environmental monitoring and its conception in the Czech Republic. *Environmental Monitoring and Assessment* 34: 105–13

Rasmussen K, H Skriver, J Tychsen, P Gudmandsen and M Olsen 1994 Environmental monitoring by remote sensing in Denmark. *Geografisk Tidsskrift* 94: 12–21

Richards L and I Biddick 1994 Sustainable economic development and environmental auditing: a local authority perspective. *Journal of Environmental Planning and Management* 37: 487–94

Sandén P and Å Danielsson 1995 Spatial properties of nutrient concentrations in the Baltic Sea. *Environmental Monitoring and Assessment* 34: 289–307

Schaeffer D J, H W Kerster, J A Perry and D K Cox 1985 Environmental audit. I. Concepts. *Environmental Management* 9: 191–8

Schmidt M G, H E Schrier and B Shah Pravakar 1995 A GIS evaluation of land use dynamic and forest soil fertility in a watershed in Nepal. *International Journal of Geographic Information Systems* 9: 317–27

Shindler B and J Neburka 1997 Public participation in forest planning: 8 attributes of success. *Journal of Forestry* 95: 17–19

Shindler B, K A Cheek and G H Stankey 1999 *Monitoring and Evaluating Citizen–Agency Interactions: A Framework Developed for Adaptive Management*. General Technical Report PNW-GTR-452, Portland, Oregon, US Department of Agriculture, Forest Service, Pacific Northwest Research Station, April

Slaats M J and R Kreutzwiser 1993 Shoreline development and regulations: do they work? *Journal of Soil and Water Conservation* 48: 158–65

Smith C 1996 Ontario's anti-pollution angels. *Toronto Star*. 6 January 1996, D6

Sokolik S L and D J Schaeffer 1986 Environmental audit. III. Improving the management of environmental information for toxic substances. *Environmental Management* 19: 311–17

Soyez D 1995 Assessing energy projects in the Saarland, Federal Republic of Germany. *Environment* 23: 82–92

Stevens D L 1994 Implementation of a national monitoring program. *Journal of Environmental Management* 42: 1–29

Thompson D and M J Wilson 1994 Environmental auditing theory and applications. *Environmental Management* 18: 605–15

Tomlinson P and S R Atkinson 1987 Environmental audits: proposed terminology. *Environmental Monitoring and Assessment* 8: 187–98

Tusa W 1990 Developing an environmental auditing program. *Risk Management* August: 24–9

Walker G and D Bayliss 1995 Environmental monitoring in urban areas: political contexts and policy problems. *Journal of Environmental Planning and Management* 38: 469–82

Walpole Island Heritage Centre and Chreod Ltd 1995 *An Environmental Audit Model for First Nations*. Wallaceburg, Ontario, Walpole Island Heritage Centre, and Ottawa, Chreod Ltd

Walton J 2000 Should monitoring be compulsory within voluntary environmental agreements? *Sustainable Development* 8: 146–54

Walz R 2000 Development of environmental indicator systems: experiences from Germany. *Environmental Management* 25: 613–23

Watzin M C and A W McIntosh 1999 Aquatic ecosystems in agricultural landscapes: a review of ecological indicators and achievable ecological outcomes. *Journal of Soil and Water Conservation* 54: 636–44

Wismer S 1999 From the ground up: quality of life indicators and sustainable community development. *Feminist Economics* 5: 109–14

Wood W W 2000 Environmental accounting: the new bottom line. *Ground Water* 38: 161

Woodley S 1999 Tourism and sustainable development in parks and protected areas. In J G Nelson, R Butler and G Wall (eds) *Tourism and Sustainable Development: Monitoring, Planning, Managing and Decision Making.* Department of Geography Publication Series No. 52 and Heritage Resources Centre Joint Publication No. 2, Waterloo, Ontario, University of Waterloo, pp. 159–74

Wooldridge C F, C McMullen and V Howe 1999 Environmental management of ports and harbours – implementation of policy through scientific monitoring. *Marine Policy* 23: 413–25

Zaleski J, R Serafin and T Palmowski 1995 Assessing planning in Poland's coastal region. *Environments* 23(1): 16–20

Zeide B 1994 Big projects, big problems. *Environmental Monitoring and Assessment* 33: 115–33

Chapter 14

Managing for Environmental Justice

14.1 Introduction

Change. Complexity. Uncertainty. Conflict. Each requires attention in resource and environmental management. The people in Bangladesh, described in Chapter 1 and who are contending with the impacts of arsenic-contaminated water, are certainly encountering each of them. The individuals and groups in Pakistan who are striving to implement the National Conservation Strategy, which will balance economic, environmental and cultural considerations, are addressing them. The *precautionary principle* was created as a guideline to handle them. *Hedging* and *flexing* have been recognized as alternative approaches toward them. *Visioning*, *backcasting* and *adaptive environmental management* are being increasingly used to anticipate and to respond to them. Techniques such as *Life Cycle Analysis* and *Environmental Audits* have been created to help managers address

---

Box 14.1 A Jeremiah* parable

I imagine spaceship earth as a kind of fortunate Titanic. On the ship's prow in the middle of the night, Jeremiah Jones peers into the dimness. Faintly perceiving some ominous shapes ahead, he cries out lustily "icebergs ahead." Unsure if he is heard, he cries out again and again. On the ship's bridge, the captain, hearing Jeremiah only after some time, turns to the navigator and asks for a course correction to avoid a collision. Ten degrees to the starboard she says. The Captain, thinking "What luck that I have already started to turn because of the bad weather ahead'" orders a five degree correction. The helmsman looks at his compass and suddenly realizes that he has been dozing for a few minutes and that the ship has actually been drifting, fortunately in the right direction. Without saying anything, he then corrects the course by two degrees. Up ahead, alone and in the cold, Jeremiah awaits a hard starboard course correction, maybe even a reversal of engines. Sensing none, he mutters to himself "they never listen to me" and prepares for the worst.

(* The prophet Jeremiah is the archetype of those who warn of future woes, hence the term *jeremiad* as a tale of woe.)

Source: Kates, 1995: 635 and 625.

---

them. And, increasing attention to *implementation*, and to *monitoring* and *evaluation*, explicitly acknowledge the need to be able to modify approaches as a result of them.

As mentioned throughout this book, it is often presumptuous to believe that humans "manage" environment and resources. More realistically, humans manage their interactions with environment and resources. For that reason, uncertainty (as a result of imperfect understanding), as well as ignorance, and conflict (as a result of many different and legitimate interests) are common. Furthermore, the complexity of biophysical and socio-economic systems is great, and is exacerbated by ongoing change, some of which is influenced or caused by human activity. Thus, to develop policies, programmes or plans for resource and environmental management, or to be able to appraise the effectiveness of initiatives, it is important to be able to recognize and deal with change, complexity, uncertainty and conflict.

In this chapter, discussion focuses on four matters deserving systematic attention: (1) developing a vision, (2) creating a process, (3) generating a product and (4) ensuring implementation and monitoring.

14.2 Developing a vision

In Chapter 3, reference was made to a comment by Alice in Wonderland – if you do not know where you want to go, any road will get you there. In other words, if there is no sense of vision or direction regarding a desirable future, then almost any choice will do. In contrast, if we have a sense of where we would like to get to, then it should be possible to take actions or intervene to try and move in the desired direction.

Given the above comments, a key task for resource and environmental managers is to help develop a vision for a desired future, as considered in Chapters 3 and 4. This task is never easy, since societies rarely if ever are homogenous (Box 14.2). Many interests exist, and conflicts frequently emerge. Thus, what may be a desirable future for one group or interest may be viewed as undesirable by another. However, by considering *desirable* futures, we should be able to move away from what too often is a preoccupation with *most probable* futures. If future search conferencing or backcasting is to be used, we need a vision of

---

Box 14.2 How far to look into the future?

It is with great trepidation that one writes about the future. After all, it hasn't happened yet and one could end up looking pretty silly, depending upon the acuity of one's vision. For that reason, and like many other forecasters, I have adopted the convention of choosing as the end of my forecast horizon a date sufficiently far in the future that there is no chance of my being around to be held accountable.

Source: Portney, 2000: 6.

> **Box 14.3**
>
> It is extremely important to approach sustainability as a learning process rather than as series of blueprints.
>
> *Source*: Brooke and Rowan, 1996: 115.

where we want to get to, so that decisions can be made which take us in the prescribed direction.

Since many different interests normally coexist in a society, determining a vision is not a task only for professional resource and environmental managers. Consultation and interaction with people living in an area are important, in order to draw upon their knowledge and understanding of conditions and processes in that area, to address their needs and to incorporate their expectations. For this reason, *partnerships* and *participatory approaches* should be important features of resource and environmental management, with particular need to draw upon *local knowledge systems*. Furthermore, in developing partnerships, greater attention needs to be given to achieving *gender balance*.

Since publication of *Our Common Future* in 1987, *sustainable development*, today more usually referred to as *sustainability*, has been viewed by many as the vision to be pursued. As explained in Chapter 4, the World Commission on Environment and Development did not present one blueprint for sustainable development, arguing instead that different countries and cultures have to custom-design an individual strategy reflecting their conditions and needs (Box 14.3).

The outcome is that different interpretations have been given to sustainability in various countries, particularly between developed and developing nations. Such differences or inconsistencies are not inherently bad. Indeed, Ralph Waldo Emerson wrote that foolish consistency is the hobgoblin of little minds. Various interpretations of sustainability are appropriate if they reflect the different situations in countries or regions. Furthermore, within a country or region, the emphasis in a sustainable development strategy could reasonably be expected to evolve over the years as conditions change. Ongoing interest in sustainable development, as a vision, can be expected to continue because it provides a way to consider how to balance economic, environmental and cultural matters. The guest statement from Walter Danjoy provides a perspective about sustainability related to developing countries.

Another concept increasingly drawing attention as a possible vision is *environmental justice*. Higgins (1993: 292) has observed that, regarding environmental justice, ". . . advocates argue their communities are disproportionately burdened with polluting facilities, the beneficiaries of which live largely outside of the minority community sites". More generally, Cutter, Holm and Clark (1996: 517) stated that "[environmental] inequity originates from three major sources of dissimilarities: social, generational and procedural. To test for outcome equity, one examines the disproportionate effect of environmental degradation on places or people arising from these dissimilarities".

Guest Statement

Sustainable development in developing countries

Walter Danjoy, Peru

Introduction

Within the context of sustainable development, developing countries, such as Peru, find sharply contradictory situations, which require a more realistic evaluation with respect to the existing relationship between the availability of natural resources, their use by the population in general and particularly by native and rural communities, and the concept of sustained development. The classic concept of sustained resource management, which is aimed at resource preservation in order to meet the needs of future generations, finds a barrier in the social element. This barrier, which limits sustainable development, is determined by the urgent needs of the population of the areas in which the resources and their area of influence are located, giving rise to the need for specific strategies for the use of these resources.

Nature of the problem

In developing countries, economic problems are frequently both a symptom and a cause of environmental degradation. Economic necessity leads to overgrazing and deforestation, thus reducing the productivity of land and soil, increasing the frequency of floods and droughts, causing greater poverty and despair, particularly in rural underdeveloped sectors of a country. The poor become trapped in a downward cycle of rapid population growth, environmental degradation and falling economic standards. They are unable to grow economically because of the pressure of population, lack of agricultural land and the inadequate use of natural resources as a source to contribute to the active national economy. Many developing countries find it difficult to overcome these controlling economic factors.

The synergistic interaction of poverty, illiteracy, malnutrition and underdevelopment make certain problems severe. Examples of this are high infant mortality, cholera and mycotoxins in food. Any improvement in this situation needs concerted action in more than one activity, particularly in the sustainable use of natural resources by the population.

Conflicts of use

The sustainable development and persistent productivity concepts may become compatible with those related to the growing need to increase food supply, minimizing the conflict among production, sustainable systems and environmental preservation. To be able to apply practices which will guarantee a sustainable production system, it is essential that they must be profitable in order to fund the investments required for the preservation of the productive environment. If adequate technological elements and strategies are not applied, combined with an element of profitability, it may turn out that the increased production may negatively affect the natural resource on which it is based, creating a vicious circle which would be hard to break.

When an economic depression hits, the rural population is forced to intensify the use of its resources. In many cases, rural inhabitants are blamed for the degradation of natural resources, mainly because of their cut-and-burn practice and migrating-type of agriculture, which is recognized as the paradigm leading to the destruction of the Amazon environment. However, there is some evidence that, in most cases and under economic and social stability conditions, rural residents may turn into a dynamic agent participating in the process, carrying out a series of strategies for the protection of their resources. When the country's economic conditions degrade, producers and rural residents change their production practices and strategies, perhaps speeding up the process of environmental degradation.

Finally, within the context of the management of natural resources and sustainable development, the above-mentioned situation creates the need for governments to adopt integrated social policies aimed at aiding the marginal population, who suffer a reduction of their only natural and cultural heritage, particularly when an economic recession alters economic levels; and consequently, the potential of natural resources.

Walter A Danjoy was born in Huancayo, Peru. Beginning at the age of 12, he came in contact with the Amazon people while accompanying his father on trips to the Upper Jungle. In 1972, he graduated from the National University of Central Peru as a forest engineer; and after graduation began his professional career at the National Office for Natural Resources Evaluation, where he worked for 20 years, carrying out activities related to the inventory and evaluation of natural resources and environmental management and monitoring, where he applied geographic information and remote sensing systems. In 1992, he completed a Master's degree in Canada, with a thesis focused on the role of remote sensing and geographic information systems for management of natural resources in the Peruvian Amazon Basin.

Currently, he is a consultant for Peruvian and international agencies, on issues related to the management of natural resources and environmental monitoring. He is also a University Professor and author of four books on natural resources. He is a member of the Latin American Society of Specialists on Remote Sensing and Geographic Information Systems (SELPER), of which he was President between 1992 and 1994.

The environmental justice movement combines values and goals from the civil rights, poor peoples, occupational safety and health, and grassroots environmental movements, and "the vision that emerges is a qualitative challenge to existing business and government agendas, as well as to the reform agendas of mainstream environmentalism" (Higgins, 1993: 292). These challenges emerged in the late 1970s and early 1980s, and focused on institutional responses related to toxic waste disposal, public facility siting (from airports to electric power transmission lines), species and wilderness preservation, nuclear power and weapons, and the political positions and growing professional orientation of mainstream environmentalism.

Those working within the framework of environmental justice do not seek only to redistribute existing and future pollution. More fundamentally, they advocate policies and practices which incorporate ecologically, economically and socially appropriate development, and argue that pollution *prevention* rather than pollution *control* should be a primary goal. The emphasis on prevention rather than control has the environmental justice movement directly opposed to strategies based on control of pollution, usually reflected in market-based approaches (permits to pollute, user pay).

Another defining characteristic of environmental justice is that it advocates *empowerment* as an end in itself, as well as a means to achieve substantive improvements. For environmental justice activists, government agencies are viewed as being too frequently uncommunicative, unresponsive, disdainful and openly hostile (Higgins, 1993: 292). As a result, environmental justice advocates promote open, transparent and accessible decision-making processes. In that context, they also believe an important task is to ensure that technical and/or scientific documents should be "deciphered and translated" so as to be understandable for citizens who wish to become informed about environmental issues, and to participate in environmental planning and decision making.

Environmental justice is a very broad concept, and incorporates numerous dimensions. Box 14.4 provides definitions for environmental justice, and several

Box 14.4 Definitions of key terms

Environmental justice: The right to a safe, healthy, productive and sustainable environment for all, in which "environment" is viewed in its totality, and includes ecological (biological), physical (natural and built), social, political, aesthetic, and economic components. Environmental justice refers to the conditions in which such a right can be freely exercised, through which individual and group identities, needs, and dignities are preserved, fulfilled, and respected in a way which provides for self actualization and personal and community empowerment.

Environmental equity: An ideal of equal treatment and protection for various racial, ethnic and income groups under environmental statutes, regulations and practices applied in a manner that yields no substantial differential impacts relative to the dominant group, and the conditions so created. It implies concepts of "fairness" and "rights", but does not necessarily address past inequities.

Environmental racism: Involves racial discrimination in environmental policy making, enforcement of laws and regulations, and targeting of communities of colour for noxious waste disposal and siting of polluting industries. Racial discrimination can be intentional or unintentional, and often reflects "institutional racism".

Environmental classism: The results of and the process by which implementation of environmental policy creates intended or unintended consequences which have disproportionate impacts (adverse or beneficial) on lower income persons, populations, or communities.

Source: Modified from: http://www-personal.umich.edu/~jrajzer/nre/definitions.html

associated key concepts. From Box 14.4, it is apparent that we need to differentiate among environmental justice, equity, racism and classism.

A decision about a land-fill site in Warren County, North Carolina, during 1982 is generally recognized as the catalysing event for the environmental justice movement in the USA. A decision was taken to create a land-fill site to receive polychlorinated biphenyl (PCB) contaminated soil, removed from 14 different places in the state, adjacent to a small, low-income, primarily African-American community. This decision highlighted that such noxious facilities were often being located in areas inhabited by minorities and/or low-income people. Civil and states rights activists organized many demonstrations, resulting in the arrest of over 500 individuals. Subsequently, the US General Accounting Office conducted a study of eight states in the US South to examine the correlation between siting of hazardous waste disposal sites and the racial and economic status of surrounding communities. The results of the study indicated that three out of every four such land fills were located in or near minority communities. Other studies confirmed these findings. From the time of the Warren County decision, the concepts of environmental justice, equity, racism and classism started to appear in public policy discussions and the research literature. These concepts can be expected to be increasingly important elements of visions in the future.

14.3 Creating a process

In the 1990s, the Ontario Ministry of Natural Resources undertook a review of its planning process. As the person responsible for the review explained, the process had to be modified so that it was not the main "issue" when resource allocation decisions were made. In other words, the official meant that too often the Ministry of Natural Resources was finding that people were not satisfied with decisions because of concerns about weaknesses in the planning process. Ministry of Natural Resources staff understood that it was not possible for every resource allocation decision to satisfy every interest, given the many conflicts. However, they did believe that an improved process should be able to result in those who did not get what they wanted being satisfied that the planning process had been fair and reasonable.

The experience in Ontario emphasizes therefore that a vision for resource and environmental management requires an accompanying process to identify issues and problems, assemble necessary information and viewpoints, determine alternative solutions, and select a course of action. And, increasingly, citizens are expecting to be involved in such planning and management processes, highlighting the importance of *partnerships* and *local knowledge systems*. And, as mentioned in Section 14.2, which focused upon development of a vision, there is growing awareness regarding the need to ensure greater *gender sensitivity* in the construction of partnerships. One form of partnership that has attracted increasing attention is *co-management*, in which specific roles and powers are delegated to local people. Co-management has been particularly effective in incorporating local knowledge systems.

The complexity and uncertainty associated with natural and human systems also have encouraged the development of what have become known as *learning organizations* and an *adaptive environmental management* process, or AEM. AEM explicitly recognizes that planners and managers deal with turbulent conditions, and that they can expect to be surprised on a regular basis. As a result, in this approach, planning and management are designed to encourage learning from mistakes, and to facilitate ongoing adjustments. AEM is realistic in its recognition of complexity and uncertainty, and of often rapidly changing conditions. However, many planners and managers still have difficulty acknowledging when they are mistaken or wrong. Furthermore, many clients or constituents are unlikely to be impressed by plans and management strategies which are continuously being changed, as that makes it difficult to know what investment decisions to make. Thus, while the concept of AEM is in many ways realistic, it also contains much idealism which often will be resisted in the "real world".

The presence of conflict has led to the emergence of *alternative dispute resolution* (ADR) processes. As indicated in Chapter 11, ADR is not a panacea for resolving all conflicts. Instead, it is an alternative to the more established judicial or legal process. The judicial system can be unduly adversarial and expensive, and lead to winners and losers. In contrast, ADR seeks to create a process in which parties in conflict come together voluntarily for joint problem solving. When parties are prepared to address their conflicts on a voluntary and joint basis, ADR offers an attractive alternative. However, if the various sides in a dispute are too entrenched in their positions and not interested in looking for mutually beneficial solutions, then ADR is not likely to be helpful.

Another process consideration relates to the role of *interdisciplinarity* in research and problem solving (Box 14.5). In this book, one message is that the perspectives and insights of many disciplines and professions are required. However, while an interdisciplinary approach is often advocated for resource and environmental management, the concept is interpreted in many different ways. When a research team or management group contains individuals using interdisciplinarity in different ways, it is unlikely that the full potential of the approach will be realized.

As Klein (1990: 56–7) noted, most activities presented as "interdisciplinary" are not, but instead are "multi-disciplinary". The distinction between the two is based on the way in which integration and synthesis are pursued. A multi-disciplinary approach is designed to take advantage of the expertise of specialists in various disciplines. With regard to sustainable development, for example, a

---

Box 14.5 Interdisciplinarity

Two claims about knowledge appear widely today. The first claim is that knowledge is increasingly interdisciplinary. . . . The second and related claim is that boundary crossing has become a defining characteristic of the age.

Source: Klein, 1996: 1.

---

multi-disciplinary approach might involve an ecologist, economist, sociologist and other disciplinary experts each posing the questions judged to be important in their discipline, and then conducting separate analyses, drawing upon the frameworks, perspectives, concepts and methods of their respective disciplines. They would work on their own and complete separate reports. Once the various reports were completed, one or more people would use the information and insights from the reports to try and obtain understanding which would be broader and deeper than from any one discipline. Thus, integration and synthesis occur at the end of the process of analysis.

In contrast, in an interdisciplinary approach, the integration and synthesis begins at the outset of the process, and continues throughout. In a sustainable development project, the various disciplinary specialists would meet before any data collection or analysis had started. The first task would be to share views about the nature of the problem to be examined, to clarify which contribution each discipline could make, and to indicate what assistance could be provided from one discipline to others. Once that task was completed, the disciplinary specialists would conduct their individual studies, but throughout the research would regularly meet with counterparts in other disciplines to share data and insights, and to discuss issues and problems for which assistance was needed. An interdisciplinary approach is more time consuming, as it requires commitment from disciplinary specialists to meet with each other on a regular basis. Furthermore, an interdisciplinary approach can be threatening to some, since specialists from other disciplines are more likely to question fundamental beliefs and assumptions that any one discipline accepts as part of its culture. However, if effective communication and sharing can occur among the team members, then one important end product of an interdisciplinary approach, beyond the systematic and continuous synthesis, is creation of respect, good will and trust among people from different disciplines. When this happens, and individuals are able and willing to acknowledge the limitations and weaknesses of their discipline, the conditions have been established for more collaborative interdisciplinary work in the future, work likely to be of even higher quality as a result of a willingness of specialists to listen to and incorporate views of specialists from other fields.

Interdisciplinarity is not a perfect approach, and indeed the obstacles to build up the necessary respect, goodwill and trust are formidable. Obstacles include the sensitivities and egos of participants, and perceived relative status or prestige of different disciplines. Sometimes an interdisciplinary team cannot overcome such barriers. When this occurs, the team may refer to itself as conducting interdisciplinary research, but in fact is engaged in multi-disciplinary work.

Another legitimate concern may be that as specialists from one discipline borrow concepts from another discipline, the result may be misuse or abuse of the concepts. In this regard, Klein (1990: 88) observed that six common problems can occur when concepts from one discipline are borrowed by another: (1) distortion and misunderstanding of the borrowed material; (2) theories, concepts, methods and data used out of context; (3) the borrowed material used in one discipline is out of favour in its home discipline; (4) phenomena treated cautiously or with scepticism in their home discipline accepted as givens or as "certain" by the borrowing discipline; (5) a particular theory or perspective is over-used or over-relied upon; and (6) the borrowing discipline over-looks or dismisses contradictory evidence, tests and explanations. These limitations are provided here, to help avoid an uncritically enthusiastic pursuit of interdisciplinarity.

For resource and environmental management, whatever process is used does not end with the development of a strategy or a plan. As noted in Chapters 12 and 13, it is also important to created processes to facilitate implementation and monitoring. These are considered in more detail in a later section.

14.4 Generating a product

Careful thought in the crafting of a vision and a process is essential. However, planners and managers should never forget that the ultimate purpose is to resolve a problem or create an opportunity for positive change. As a result, a vision and process should lead to an output, which could be a strategy or plan, and then ultimately to outcomes and impacts, or new desirable conditions as a result of initiatives flowing from a strategy or plan. As we became aware when examining *learning* organizations and *adaptive environmental management*, a strategy or plan should not be viewed as a static product. In other words, the preparation of a strategy or plan does not mean that the work of resource and environmental planners and managers is completed. Due to changing conditions, turbulence and surprises, it may become necessary to modify the product. Furthermore, as a result of new understanding and knowledge, planners and managers may become aware of opportunities for improvement. A key lesson is that the product should never be viewed as fixed and unchangeable.

The design of the strategy should incorporate some basic ideas. First, it should reflect the spirit of the *precautionary principle*. In other words, lack of knowledge or understanding should not be used as a reason for not taking action. Second, in situations characterized by high uncertainty, it is appropriate to base the approach on *flexing*. In other words, the approach should strive to achieve what is considered to be the option providing the most benefits, but also recognize that there may be a need to make adjustments as time unfolds.

In developing a product which can guide actions to resolve a problem or to create opportunities, planners and managers should draw upon a mix of techniques or tools. In this book, it has been argued that in the early stages we should use a combination of *forecasting* and *backcasting* to ensure we consider the most probable, desirable and feasible futures. Having outlined some options, we then need to analyse them systematically. Such analysis should use various techniques. *Environmental impact assessment* should draw attention to the environmental gains and losses of possible choices. And, if impact assessment is broadened to incorporate *social impact assessment*, then sociocultural issues also will be systematically considered. *Life-cycle analysis* emphasizes that we should consider the implications of options from "cradle to grave". Finally, *environmental audits* provide another way to determine performance relative to some generally accepted standards or guidelines.

Resource and environmental planners and managers have a broad suite of techniques from which to choose when analysing options along the way to developing a strategy or plan. As in all good analysis, it is important not to over-rely on any one method or technique, but to use a combination to ensure that the strengths of one offset the weaknesses of another. Judgement has to be exercised about the number of techniques to be used, since the addition of each extra technique implies more time and resources to complete the task. At some point, diminishing returns set in, and the advantages of conducting one more type of analysis becomes marginal. There is no recipe book which can identify the most appropriate mix of methods to use. Analysts have to make choices, based on their judgement about conditions and needs.

A continuing theme in this book is that the resource and environmental management process never ends. Nothing highlights that more than the concepts of implementation and monitoring. Visions, processes and products are unlikely to have an impact if action does not occur, and there is the capacity to make adjustments. It is these two aspects, implementation and monitoring, which are considered in the next section.

14.5 Ensuring implementation and monitoring

In Chapter 12, it was noted that many obstacles can thwart effective implementation of policies, programmes or plans. Indeed, it was commented that implementation failure seems to be similar to "original sin". Both appear to be everywhere and to be ineradicable! And, evidence would seem to support this conclusion, since there often seem to be more plans generated than action taken.

The discussion in Chapter 12 identified many obstacles which can hinder implementation. We also became aware that there are *programmed* and *adaptive* approaches to implementation. The conclusion in that chapter was not that one approach was correct and the other was wrong, but rather that each offers advantages in different situations. As planners and managers, we should be aware of the strengths and weaknesses of programmed and adaptive approaches, and be able to determine which one is most likely to be effective for a given situation. In Chapter 2, we also reviewed the merits of different *planning models* or *schools*. These make different assumptions about problem situations. Regarding implementation issues, it would seem that more use of *incremental* and *transactive* planning models might lead to improvements.

A wise person learns from his or her, or other people's, mistakes. To be able to learn from experience requires a willingness and capacity to track or monitor. Certainly, monitoring is one of the cornerstones of *learning organizations* and *adaptive environmental management* which approach problem solving as a social learning experiment, and deliberately seek information which will allow improvements to be made. In Chapter 13 we saw that many agencies and countries have introduced *state of environment* or *state of sustainability reporting* as a way to monitor policies, programmes and plans. *Environmental auditing* is also used, often at a site level, to measure performance relative to some pre-determined

Figure 14.1 Children in the village of Tumba, Sokoto State, in northwestern Nigeria. Decisions taken by resource and environmental planners and managers will affect their lives, as well as generations to follow (Bruce Mitchell)

standards. Much monitoring is still at a fairly basic level, with attention being given more to *inputs* than *outputs*, mainly for the pragmatic reason that inputs usually are easier to measure and collect data on. Hardly any attention is given to monitoring outcomes and impacts, which occur in the medium to long term.

It is difficult to conceive how we can learn from experience or mistakes if we do not systematically monitor our initiatives and activities. Yet, the reality is that often there is considerable resistance to monitoring. People who have invested professional reputations or egos in policies or plans often are not anxious to see them monitored, if there is significant likelihood that errors will then become public knowledge. This reluctance and inertia in many instances still has to be overcome, and this will not be a minor task.

14.6 Anticipating the future

Will we encounter the future like a Captain Ahab travelling on the *Pequod* in pursuit of Moby Dick, or like a Jeremiah Jones travelling on a "fortunate" *Titanic* (Box 14.1)? A main message from this book is that we do have choices; it is up to us to identify what those are and how we might act on them. The future may appear unclear and fuzzy, and conflicting signals may be seen regarding the most appropriate course of action to pursue. Furthermore, we must be mindful that the decisions we take today will have implications for our children, and future generations (Figure 14.1).

While the "right" course of action may not be obvious, decisions still must be taken, and commitments made since needs exist, problems emerge or persist, and opportunities present themselves. In that context, we should heed the words of Emil Salim (1988: v), the first Minister of Population and Environment for Indonesia and a member of the World Commission on Environment and Development. While he was speaking about his own country, his comments have much more general relevance:

> As a developing country, Indonesia faces the necessity of having to start sailing while still building the ship. We don't have the time to wait until all concepts are well established; until the theories are completed. The problems cannot wait until we can think the problems through.

References and further reading

Adeola F O 2000 Endangered community, enduring people: toxic contamination, health, and adaptive responses in a local context. *Environment and Behavior* 32: 209–49

Agyeman J and B Evans 1999 Sustainability, equity and environmental justice. *Local Environment* 4: 3–4

Albrecht S L 1995 Equity and justice in environmental decision-making: a proposed research agenda. *Society and Natural Resources* 8: 67–72

Anderson D L, A B Anderson, J M Oakes and M R Fraser 1994 Environmental equity: the demographics of dumping. *Demography* 31: 229–48

Aplin G 2000 Environmental rationalism and beyond: toward a more just sharing of power and influence. *Australian Geographer* 31: 273–87

Armour A 1992 The co-operative process: facility siting the democratic way. *Plan Canada* March: 29–34

Baland J-M and J-P Platteau 1999 The ambiguous impact of inequality on local resource management. *World Development* 27: 773–88

Ballard K R and R G Kuhn 1996 Developing and testing a facility location model for Canadian nuclear fuel waste. *Risk Analysis* 16: 821–32

Baxter B H 2000 Ecological justice and justice as impartiality. *Environmental Politics* 9: 43–64

Baxter J W, J D Eyles and S J Elliott 1999 From siting principles to siting practices: a case study of discord among trust, equity and community participation. *Journal of Environmental Planning and Management* 42: 501–25

Been V 1994 Locally undesirable land uses in minority neighborhoods: disproportionate siting or market dynamics. *Yale Law Journal* 103: 1383–422

Boerner C and T Lambert 1994 *Environmental Justice?* Policy Study No. 121, St Louis Missouri, Washington University, Center for the Study of American Business, April

Boerner C and T Lambert 1995 Environmental injustice. *Public Interest* 118: 61–82

Bond M 2000 Culture: striking a fair deal. *Geographical* Magazine 72(7): 35–8

Bowen W, M Salling, K Haynes and E Cyran 1995 Toward environmental justice: spatial equity in Ohio and Cleveland. *Annals of the Association of American Geographers* 85: 641–63

Boyce J K 1995 Equity and the environment: social justice today as a prerequisite for sustainability in the future. *Alternatives* 21(1): 12–17

Brooke J and L Rowan 1996 Professionalism, practice and possibilities. *Local Environment* 1: 113–18

Bryant B (ed.) 1995 *Environmental Justice: Issues, Policies, and Solutions.* Washington, DC, Island Press

Bryant B and P Mohai (eds) 1992 *Race and the Incidence of Environmental Hazards: A Time for Discourse.* Boulder, Colorado, Westview Press

Bullard R D (ed.) 1993 *Confronting Environmental Racism: Voices from the Grassroots.* Boston, Massachusetts, South End Press

Bullard R D 1994a *Dumping in Dixie: Race, Class, and Environmental Quality.* Second edition. Boulder, Colorado, Westview Press

Bullard R D (ed.) 1994b *Unequal Protection: Environmental Justice and Communities of Color.* San Francisco, Sierra Club Books

Burningham K 2000 Using the language of NIMBY: a topic for research, not an activity for researchers. *Local Environment* 5: 55–67

Cable S and M Benson 1993 Acting locally: environmental injustice and the emergence of grass-roots environmental organizations. *Social Problems* 49: 464–77

Capek S 1993 The environment justice frame: a conceptual discussion and an application. *Social Problems* 40: 5–24

Clapp J 1998 Foreign direct investment in hazardous industries in developing countries: rethinking the debate. *Environmental Politics* 7: 92–113

Clark T W 1999 Interdisciplinary problem-solving: next steps in the Greater Yellowstone ecosystem. *Policy Sciences* 32: 393–414

Clinton W J, President 1994 Memorandum on Environmental Justice (11 February 1994). *Public Papers of the President.* Washington, DC, Government Printing Office, pp. 241–2

Cole L 1994 Environmental justice litigation: another stone in David's sling. *Fordham Urban Law Journal* 21: 523–45

Concepción C M 1993 Environment and industrialization in Puerto Rico: disenfranchising the people. *Journal of Environmental Planning and Management* 36: 269–82

Cooper D E and J A Palmer 1995 *Just Environments: Intergenerational, International and Interspecies.* London, Routledge

Cutter S L 1995 Race, class and environmental justice. *Progress in Human Geography* 19: 111–22

Cutter S L, D Holm and L Clark 1996 The role of geographic scale in monitoring environmental justice. *Risk Analysis* 16: 517–26

Dahlberg K and J Bennett (eds) 1986 *Natural Resources and People: Conceptual Issues in Interdisciplinary Research*. Boulder, Colorado, Westview Press

Daniere A G and L M Takahashi 1999 Public policy and human dignity in Thailand: environmental policies and human values. *Policy Sciences* 32: 247–68

Dawson J I 2000 The two faces of environmental justice: lessons from the eco-nationalist phenomenon. *Environmental Politics* 9: 22–60

Dwivedi R 1999 Displacement, risks and resistance: local perceptions and actions in the Sardar Sarovar. *Development and Change* 30: 43–78

Forsyth T 1999 Environmental activism and the construction of risk: implications for NGO alliances. *Journal of International Development* 11: 687–700

Godsil R D 1994 The question of risk: incorporating community perceptions into environmental risk assessments. *Fordham Urban Law Journal* 21: 547–76

Goldman B A 1993 *Not Just Prosperity: Achieving Sustainability with Environmental Justice*. Vienna, Virginia, National Wildlife Federation

Goldman B A 1996 What is the future of environmental justice? *Antipode* 28: 122–41

Graham J D, N D Beaulieu, D Sussman, M Sadowitz and Y-C Li 1999 Who lives near coke plants and oil refineries? An exploration of the environmental inequity hypothesis. *Risk Analysis* 19: 171–86

Greenberg D 2000 Reconstructing race and protest: environmental justice in New York city. *Environmental History* 5: 223–50

Groothuis P and G Miller 1997 The role of social distrust in risk–benefit analysis: a study of the siting of a hazardous waste disposal facility. *Journal of Risk and Uncertainty* 15: 241–57

Gugliotta A 2000 Class, gender and coal smoke: gender ideology and environmental justice in Pittsburgh, 1868–1914. *Environmental History* 5: 165–93

Hart D 2000 The impact of the European Convention on Human Rights on planning and environmental law. *Journal of Planning and Environmental Law* February: 117–33

Hart S 1995 A survey of environmental justice legislation in the States. *Washington University Law Quarterly* 73: 1459–75

Harvey D 1996 *Justice, Nature and the Geography of Difference*. Oxford, Blackwell

Heiman M K 1996 Race, waste, and class: new perspectives on environmental justice. *Antipode* 28: 111–21

Higgins R R 1993 Race and environmental equity: an overview of the environmental justice issue in the policy process. *Polity* 26: 281–300

Hine D W, C Summers, M Prystupa and A McKenzie-Richards 1997 Public opposition to a nuclear waste repository in Canada: an investigation of cultural and economic effects. *Risk Analysis* 17: 293–302

Hird J A 1993 Environmental policy and equity: the case of Superfund. *Journal of Policy Analysis and Management* 12: 323–43

Hofrichter R (ed.) 1993 *Toxic Struggle: The Theory and Practice of Environmental Justice*. Philadelphia, New Society Publishers

Ibitayo O O and K D Pijawka 1999 Reversing NIMBY: an assessment of state strategies for siting hazardous-waste facilities. *Environment and Planning C* 17: 379–89

Jerrett M, J Eyles, D Cole and S Reader 1997 Environmental equity in Canada: an empirical investigation into the income distribution of pollution in Ontario. *Environment and Planning A* 29: 1777–1800

Kasperson R and K Dow 1991 Developmental and geographical equity in global change. *Evaluation Review* 15: 149–71

Kasperson R, D Golding and S Tuler 1992 Social distrust as a factor in siting hazardous facilities and communicating risks. *Journal of Social Issues* 48: 161–87

Kates R W 1995 Labnotes from the Jeremiah experiment: hope for a sustainable transition. *Annals of the Association of American Geographers* 85: 623–40

Klein J T 1990 *Interdisciplinarity: History, Theory and Practice.* Detroit, Wayne State University

Klein J T 1996 *Crossing Boundaries: Knowledge, Disciplinarities, and Interdisciplinarities.* Charlottesville and London, University Press of Virginia

Klein J T and W G Doty (eds) 1994 *Interdisciplinary Studies Today.* San Francisco, Jossey-Bass

Kockelmans J (ed.) 1979 *Interdisciplinarity and Higher Education.* University Park, Pennsylvania State University Press

Kunreuther H, K Fitzgerald and T Aarts 1993 Siting noxious facilities: a test of the facility siting credo. *Risk Analysis* 13: 301–15

Laituri M and A Kirby 1994 Finding fairness in America's cities? The search for environmental equity in everyday life. *Journal of Social Issues* 50: 121–39

Lake R W 1996a Rethinking NIMBY. *Journal of the American Planning Association* 59: 87–93

Lake R 1996b Volunteers, NIMBYs, and environmental justice: dilemmas of democratic practice. *Antipode* 28: 160–74

Lawrence D 1996 Approaches and methods of siting locally unwanted waste facilities. *Journal of Environmental Planning and Management* 39: 165–87

Lawrence R L, S E Daniels and G H Stankey 1997 Procedural justice and public involvement in natural resource decision making. *Society and Natural Resources* 10: 577–89

Lazarus R J and S Tai 1999 Integrating environmental justice into EPA permitting authority. *Ecological Law Quarterly* 26: 617–78

Low N P and B J Gleeson 1997 Justice in and to the environment: ethical uncertainties and political practices. *Environment and Planning A* 29: 21–42

Low N P and B Gleeson 1998 Situating justice in the environment: the case of BHP at the Ok Tedi Copper Mine. *Antipode* 30: 201–26

McCaull J 1976 Discriminatory air pollution: if poor, don't breathe. *Environments* 2: 26–31

McGee T K 1999 Private response and individual action: community responses to chronic environmental lead contamination. *Environment and Behavior* 31: 66–83

McGlinn L 2000 Spatial patterns of hazardous waste generation and management in the United States. *Professional Geographer* 52: 11–22

Meyer P B and T S Lyons 2000 Lessons from private sector brownfield redevelopers: planning public support for urban regeneration. *Journal of the American Planning Association* 66: 46–57

Nabalamba A and G K Warriner 1998 Environmental equality: pollution, race and socio-economic status in Michigan. *Environments* 26: 58–75

Okeke C U and A Armour 2000 Post landfill-siting perceptions of nearby residents: a case study of Halton landfill. *Applied Geography* 20: 137–54

Organization for Economic Cooperation and Development 1972 *Interdisciplinarity: Problems of Teaching and Research in Universities.* Paris, Organization for Economic Cooperation and Development

Padgett D A and N O Imani 1999 Qualitative and quantitative assessment of land-use managers' attitudes toward environmental justice. *Environmental Management* 24: 509–15

Perhac R M 1999 Environmental justice: the issue of disproportionality. *Environmental Ethics* 21: 81–92

Portney PR 2000 Environmental problems and policy: 2000–2050. *Resources* 138: 6–10

Pulido L 2000 Rethinking environmental racism: white privilege and urban development in Southern California. *Annals of the Association of American Geographers* 90: 12–40

Rabe B 1992 When siting works, Canada-style. *Journal of Health Politics, Policy and Law* 17: 119–42

Rephann T J 2000 The economic impacts of LULUs. *Environment and Planning C* 18: 393–407

Ross H 1994 Using NEPA in the fight for environmental justice. *William and Mary Journal of Environmental Law* 18: 353–74

Rowlands I H 1997 International fairness and justice in addressing global climate change. *Environmental Politics* 6: 1–30

Salim E 1988 Foreword. In M Soerjani (ed.) *Enhancing the role of Environmental Study Centres for Sustainable Development in Indonesia*. Jakarta, UNDP/World Bank/Government of Indonesia, Development of Environmental Study Centres Project, p. *v*

Salter L and A Hearn (eds) 1996 *Outside the Lines: Issues and Problems in Interdisciplinary Research*. Montreal, McGill-Queen's Press

Symes G J, B E Nancarrow and J A NcCreddin 1999 Defining the components of fairness in the allocation of water to environmental and human uses. *Journal of Environmental Management* 57: 51–70

Takahashi L M and S L Gaber 1998 Controversial facility siting in the urban environment: resident and planner perceptions in the United States. *Environment and Behavior* 30: 184–215

Tarrant M A and R Porter 1999 Environmental justice and the spatial distribution of fish consumption advisory areas in the Southern Appalachians: a geographic information systems approach. *Human Dimensions of Wildlife* 4: 1–17

Towers G 2000 Applying the political geography of scale: grassroots strategies and environmental justice. *Professional Geographer* 52: 23–36

United Church of Christ Commission for Racial Justice 1991 *Proceedings of the First National People of Color Environmental Leadership Summit*. New York, United Church of Christ

US General Accounting Office 1993 *Siting of Hazardous Waste Landfills and their Correlation with Racial and Economic Status of Surrounding Communities*. Washington, DC, General Accounting Office

Walker G 2000 Urban planning, hazardous installations, and blight: an evaluation of responses to hazard-development conflict. *Environment and Planning C* 18: 127–43

Warhurst A and P Mitchell 2000 Corporate social responsibility and the case of Summitville mine. *Resources Policy* 26: 91–102

Warren K J 1999 Environmental justice. *Environmental Ethics* 21: 151–61

Watters L and C Dugger 1997 The hunt for Gray whales: the dilemma of Native American treaty rights and the International Moratorium on whaling. *Columbia Journal of Environmental Law* 22: 319–52

Weingart P and N Stehr (eds) 2000 *Practising Interdisciplinarity*. Toronto, University of Toronto Press

Wenz P S 2000 Environmental justice through improved efficiency. *Environmental Values* 9: 173–88

Westing A H 1993 Human rights and the environment. *Environmental Conservation* 20: 99–100

Westing A H 1999 Towards a universal recognition of environmental responsibilities. *Environmental Conservation* 26: 157–8

Williams B L, S Brown and M Greenberg 1999 Determinants of trust perceptions among residents surrounding the Savannah River Nuclear Weapons Site. *Environment and Behavior* 31: 354–71

Zimmerman R 1993 Social equity and environmental risk. *Risk Analysis* 13: 649–66

Index

' ' U.W.E.L. LEARNING RESOURCES